INTRODUCTION TO UNMANNED AIRCRAFT SYSTEMS

SECOND EDITION

INTRODUCTION TO UNMANNED AIRCRAFT SYSTEMS

SECOND EDITION

EDITED BY

DOUGLAS M. MARSHALL · R. KURT BARNHART

ERIC SHAPPEE · MICHAEL MOST

CRC Press is an imprint of the
Taylor & Francis Group, an **informa** business

CRC Press
Taylor & Francis Group
6000 Broken Sound Parkway NW, Suite 300
Boca Raton, FL 33487-2742

Printed on acid-free paper
Version Date: 20151207

International Standard Book Number-13: 978-1-4822-6393-0 (Hardback)

Library of Congress Cataloging-in-Publication Data

Names: Marshall, Douglas M., 1947- editor, author. | Barnhart, Richard K., editor, author. | Shappee, Eric, 1968- editor, author. | Most, Michael, editor, author.
Title: Introduction to unmanned aircraft systems / [compiled by] Douglas M. Marshall, R. Kurt Barnhart, Eric Shappee, Michael Most.
Description: Second edition. | Boca Raton : Taylor & Francis, CRC Press, 2016. | Includes bibliographical references and index.
Identifiers: LCCN 2015036667 | ISBN 9781482263930
Subjects: LCSH: Drone aircraft. | Drone aircraft--Automatic control.
Classification: LCC TL589.4 .I68 2016 | DDC 629.133/3--dc23
LC record available at http://lccn.loc.gov/2015036667

Visit the Taylor & Francis Web site at
http://www.taylorandfrancis.com

and the CRC Press Web site at
http://www.crcpress.com

Contents

Preface

It is our pleasure to present the second edition of *Introduction to Unmanned Aircraft Systems*. It is well understood that the unmanned aircraft systems (UAS) industry is highly dynamic and constantly evolving with the advancement of science and technological enablement. As such, the aim of this book is to identify and survey the fundamentals of UAS operations, which will serve as either a basic orientation to UAS or as a foundation for further study. The first edition was birthed out of an unsuccessful search for suitable texts for such a course. The second edition has been expanded to cover additional topics to a greater depth. Both are suitable for survey and introductory collegiate courses in UAS, but the second edition also offers content to support more advanced coursework. The chapters have been individually contributed by some of the nation's foremost experts in UAS operations at the collegiate level; therefore, the reader may note some variation in writing style. It was decided to leave the contributions in this form in the interest of preserving the author's intent, thereby improving the quality of information contained herein. This book is written from a non-engineering, civilian, operational perspective aimed at those who will operate or employ UAS for a variety of future missions.

This publication would not have been possible without the close cooperation of all the editors and contributors; a heartfelt thank you to all who gave of your time to make this possible.

Your feedback is welcomed as a basis for future editions of this book as the industry continues to advance.

Acknowledgments

The editors would like to personally thank all who contributed to this work and their families for allowing them the time to make this sacrifice. The end result is worth it. For all your services we are eternally grateful.

Editors

Richard Kurt Barnhart, PhD, is a professor and currently the associate dean of research at Kansas State University Salina in addition to serving as the executive director of the Applied Aviation Research Center, which established and now oversees the Unmanned Aerial Systems Program office. Most recently, Dr. Barnhart was head of the Aviation Department at Kansas State University. He is a member of the graduate faculty at K-State and holds a commercial pilot certificate with instrument, multiengine, seaplane, and glider ratings. He also is a certified flight instructor with instrument and multiengine ratings. Dr. Barnhart also holds an airframe and power plant certificate with inspection authorization and is a former FAA (Federal Aviation Administration) designated examiner for aircraft maintenance technicians. He earned an AS in aviation maintenance technology from Vincennes University, a BS in aviation administration from Purdue University, an MBAA from Embry-Riddle Aeronautical University, and a PhD in educational administration from Indiana State University. Dr. Barnhart's research agenda has focused on aviation psychology and human factors in the past and more recently in the area of integration of UAS into the National Airspace System. His industry experience includes work as an R&D inspector with Rolls Royce Engine Company where he worked on the RQ-4 Unmanned Reconnaissance Aircraft Development Program as well as other development projects for the Cessna Citation X, V-22 Osprey, Saab 2000, C-130. He also served as an aircraft systems instructor for American Trans-Air airlines. Formerly, Dr. Barnhart was an associate professor and acting department chair of the Aerospace Technology Department at Indiana State University, where he was responsible for teaching flight and upper division administrative classes with an emphasis on aviation management, air carrier safety, and aircraft accident investigation. Courses taught include Aviation Risk Analysis, Citation II Ground School, King Air 200 Flight, Air Navigation, Air Transportation, Instrument Ground School, and many others.

Douglas M. Marshall, J.D., is the owner of TrueNorth Consulting LLC, a UAS support and service organization founded in 2007. Previously, he was a division manager, UAS Regulations & Standards Development at the Physical Science Laboratory, New Mexico State University, and professor of aviation at the University of North Dakota. He has been engaged full time on UAS-related activities for over 9 years, is the coeditor of two books related to aviation, is the coauthor and editor of *Introduction to Unmanned Aircraft Systems*, and is the author of numerous published articles on aviation law, regulations, and remotely piloted aircraft. He has served on RTCA SC-203, ASTM F-38, and SAE G-10 Committees; the AUVSI Advocacy Committee; the Arctic Monitoring and Assessment Program UAS Expert Group; the Small UAS Aviation Rulemaking Committee; and the Part 91 Working Group serving the current UAS Aviation Rulemaking Committee. Currently, he chairs the ASTM F38.02.01 Task Group on Standards for Operations Over People. He has also served on the Steering Committee, Civil Applications of Unmanned Aerial Systems Conference, Boulder, Colorado, and several other committees dedicated to the development of UAS and has delivered presentations on international aviation regulations and airspace issues at conferences around the world.

Michael T. Most, PhD, is an academic lead of the Unmanned Aircraft Systems Program at Kansas State University, prior to which he was an associate professor and chair of the Department of Aviation Technologies at Southern Illinois University. He has authored numerous articles for technical and refereed journals on aviation, aircraft design, and the use of GIS to investigate aviation-related environmental externalities and delivered several peer-reviewed papers on these same topics. Dr. Most holds FAA private pilot and A&P technician certificates, ASTM National Center for Aerospace and Transportation Technologies avionics certification, and a PhD in environmental resources and policy with an emphasis in remote sensing and geographic information systems.

Eric J. Shappee is a professor of aviation at Kansas State University Salina in the Professional Pilot Program. He teaches numerous aviation courses, which include Introduction to Aviation, System Safety, Safety Management, and Aviation Accident Investigation. Shappee holds a commercial pilot certificate with instrument, multiengine, and glider ratings. He is also a certified flight instructor with gold seal. Shappee earned two associate degrees from Antelope Valley College (Lancaster, California), a bachelor's in aeronautical science, and a master's in aeronautical science and safety from Embry-Riddle Aeronautical University. Shappee's main area of focus in aviation is safety. He has developed several risk assessment tools for K-State and other aviation organizations. Further, he is a member of the International Society of Air Safety Investigators and has attended the NTSB Academy. Shappee has been active in the field of aviation since 1986, and teaching since 1995. During his career in aviation, Shappee has also spent time working with unmanned aerial systems including the Predator and Aerosonde.

Contributors

Brian Argrow is a professor of aerospace engineering sciences and founding director (emeritus) of the Research and Engineering Center for Unmanned Vehicles (RECUV) at the University of Colorado Boulder. Professor Argrow is an author or coauthor of more than 100 journals and conference proceedings and he received numerous teaching and education awards. He is chair-emeritus of the AIAA Unmanned Systems Program Committee. He received the Air Force Exemplary Civilian Service Award for service on the Air Force Scientific Advisory Board (2003, 2005–2009) where he coauthored two reports on UAS and served on the NASA Advisory Council's UAS Subcommittee from 2011 to 2013. In 2014, he cofounded and now codirects the Unmanned Aircraft System and Severe Storms Research Group (USSRG). Professor Argrow currently serves on the ASTM F38 Subcommittee for Small VTOL UAS Operations over People.

Mark Blanks has held a variety of positions in the aviation industry including aircraft maintenance, flight test, and aircraft certification. He joined the faculty of Middle Tennessee State University in 2009 as an aviation maintenance instructor, eventually transitioning into a full-time UAS position. Blanks accepted the position of UAS program manager for Kansas State University in January 2013. In his current role, he oversees the growth and development of the K-State UAS academic and research programs. Blanks serves as the chairman for the ASTM F38-02 Subcommittee on UAS Flight Operations and is a member of the AUVSI Board of Directors. He is also actively involved in research to support the integration of UAS into the National Airspace System. He is a licensed Airframe and Powerplant technician and holds a private pilot license with an instrument rating. He is currently the associate director of the Mid-Atlantic Aviation Partnership and, in this capacity, is responsible for the operation of the Virginia Tech test site.

Dallas Brooks is the director of UAS Research and Development for New Mexico State University's Physical Science Laboratory (PSL). He is responsible for NMSU's broad spectrum of government and commercial UAS research, development, test, and engineering (RDT&E) programs. A recognized national leader in unmanned systems integration, he engages and coordinates with national and international regulatory, support, and administrative agencies to help ensure that the tremendous capabilities of unmanned systems are realized. Brooks' aviation and technical experience spans over 30 years, primarily in service to America's armed forces at home and overseas. He serves on multiple national-level boards, committees, and steering groups. He is the executive vice chairman and the strategic planning committee chairman of the National Board of Directors for the Association of Unmanned Vehicle Systems International (AUVSI), the world's largest nonprofit organization devoted exclusively to advancing the unmanned systems and robotics community. He cochairs the Federal multiagency UAS Sense and Avoid Science and Research Panel, supporting the FAA, the DoD, NASA, and DHS to identify and solve key sense-and-avoid challenges. He is also a member of the FAA's UAS Aviation Rulemaking Committee and serves on the Executive Council of the FAA's UAS Center of Excellence.

Joshua Brungardt is the executive vice president for PARADIGMisr in Bend, Oregon. He is also a courtesy faculty member of Oregon State University's College of Forestry. He has served as the Unmanned Aircraft System (UAS) director for Kansas State University helping to grow one of the largest and top renowned UAS Academic programs in the United States. Brungardt served on the Kansas Governor's Aviation Advisory Council from 2010 to 2012 and is currently a member of the board of directors for the Oregon Aviation Industries Board. In 2010, Brungardt attended senior officer training on the Predator UAS at Creech AFB with the 11th Reconnaissance SQ. In his positions, and as a consultant to numerous UAS companies, Brungardt has written over 80 approved UAS Certificates of Authorization from the FAA. Brungardt has also been chief pilot for High Performance Aircraft Training, EFIS Training, and Lancair Aircraft. He holds ATP & CFII ratings with over 4000 hours, as well as having raced at the Reno National Air Races. In addition to completing over 75 first flights on experimental aircraft, he has served as an instructor and test pilot to the U.S. Air Force. In 2006, Brungardt founded the pilot training company EFIS Training, which focused on pilots transitioning to glass cockpits. Brungardt received a bachelor's degree in airway science and an associate's degree in professional pilot from Kansas State University in 2002. In 2013, Brungardt was awarded the Alumni Fellow of the year award from KSU's College of Technology and Aviation.

Stephen P. Cook is a principal safety engineer in the Navigation and Unmanned Aircraft Systems (UAS) Department at the MITRE Center for Advanced Aviation System Development in McLean, Virginia. In this role he supports multiple efforts to integrate civil and military unmanned aircraft into the National Airspace System (NAS). Dr. Cook currently cochairs the UAS Science and Research Panel (SARP). The SARP is responsible for identifying research gaps related to integrating UAS into the NAS, recommending solutions, and promoting alignment of research efforts across the government agencies represented in the UAS Executive Committee. Before joining MITRE, Dr. Cook served as the UAS airworthiness certification lead at the Naval Air Systems Command and led the NATO Technical Working Group charged with developing an airworthiness standard for military fixed-wing UAS. Dr. Cook earned his PhD in aerospace engineering from the University of Maryland in 2003, having previously earned an MS and a BS in aerospace engineering from North Carolina State University in 1993 and 1990, respectively.

Charles Jarnot, USAR, LTC (Ret.), is a graduate of the Aviation Department of Western Michigan, and holds a master's degree in aeronautical science from Embry-Riddle Aeronautical University. Jarnot spent his career as a U.S. Army rotary wing aviator and aviation advisor, serving in numerous locations throughout the United States and the world, including Korea and Afghanistan, where he most recently served in a UAS support role helping to field the MMist CQ-10A Snowgoose™ Jarnot has also served as an adjunct faculty member with Kansas State University teaching an introductory UAS course. In addition, he has field experience on other UAS platforms including Insitu's ScanEagle™ where he is fielding this asset in his current position in the Middle East. One of Jarnot's passions is aviation history, and he has a particular interest in rotary wing history.

Warren Jensen received his medical degree from the University of California, San Francisco. He was engaged in private practice for 8 years before his residency and board certification in aerospace medicine. Dr. Jensen joined the University of North Dakota in 1993, where he teaches human factors, aerospace physiology, and human performance aspects of aviation and space flight. He has served as the state air surgeon for the North Dakota Air National Guard, is an active pilot, and FAA designated senior aviation medical examiner. He currently consults with the Customs and Border Protection Agency to provide human factors coursework for Predator operations.

Saeed M. Khan is a professor of engineering technology at Kansas State University Salina. He earned his MS and PhD in electrical engineering from the University of Connecticut and a BS in electrical engineering from the Bangladesh University of Engineering and Technology. Dr. Khan's specialization is in antennas and electromagnetic wave propagation, and he has been working in stealth technology since he was a graduate student and on smart skins for radar evasive aircraft since 1994. Since 2008, Professor Khan has been active in various aspects of UAS research, which include sense-and-avoid systems, multipath mitigation for UAVs in urban canyons, radar cloaking materials, and wireless power transfer. He has over 40 papers, book chapters, and invention disclosures and has received over U.S.$3.5 M in funding as principal investigator (PI) or co-PI. Professor Khan was recently honored with the 2013–2014 Rex McArthur Family Fellow for demonstrating teaching excellence, commitment to research, and honorable service to the University.

Gabriel B. Ladd earned a bachelor's degree in aerospace engineering from Boston University and a master's degree in environmental science from the University of Maryland. He conducted his undergraduate research at NASA Goddard Space Flight center, focussing on sensor design, testing, and building data for a UAS-based collection system for CO_2 and water vapor eddy correlation. Continuing at NASA Goddard while working on his master's thesis research on UAS applications in precision agriculture, he collaborated with the AeroScience lab to design and build data collection and processing systems for small UAS as well as ground sampling equipment.

After graduate school Ladd worked at the Washington DC area consulting firm Enegis where he led the UAS consulting efforts and developed geospatial computer models to answer national level policy questions for the U.S. and foreign governments. He is currently an independent consultant, working with industry leading UAS operators to create camera systems and data workflows to help them collect, process, and analyze UAS data. Among his clients are industry leaders such as American Aerospace Technologies Inc. (AATI). Ladd's work with AATI is pioneering applications, data workflows, and automation processes for data collected from Medium Altitude Long Endurance (MALE) UAS Beyond Line of Sight (BLOS) operations in the National Air Space (NAS).

Eric McClafferty is a partner and the head of the 35-person International Trade Practice Group at the international law firm Kelley Drye & Warren LLP in Washington, DC. The group was recognized by Law 360 as U.S. International Trade Practice of the Year in 2013 and 2014. He helps universities, large and small companies, individuals, and industry associations comply with U.S. regulations on international trade, particularly those relating to exports of unmanned systems. He earned a BA and an MA from the University of Michigan and JD from the University of Virginia.

Nathan Maresch joined the Kansas State University UAS initiative in 2009 to research avionics development and miniaturization. He develops, integrates, and repairs unmanned aircraft electronics systems and components. He also develops and maintains unmanned simulator systems.

During his time in K-State Salina's UAS lab, Maresch has been involved with a variety of research projects, including a sense-and-avoid project with the creation of a two-dimensional obstacle avoidance system and a wireless power transmission project in conjunction with engineering technology faculty.

Maresch is a K-State alumnus, graduating summa cum laude with a bachelor of science in electronic and computer engineering technology. He served as an instructor for various engineering technology labs at K-State Salina. He holds a private pilot's license and has industry experience in industrial controls and automation. He is a member of the Institute of Electrical and Electronics Engineers and belongs to the Phi Kappa Phi honor society.

Ben Miller is the founder and Unmanned Aircraft Program director for the Mesa County Sheriff's Office. Considered a thought leader on the applications of small-unmanned aircraft by numerous organizations and agencies, Miller's perspective has been sought in presentations at UAS gatherings across the country and Canada, including testimony to the U.S. Senate Judiciary Committee and other members of Congress at both the state and national level. His expertise has been requested by prestigious organizations such as NASA, the FAA's NextGen Institute, U.S. Department of Defense, the U.S. Secret Service, and the Center for Strategic and International Studies.

Miller has assisted the U.S. Department of Justice and its coordinating effort to assist the FAA with guidelines regarding the use of UAS by public safety organizations. He has offered his expertise to public safety departments across the country and continues his focus on the integration of UAS. He now manages an operational team that has flown more public safety-related missions than any other state or local law enforcement agency in the United States and is focused on developing a defensible process for documenting and reconstructing crime scenes in highly accurate, detailed three-dimensional models, and orthomosaic imagery constructed from small UAS aerial data. He is the recipient of the Mesa County Sheriff's 2011 Excellence Award as well as the 2012 AUVSI Member of the year award. He participates in committees throughout the nation including ASTM's F38 Committee on sUAS and serves on the AUVSI Board of Directors.

Rose Mooney is an executive director of the Mid-Atlantic Aviation Partnership (MAAP), which is the FAA awarded UAS Test site in Virginia, New Jersey, and Maryland. Mooney's career started with developing manufacturing robots, then progressed to medical devices, wireless communications, and UAS. Prior to her current position, she spent over 30 years working in industry in engineering. Mooney's interest in aviation began as a young child living in Queens NY in the approach path of LaGuardia Airport. She spent her nights watching the aircraft and being intrigued by flight. It seemed a natural fit when she began working in UAS. She earned her BA in information systems management from Notre Dame University of Maryland after completing studies in computer science at University of Baltimore County.

William H. Semke earned a PhD in mechanical engineering in 1999, an MS in engineering mechanics in 1993 at the University of Wisconsin-Madison, and a BS in physics at Bemidji State University in 1991. He came to the University of North Dakota in 2000.

He is currently a professor of mechanical engineering at the University of North Dakota where he conducts contemporary research in precision motion, vibration control, and aerospace hardware design, along with instruction in the areas of mechanical design and experimental methods. He also serves as the director of the Unmanned Aircraft Systems Engineering (UASE) Laboratory within the UND Center of Excellence for UAS Research and Education.

Benjamin Trapnell is an associate professor of aeronautics at the University of North Dakota. He holds a Bachelor of Science in physical sciences from the United States Naval Academy and a Master of Aeronautical Sciences degree from Embry-Riddle Aeronautical University. He studied Aviation Safety Programs Management at the Naval Postgraduate School in Monterey, California and has completed the Introduction to UAS Flight Test Short Course at the U.S. Naval Test Pilot School.

Professor Trapnell has researched unmanned aircraft systems, sense and avoid technology, and the UAS regulatory environment for the Federal Aviation Administration. He led an interdisciplinary team that developed a Ganged Phased Array Radar risk mitigation system for the US Air Force. In his current assignment, he developed and implemented the first undergraduate Unmanned Aircraft Systems Operations degree program. Professor Trapnell has also developed and implemented the first undergraduate degree worldwide in unmanned aircraft systems operations. He has also worked closely with the FAA and the United States Air Force via Joint Unmanned Aircraft Systems Center of Excellence.

Professor Trapnell served as a carrier aircraft plane commander and a flight instructor for the United States Navy. He holds commercial airplane and glider ratings and is a Life Member of the Academy of Model Aeronautics where he is a Leader Member as well as a Contest Director. Professor Trapnell serves as chairman of the Unmanned Aircraft Systems Committee for the University Aviation Association.

1

History

Charles Jarnot, Edited by Benjamin Trapnell

CONTENTS

1.1 The Beginning

Predating that of manned aircraft, the developmental history of the unmanned version can be argued to begin the real movement forward in man's age-old desire to rise above the limitations imposed by gravity. To soar with the birds, to see from a vantage known only to the avian world, has been throughout history one of the strongest motivational forces in mankind's history. Whether in its mythology or in the earliest texts of the renaissance, visionaries have provided glimpses of what might be possible and, in their own ways, began sketching out a road map for future generations of imaginatives to explore and clear even greater paths to their success. Indeed, from centuries past when Chinese kites graced the skies, to the first hot air balloons, unmanned flying craft utilized the technologies of the day to pave a way forward for the development of manned aircraft. And with the development of manned aircraft came the realization that unmanned aircraft were not rendered obsolete. On the contrary, the advancement of systems development of manned aircraft, coupled with advancements in electronic systems, enabled the integration of automation that refined, if not defined, the capabilities of both.

In modern times, unmanned aircraft have come to mean an autonomous or remotely piloted air vehicle that is used to navigate in the air. Even the name assigned to unoccupied aircraft has changed over the years, as viewed by aircraft manufacturers, civil aviation authorities, and the military. Aerial torpedoes, drones, pilotless vehicle, radio-controlled aircraft, remotely controlled aircraft, remotely piloted aircraft, autonomous aircraft, unmanned aerial vehicles (UAVs), and others, are but a few of the names used to describe a flying machine operated without an onboard pilot.

As they progress through this chapter, observant readers will discover that all aircraft, manned or unmanned, followed essentially the same developmental progress involving the development of aerodynamic forces by wings or rotors that offset the weight of the craft, allowing it to fly. This progression involved the development of aircraft control, allowing the pilot to maneuver the aircraft in pitch and bank, effecting safe, aerodynamic control. When more than gliding was desired, the development of the aircraft meant creating suitable propulsion systems; lightweight and powerful enough to propel the craft through the air. With the ability to fly greater distances, the need arose for proper navigational systems, while the development of flight and navigation automation systems reduced the pilot workloads in flight. None of these were trivial matters, as each relied upon the unique adaptation of immature existing technologies to create the new ones that were needed. Advancements in the sciences of aerodynamics, structures, propulsion, flight control systems, stabilization systems, navigation systems, and the integration of all in flight automation systems made the nearly parallel development of manned and unmanned aircraft systems possible. It continues today with refinements made feasible by the advancements in computer technologies and potential energy systems.

In the early years of aviation, the idea of flying an aircraft with no one onboard had the obvious advantage of removing the risk to life and limb of these highly experimental contraptions.* As a result, several mishaps are recorded where advances were made without injury to an onboard pilot. Although such approaches to remove people from the equation were used, the lack of a satisfactory method to affect control limited the use of these early unmanned aircraft. Early aviation developmental efforts quickly turned to the use of the first "test pilots" to fly these groundbreaking craft. Further advances beyond unmanned gliders proved painful, as even pioneer Otto Lilienthal was killed flying an experimental glider in 1896.

As seen in the modern use of unmanned aircraft, historically unmanned aircraft often followed a consistent operational pattern, described today as the three D's, which stands for *dangerous, dirty,* and *dull. Dangerous* means that someone is either trying to bring down the aircraft or where the life of the pilot may be at undue risk operationally. *Dirty* is where the environment may be contaminated by chemical, biological, or radiological hazards precluding human exposure. Finally, *dull* is where the task requires long hours in the air, making manned flight fatiguing, stressful, and therefore not desirable.

* The German aviation pioneer Otto Lilienthal, circa 1890s, employed unmanned gliders as experimental test beds for main lifting wing designs and the development of lightweight aero-structures. So, too, did the Wright Brothers, flying their first gliders as kites, to unlock the mathematics of lift and drag, and working out the details of aircraft control, all the while remaining safely on the ground.

1.2 The Need for Effective Control

The Wright Brothers' success in flying the first airplane is more of a technical success story in solving the ability to control a piloted, heavier-than-air craft. Many aviation pioneers either used weight shifting to control their inventions, or aerodynamic design (i.e., dihedral) to stabilize their craft, hoping that a solution would evolve during testing. Dr. Samuel P. Langley, the heavily government-financed early airplane designer competing with those two bicycle mechanics from Ohio, also wrestled with the problem of how to control an airplane in flight. Dr. Langley's attempts with a far more sophisticated and better powered airplane, however, ended up headfirst in the Potomac river; not once, but twice, over the issue of adequate flight control. After the Wright Brothers taught the fledgling aviation world the secrets of controlled flight, namely, their wing-warping approach to roll control, and a movable horizontal "rudder" for pitch control, aircraft development experienced a burst of technical advancement. Yet it took the tragedy of World War I and the military demands of the 1914–1918 conflict to stimulate the rapid development of a useful tool. All aspects of aircraft design, from relatively advanced power plants, fuselage structures, lifting wing configurations, and control surface arrangements, began to mature into what we see today as the "airplane." It was in the crucible of "the war to end all wars" that aviation came of age and, along with this wave of technological advancement came the critical but little recognized necessity of achieving effective flight control.

1.3 The Radio and the Autopilot

As is often the case with many game-changing technological advances, inventions of seemingly unrelated items combined in new arrangements to serve as the catalyst for new concepts. Such is the case with unmanned aircraft. Even before the first Wright Brothers' flight in 1903, the famous electrical inventor Nicola Tesla promoted the idea of a remotely piloted aircraft in the late 1890s to act as a flying guided bomb. His concept appears to have been an outgrowth of his work building the world's first guided underwater torpedoes, controlled by what was then called "teleautomation," in 1898. Tesla preceded the invention of the radio in 1893 by demonstrating one of the first practical applications of a device known as a full spectrum spark-gap transmitter. Tesla went on to help develop frequency separation and is recognized by many as the real inventor of the modern radio.

While the electrical genius Tesla was busy designing the first electric architecture for the City of New York, another inventor, Elmer Sperry, the founder of the famed flight control firm that today bears his name, was developing the first practical *gyro-control* system. Sperry's work, like Tesla's, focused initially on underwater torpedoes for the Navy. He developed a three-axis mechanical gyroscope system that took inputs from the gyros and converted them to simple magnetic signals, which in turn were used to affect actuators. The slow speeds of water travel, and weight not being as critical an issue for sea craft, allowed Sperry to perfect his design of the world's first practical mechanical autopilot. Next, Sperry turned his attention to the growing new aircraft industry as a possible market for his maritime invention, not for the purpose of operating an aircraft unmanned, but as a safety device to help tame unstable manned aircraft, and to assist the pilot in maintaining their bearings in bad weather. Sperry began adopting his system of control on early aircraft with the help of airframe designer Glenn Curtiss. Together they made a perfect team of flyer–designer and automation inventor. Following excellent prewar progress on the idea, the demand during World War I to find new weapons to combat Germany's battleships combined the inventions of the radio, airplane, and mechanical autopilot to field the world's first practical unmanned aircraft, an aerial torpedo.

1.4 The Aerial Torpedo: The First Modern Unmanned Aircraft (March 6, 1918)

In late 1916, with war raging in Europe, the U.S. Navy, a military arm of a still neutral country, funded Sperry to begin developing an unmanned aerial torpedo. Elmer Sperry put together a team to tackle the most daunting aerospace endeavor of the time. The Navy contract directed Sperry to build a small, lightweight airplane that could be self-launched without a pilot, fly unmanned out to 1000 yards guided to a target and detonate its warhead at a point close enough to be effective against a warship. Considering that the airplane had just been invented 13 years earlier, the ability to even build an airframe capable of carrying a large warhead against an armored ship, a sizable radio with batteries, heavy electrical actuators, and a large mechanical three-axis gyrostabilization unit, was by itself incredible, but then integrating these primitive technologies into an effective flight profile—spectacular.

Sperry tapped his son Lawrence to lead the flight testing conducted on Long Island, New York. As the United States entered World War I in mid-1917, these various technologies were brought together to begin testing. It is a credit to the substantial funding provided by the U.S. Navy that the project was able to weather a long series of setbacks, crashes, and outright failures of the various pieces that were to make up the Curtis N-9 Aerial Torpedo. Everything that could go wrong did. Catapults failed; engines died; airframe after airframe crashed in stalls, rollovers, and crosswind shifts. The Sperry team persevered and finally on March 6, 1918, the Curtis prototype successfully launched unmanned, flew its 1000-yard course in stable flight and dived on its target at the intended time and place, then recovered and landed. Thus was born the world's first true unmanned system, or "drone."

Not to be outdone by the Navy, the Army invested in an aerial bomb concept similar to the aerial torpedo. This effort continued to leverage Sperry's mechanical gyrostabilization technology and ran nearly concurrent with the Navy program. Charles Kettering designed a lightweight biplane that incorporated aerodynamic static stability features not emphasized on manned aircraft, such as exaggerated main wing dihedral, which increases an airplane's roll stability at the price of complexity and some loss in maneuverability. The Ford Motor Company was tapped to design a new lightweight V-4 engine that developed 41 horsepower and weighed 151 pounds. The landing gear had a very wide stance to reduce ground roll on landings. To further reduce cost and to highlight the disposable nature of the aircraft, the frame incorporated pasteboard and paper skin alongside traditional cloth. The aircraft employed a catapult system with a nonadjustable full throttle setting.

The Kettering aerial bomb, dubbed the *Bug* (Figure 1.1), demonstrated impressive distance and altitude performance, having flown some tests at 100 miles distance and 10,000-ft altitudes. To prove the validity of the airframe components, a model was built with a manned cockpit so that a test pilot could fly the aircraft. Unlike the Navy aerial torpedo, which was never put into service production, the aerial bomb was the first mass-produced

FIGURE 1.1
U.S. Army Liberty Eagle (Kettering Bug).

unmanned aircraft. While too late to see combat in World War I, the aircraft served in testing roles for some 12 to 18 months after the war. The aerial bomb had a supporter in the form of then Colonel Henry "Hap" Arnold, who later became a five-star general in charge of the entire U.S. Army Air Forces in World War II (WWII). The program garnered significant attention when Secretary of War Newton Baker observed a test flight in October 1918. After the war, some 12 Bugs were used alongside several aerial torpedoes for continued test flights at Calstrom Field in Florida.

1.5 The Target Drone

Surprisingly, most of the world's aviation efforts in unmanned aircraft after World War I did not pursue weapon platforms like the wartime aerial torpedo and bomb. Instead, work focused primarily on employing unmanned aircraft technology as target drones. In the interwar years (1919–1939) the manned aircraft's ability to influence the outcome of ground and naval warfare was recognized, and militaries around the world invested more in antiaircraft weaponry. This in turn created a demand for realistic targets, and the unmanned target drone was born. Target drones also played a key role in testing air war doctrine. The British Royal Air Force was in a debate with the Royal Navy over the ability of an airplane to sink a ship. In the early 1920s, General Billy Mitchell of the Army Air Corps sunk a war prize German battleship and subsequent older target warships to the dismay of the U.S. Navy. The counterargument to these demonstrations was that a fully manned ship armed with antiaircraft guns would easily shoot down attacking aircraft. The British used unmanned target drones flown over such armed warships to test the validity of the argument. In 1933, to the surprise of all, a target drone flew over 40 missions above Royal Navy warships armed with the latest antiaircraft guns without being shot down. Unmanned aircraft technology played a key role in formulating air power doctrine and provided key data that contributed to America, England, and Japan concluding that aircraft carriers, which played such a vital role in upcoming WWII, were a good investment.

In the United States, the target drone effort was influenced by the development of the Sperry Messenger, a lightweight biplane built in both manned and unmanned versions as a courier for military applications and as a possible torpedo carrier. Some 20 of these aircraft were ordered. The U.S. Army identified this class of aircraft in 1920 as a Messenger

Aerial Torpedo (MAT). The effort waned in the early 1920s, however, and Sperry Aircraft Corporation withdrew from active unmanned aircraft design with the untimely death of Lawrence Sperry, the son of the founder and victim of an aircraft accident.

As the U.S. Army lost interest in the MAT program, the service turned its attention to target drones. By 1933, Reginald Denny, a British-born actor and an avid model aviation enthusiast, using an aircraft obtained from aviation modeler Walter Righter, perfected a radio-controlled airplane only 10 ft long and powered by a single-cylinder 8 hp engine. Having been an observer/gunner in the Royal Flying Corps in World War I, Denney saw the possibility of creating a radio-controlled target drone for the United States Army to allow gunners to actually shoot at an airborne target known as the "Denneyplane." With this aircraft Denny won an Army contract and produced the target drone, as well as later models, in a shop located in Southern California. The Army designated this craft the OQ-1 Radioplane, and subsequent versions continued with the OQ (subscale target) designation. The Navy bought the aircraft, designated Target Drone Denny 1 (TDD-1). Some 15,000 of all variants were produced and they served throughout WWII as the world's most popular target drone. The company was eventually sold to Northrop in 1952.

In the late 1930s, the U.S. Navy returned to the unmanned aircraft arena with the development of the Navy Research Lab N2C-2 Target Drone (Figure 1.2). This 2500-pound radial engine biplane was instrumental in identifying the deficiencies in Naval antiaircraft prowess. Much like the earlier Royal Air Force experience with the Royal Navy, where drones survived numerous passes on well-armed warships, the U.S. Navy battleship *Utah* failed to shoot down any N2C-2 drones that were making mock attacks on the ship. Curiously enough, the U.S. Navy added yet another title by describing this class of unmanned drones as No Live Operator Onboard (NOLO).*

During the same interwar years the British Royal Navy attempted to develop an unmanned aerial torpedo and an unmanned target drone, both utilizing the same fuselage. Several attempts were made to launch these aircraft from ships with little success.

FIGURE 1.2
Curtiss N2C-2 target drone.

* The Navy target drone program of the late 1930s developed the technique of controlling an unmanned aircraft from a manned aircraft in flight. Used with some success during WWII, the technique was rediscovered and used in Vietnam and to much greater effect in the Iraq conflict.

The Royal Aircraft Establishment (RAE) finally gained a measure of success with a "Long Range Gun with Lynx engine" called together the "*Larynx*." This program was followed by the Royal Air Force automating an existing manned aircraft as its first practical target drone. This effort involved utilizing the Fairey Scout 111F manned aircraft converted as a gyrostabilized radio-controlled plane, now referred to as the *Queen*. Of the five built, the first four crashed on their first flight. The fifth aircraft, however, proved more successful and succeeded in subsequent gunnery trials.

The next evolution was to take the Fairey flight control system and combine it with the excellent, and highly stable, DeHavilland Gypsy Moth. Dubbed the *Queen Bee*, this aircraft proved much more reliable than the earlier Queen, with the Royal Air Force placing an order for 420 target drones. This led to the designation of an unmanned aircraft being described by the letter *Q* to denote unmanned operation. This protocol was adopted as well by the U.S. Military. Although unverified by research, the term *drone* is believed by some to have originated with the *Queen* name as meaning "a bee or drone." Between WWI and WWII, almost all nations with an aviation industry embarked on some form of unmanned aircraft research. These efforts were mainly in the form of target drones.*

1.6 WWII U.S. Navy Assault Drone

The U.S. Navy leveraged its experience with the 1930s N2C-2 Target Drone, which was controlled by an operator from a nearby manned aircraft in flight, to develop a large-scale aerial torpedo now reclassified as an assault drone. Initially, the assault drone effort took the form of the TDN-1 built in a 200-unit production run in early 1940. This aircraft had a wingspan of 48 ft and was powered by twin six-cylinder O-435 Lycoming engines with 220 hp each in a high wing configuration (Figure 1.3). The aircraft was intended to be employed as a bomb or torpedo carrier, in high-risk environments, to mitigate the risk to

FIGURE 1.3
Naval aircraft factory TDN-1 "Assault Drone."

* Germany, however, was an exception. Paul Schmidt, who pioneered the pulse jet as a low-cost, simple, high-performance thrusting device in 1935, found his work being considered by Luftwaffe General Erhard Milch, who recommended the new pulse jet be adapted to unmanned aircraft, which later took the form of the Fieseler Fi 103 flying bomb; better known to the Allies as the "Buzz Bomb."

aircrews. The groundbreaking advancement of this unmanned aircraft was the first use of a detection sensor in the form of a primitive 75-pound RCA television camera in the nose to provide a remote pilot better terminal guidance from standoff distances. Given the relatively poor reliability and resolution of the first TVs, this was indeed a remarkable feat of new technology integration. The TDN-1 was superseded by a more advanced model called the Navy/Interstate TDR-1 Assault Drone. Some 140 examples were built. A Special Air Task Force (SATFOR) was organized and sent to the Pacific Theater. It used the technology against the Japanese during the Bougainville Island Campaign in 1944 with limited, but definable, success. Operationally, a Navy Avenger Torpedo Bomber was flown as the guiding aircraft. Equipped with radio transmitters to affect radio control, a television receiver was also installed which enabled an operator to guide the drone to its target from as much as 25 miles away. Approximately, 50 aircraft were thus employed against various targets attaining roughly a 33% success rate.

The U.S. Navy and Army Air Forces then turned to outfitting older four-engine bombers into unmanned aircraft to be deployed in the European Theater to destroy highly defended, high-priority targets such as V-1 Buzz Bomb bunkers in Siracourt and the heavily fortified U-boat pens in Heligoland. Operation Aphrodite, as it was called, consisted of stripping out Navy PB4Y-2 Privateers (the Navy version of the Consolidated B-24) and B-17s and packing them with high explosives. They were equipped with a Sperry-designed, three-axis autopilot for stabilization, radio control links for remote control, and RCA television cameras in the cockpit. The concept of the operation was for pilots in the aircraft to control it during takeoff. Once established in remote-controlled cruising flights, these pilots would arm the explosives and parachute from the "flying bomb" over friendly England while the aircraft, controlled by an operator in a nearby manned bomber, would be guided to its target. Operations commenced in August 1944, with rather dubious results. On the first mission, the aircraft spun out of control after the pilot left the plane. Subsequent flights ended similarly with loss of control of the aircraft or lack of suitable visibility to fly the aircraft accurately to the target. On August 12, 1944, an aircraft with two pilots at the controls detonated prematurely, killing Navy Lieutenants Wilford J. Wiley and Joseph P. Kennedy. The latter was President John F. Kennedy's older brother and son of the former U.S. Ambassador to England, Joseph Kennedy. Continued failures with equipment and/or operational weather-related incidents, combined with the rapid advancement of Allied forces in Europe, forced the cancellation of the program. In retrospect, one might consider this the first use of unmanned aircraft as an offensive weapon.

1.7 WWII German V-1 Buzz Bomb

The most significant unmanned aircraft of WWII was Nazi Germany's V-1 Buzz Bomb (Vengeance Weapon-1). Based on the earlier 1930s work by inventor Paul Schmidt in developing a practical pulse jet, the aircraft integrated an advanced, lightweight, and reliable three-axis gyrostabilized autopilot, a radio signal baseline system for accurate launch point data, and a robust steel fuselage that was resistant to battle damage. The V-1 represented the first successful, mass-produced, cruise-missile-type unmanned aircraft, and its configuration influenced many postwar follow-on unmanned aircraft designs (Figure 1.4).

The V-1 was manufactured by Fieseler Aircraft Company in large numbers, with over 25,000 built. This high number makes the V-1 the most numerous combat unmanned

FIGURE 1.4
Fieseler FI 103 (V-1) German Buzz Bomb.

aircraft in history, excluding modern hand-launched platforms. The aircraft was flexible in being capable of both ground and air-launching. It utilized a powerful pneumatic catapult system, which is a familiar feature on many modern-day unmanned aircraft systems. The pulse jet was a simple, lightweight, high-thrust device that operated on the principle of cyclic compressions/explosions at about 50 times a second. Employing closing veins to direct the gas toward the exhaust, these cycles created the hallmark "buzz" sound made by the engines in flight. Although not fuel efficient by traditional jet engine standards, the pulse jet was inexpensive to produce, provided high thrust, was reliable, and could operate with significant battle damage. The V-1 was also the world's first jet-powered unmanned aircraft, weighing about 5000 pounds, with an impressive 1800-pound warhead.

Operationally the V-1 was primarily employed from ground-launch rail systems. A small number were air launched from Heinkel 111 bombers, making the V-1 the world's first air-launched unmanned aircraft as well. Some 10,000 V-1s were launched against Allied cities and military targets, killing some 7000 people.

Though the V-1's guidance system allowed it to maintain heading and altitude, it was unable to provide the capability of in-flight navigation. Accurate weather forecasts, primarily wind direction and speed, were necessary to allow the operators to launch the aircraft in the right direction. At a pre-determined time in the flight, a device would close the fuel valve thereby terminating powered flight. The aircraft would then assume a nose-down attitude, with the warhead detonating on contact with the ground. It was necessary, therefore, to have an accurate intelligence to determine where these bombs landed so that proper preflight preparation would allow for some modicum of accuracy. Though a meager 25% were considered "successful," when compared to its fairly low cost, and the reportedly devastating effect it had on public morale, it has been argued that the V-1 was an effective weaponized unmanned aircraft. Mass produced, and employing many firsts for autonomously flown aircraft, it influenced future designs and provided the historical context to fund many more sophisticated unmanned programs during the following Cold War.* The U.S. Navy built a reverse-engineered copy for use in the invasion of Japan and

* The operational capabilities of the V-1 and its ability to carry a large weapon across international borders unmanned is likely the impetus for the inclusion of the following statement in the initial charter of the International Civil Aviation Organization's Chicago Convention of 1944: "No aircraft capable of being flown without a pilot shall be flown without a pilot over the territory of a contracting State without special authorization by that State and in accordance with the terms of such authorization. Each contracting State undertakes to insure that the flight of such aircraft without a pilot in regions open to civil aircraft shall be so controlled as to obviate danger to civil aircraft." Convention on International Civil Aviation, Art. 8. Chicago, 1944.

launched improved versions from submarines on the surface, gaining yet another title as the world's first naval-launched, jet-powered, unmanned cruise missile.

1.8 WWII German Mistletoe

The teaming of manned and unmanned aircraft was not the exclusive domain of the Allies in WWII. The Germans, in addition to the V-1, built a significant number of piggyback aircraft configurations known as Mistletoe Bombers. The main issue with the effectiveness of the V-1 was that it was not very accurate in flying to its desired target. Mistel (Mistletoe) was an attempt by the Germans to deal with this problem. The concept was for an unmanned bomber, usually a twin-engine JU-88, being modified to carry a manned fighter, supported by struts, on its upper surface. The pilot of the manned fighter would guide the bomber to its target and then release it allowing an on-board stabilization system to allow the explosive-laden bomber to glide to the target. Though about 250 such examples were built, the concept had marginal success, due primarily to operational challenges rather than technical issues.

The German Mistletoe concept could be termed more of a guided bomb than an unmanned aircraft, and several gliding guided bombs were developed by the Germans with limited success. The lines between guided missile and unmanned aircraft are not always clear, and in WWII, the V-1 assault drones, explosive-packed, radio-controlled bombers, and the piggyback Mistletoe configuration all involved forms of an airplane, which places them in the category of unmanned aircraft. This distinction is far less clear in the view of future cruise missiles, which are more closely related to their ballistic cousins than airplanes.

1.9 Early Unmanned Reconnaissance Aircraft

As we have seen from the beginning of the first successful unmanned aircraft flight in 1918 through to WWII, unmanned aircraft have been employed mainly in the target drone and weapon delivery roles. Unmanned aircraft development in the follow-on Cold War years shifted dramatically toward reconnaissance and decoy missions. This trend has continued today, where nearly 90% of unmanned aircraft are involved in some form of data gathering in the military, law enforcement, and environmental monitoring arenas. The main reasons unmanned aircraft were not employed in WWII for reconnaissance had more to do with the imagery technology and navigation requirements than the aircraft platforms themselves. Cameras in the 1940s required relatively accurate navigation to gain the desired areas of interest, and navigation technology of the day could not compete as well as a trained pilot with a map. This changed in the postwar years with the advent of radar mapping, better radio navigation, Loran-type networks, and inertial navigation systems, all enabling an unmanned aircraft to fly autonomously to and from the target area with sufficient accuracy.

One of the first reconnaissance high-performance unmanned aircraft to be evaluated was the Radio Plane YQ-1B high-altitude target drone modified to carry cameras, subsequently GAM-67. This turbojet-powered aircraft was primarily air launched from B-47 aircraft and was proposed to be used in the suppression of enemy antiaircraft destruction (SEAD) role. Cameras were also proposed, but the program was canceled after only about 20 were built. Poor range and high cost were given as the reasons for cancelation.

1.10 Radar Decoys: 1950s–1970s

The Vietnam War of the 1960s and early 1970s created a high demand for countermeasures to Soviet-built surface-to-air missiles (SAMs) used by the North Vietnamese. The missile threat relied extensively on radar detection of American aircraft. Jamming of these radars was attempted with mixed results. However, even under the best of circumstances, jamming ground-based radars with airborne systems was problematic in that the ground system probably has access to more power, enabling the radar to overcome the jamming emitter. A more effective solution is to fool the radar into believing it has locked on to a real aircraft and having it waste its expensive missiles on a false target. The U.S. Air Force embarked on such a solution by developing a series of unmanned aircraft to decoy enemy SAM batteries.

To fool a radar operator into believing a decoy resembles an American B-52 Bomber, for example, the aircraft does not need to be built to physically resemble the actual aircraft. Only minor radar reflectors are needed to create a return radar signal that mimics the actual bomber. The additions of radios that mimic the electronic signatures of such aircraft enhance the illusion. As a result, the unmanned Air Force decoys were small in size but had the desired effect. The most frequently used radar decoy was the McDonnell ADM-20 Quail, which could be carried inside the bomb bay of a B-52 and air launched prior to the bombing run. The Quail was about 1000 pounds, had a range of 400 miles, and could mimic the speed and maneuvers of a B-52. As radar resolution improved, the decoys became less effective and most were out of service by the 1970s.

1.11 Long-Range Reconnaissance Unmanned Aircraft Systems: 1960s–1970s

The U.S. Air Force pioneered the first mass-produced, long-range, high-speed unmanned aircraft designed to conduct primarily reconnaissance missions, but these systems evolved into supporting a wide array of tasks, from suppression of enemy air defenses to weapons delivery. The Ryan model 147, later renamed the AQM-34 Lightning Bug and Firefly series, has the longest service record for an unmanned aircraft. Designed as an initiative of the Ryan Aircraft Company in the late 1950s from an earlier target drone, the aircraft was powered by a turbojet, employed low drag wing and fuselage configuration and could reach altitudes in excess of 50,000 ft and speeds of 600 knots (high subsonic).

The Bug, as it was called by its operators, had a long career and flew in a wide range of high- and low-altitude profiles performing electronic signal-gathering intelligence, camera reconnaissance, and various decoy radar signal transmissions. A frequent violator of Communist airspace, many were shot down, but enough successfully completed their missions to justify their use. The aircraft underwent many modifications over its operational use spanning the

FIGURE 1.5
AQM Lightning Bug.

early 1960s to 2003. Many unique and groundbreaking technologies were employed in the Bug unmanned aircraft, including air launch from the wing store of modified DC-130 aircraft to midair parachute snag recovery from H-2 "Jolly Green Giant" helicopters. The AQM-34, as it was renamed late in its career, performed high-priority missions of great national importance, such as reconnaissance missions during the 1960s Cuban Missile Crisis, to relatively mundane tasks as a target drone for fighter aircraft air-to-air missiles (Figure 1.5).

1.12 First Helicopter Unmanned Aircraft Systems: 1960s–1970s

The U.S. Navy's QH-50 DASH, fielded in the early 1960s, established several firsts for unmanned aircraft. This unusual stacked, counter-rotating rotary wing aircraft was the first unmanned helicopter and the first unmanned aircraft to take off and land on a ship at sea. The requirement for the Drone Anti-Submarine Helicopter (DASH) was to extend the delivery range of antisubmarine homing torpedoes. A typical Destroyer in the early 1960s could detect a submarine to ranges over 20 miles, but could only launch weapons at less than 5 miles. This small, compact, unmanned helicopter only needed to fly off to the maximum detection range and drop its homing torpedoes over the submerged submarine. The QH-50 DASH used remote control via a pilot on the fantail of a ship to take off and land, and then employed a gyrostabilizer autopilot to direct the craft to a location that was tracked by the launching ship's radar. Over 700 were built and were used from 1960 to the mid-1970s, where they finished up their career as towing targets for antiaircraft gunnery. Several countries operated the aircraft, including France and Japan.

1.13 The Hunt for Autonomous Operation

From the very first unmanned aircraft, designers strived to gain as much independent flight operation from manned ground control as possible. Military requirements called for maximum standoff distance, long endurance, and significant data streams from onboard sensors. The demand for data competed with bandwidth for flight control transmission, further driving the need for self-flight or autonomous operation. Enemy jamming may delay sensor transmission, but disrupting required flight control information might cause the loss of the aircraft. Cognizant of the British ability to jam its signals, the German V-1 Flying Bomb of WWII intentionally employed a crude, fully autonomous flight control and navigation system based on mechanical gyros, timers, and some primitive preprogramming involving fuel shutoff to initiate the termination dive. It was not until the advent of small, lightweight digital computers, inertial navigation technology, and finally the global positioning system (GPS) satellite network, that autonomous unmanned aircraft operation gained flight autonomy on par with a human-piloted vehicle.

Lightweight computer technology developed in the 1970s, which led to the worldwide explosion in personal computers and the digitalization of everyday items from wristwatches to kitchen appliances, played the most significant role in unmanned aircraft autonomy. With each advance in computing power and cache memory retrieval, unmanned aircraft gained greater flexibility in addressing changes in winds and weather conditions as well as new variables affecting the mission equipment payloads. Mapping data could now be stored aboard the aircraft, which not only improved navigation but also enabled more accurate sensor camera imagery.

1.14 The Birth of the Twin Boom Pushers

The U.S. Marine Corps' groundbreaking work in the late 1960s with the Bikini built by Republic laid the foundation of what was to become the most popular UAS configuration, leading to today's RQ-7 Shadow, which is the most abundant UAS outside of the hand-launched Raven. The Marine Corps Bikini fuselage focused on providing the camera payload with a nearly unobstructed field of view attained by placement in the nose section. This led to a pusher engine arrangement further simplified by a twin boom tail. Although delta pushers were attempted, most notably the Aquila UAS, this aerodynamic configuration made weight and balance a more challenging proposition, since the elevator moment arms were generally fixed, whereas the twin booms could be easily extended.

In the late 1970s, capitalizing on the Marine Corps Bikini configuration, the Israelis developed a small tactical battlefield surveillance UAS called the Scout, built by Israel Aerospace Industries (IAI). The Scout was accompanied by IAI Decoy UAV-A and the Ryan-built Mabat. The decoys were designed to be flown against SAM batteries so as to fool their radar into activating early or even firing a missile on the drone itself. The Mabat was designed to collect antiaircraft radar signals associated with SAM batteries. Finally, the Scout was designed to exploit the actions of the other two in order to put eyes on the SAM batteries for targeting information and damage assessment after a strike. In addition, the Scout provided close-up battlefield imagery to maneuvering ground commanders, a first for unmanned aircraft. This approach differed greatly from all the previous reconnaissance UAS platforms, in that their imagery was more operational and strategic, with film being developed afterward or even electronically transmitted to a collection center for analysis. The advances in small-sized computers enabled this real-time bird's-eye view to a maneuvering leader on the ground to directly influence the decision process on small groups of soldiers or even individual tank movement.

Israeli forces made significant advances in battlefield situational awareness during the June 1982 Bekaa Valley conflict between Israeli and Syrian forces. Operation Peace for Galilee, as it was called by Israel, involved an Israeli ground offensive against Hezbollah terrorists occupying southern Lebanon. Syria, allied with Hezbollah, occupied a large portion of the Bekaa Valley with a sizable ground force consisting of large numbers of new Soviet tanks and heavy artillery. Syrian forces were supported by sophisticated Soviet-built SAM batteries. Israel used a combination of jet-powered decoys and Mabat signal-gathering UASs to detect and identify the Syria SAM battery operating frequencies, and then using the Scout with other manned assets quickly destroyed most of the SAM threat, enabling the Israeli ground forces to maneuver with close air support. The Scout UAS, with its twin boom pusher configuration, flew along the sand dunes of the Bekaa Valley and identified Syrian tanks with near real-time data feed to maneuvering Israeli small-unit commanders. This eye-in-the-sky advantage enabled a smaller force to move with greater speed, provided excellent targeting data to Cobra attack helicopters, and directed very effective artillery fires. The Scout UAS was too small to be picked up and tracked by Syrian Soviet-designed radar and proved to be too difficult to observe by fast-flying Syrian jet fighters. The 1982 Bekaa Valley experience initiated a worldwide race to develop close-battle unmanned aircraft.

1.15 Desert Storm: 1991

Whereas the short 1982 Israeli–Syrian Bekaa Valley campaign represented the first use of close battle UASs, Desert Storm in 1991 represented the first wide-scale employment. The United States and its allies used unmanned aircraft continuously from Desert Shield through Desert Storm. The most frequently employed system was the now-familiar twin boom pusher configuration of the Pointer and Pioneer (Figure 1.6). The aircraft was a joint Israeli–U.S. effort that used a 27 hp snowmobile engine, flew via a remote control joystick on the ground, had a range of about 100 miles, and required an altitude of 2000 ft to maintain a line-of-sight transmission data link. Fully autonomous flight was technically possible, but users of these aircraft opted to have a manned pilot remotely operating the aircraft to achieve more responsive battlefield maneuvering at a desired point of interest. GPS and computer power were not yet sufficiently integrated to enable ground operators

FIGURE 1.6
AAI RQ2 Pioneer.

to simply designate waypoints on short notice. Also, imagery feeds via satellite links were not sufficiently developed at that small size to affect transmission of data. During Desert Storm, U.S. forces flew some 500 UAS sorties. The Pointer and Pioneers guided artillery, even directing the heavy 16-inch gunfire from the battleship *Iowa*. There is a documented case in which a group of Iraqi soldiers attempted to surrender to a Pointer flying low over the desert.

Most militaries around the world concluded after the Desert Storm experience that UAS platforms did indeed have a role to play in spotting enemy locations and directing artillery fires. Conversely, most military analysts concluded that the vulnerable data links precluded UAS use across the board as a replacement for many manned aircraft missions and roles. This opinion was based in part on the limitations of the line-of-sight data link of the Pointer and Pioneer, and a deep-rooted cultural opposition by manned aircraft pilots and their leadership. A large segment of a nation's defense budget is dedicated to the procurement of military aircraft and the training and employment of large numbers of pilots, navigators, and other crew members. Most air forces choose their senior leaders after years of having proved themselves in the cockpit flying tactical aircraft. The very idea of cheaper, unmanned aircraft replacing manned platforms ran against what President Eisenhower warned as the self-fulfilling "military–industrial complex."

1.16 Overcoming the Manned Pilot Bias

From the 1990s to the terrorist attacks of 9/11, unmanned aircraft made slow progress, leveraging the increases in small, compact, low-cost computers and the miniaturization of a more accurate GPS signal. However, the barrier to widespread acceptance lay with manned aircraft platforms and the pilots who saw UAS technology as replacing their livelihoods. When 9/11 occurred, the U.S. Army had only 30 unmanned aircraft. In 2010, that number was over 2000. The argument against unmanned aircraft had finally given way to the low cost, the reduced risk, and the practicality of a drone, as the press still calls them today, performing the long, boring missions of countless hours of surveillance in both Iraq and Afghanistan. With a person still in the loop of any lethal missile leaving the rails of an Air Force Predator UAS, the "responsibility" argument has for the time being been addressed.

1.17 Amateur-Built Unmanned Aircraft

As mentioned earlier in this chapter, Nicola Tesla pioneered the development of a means for successfully controlling an object from a remote location. It shouldn't be difficult, then, to see from a militaristic standpoint that such a device would have a significant impact in the nature of armed conflict. It should be obvious, as well, that inquisitive amateurs would also be interested in investigating the possibilities of utilizing this technology for controlling model aircraft that had heretofore been flown by line control or as free-flying models. In the 1930s, the British developed the Queen Bee remotely piloted aerial drone. Reginald Denny, a British actor and avid aeromodeler, like many enthusiasts who look to make a career out of their hobby, used this aircraft modeling passion, and a desire to incorporate the new methods for radio control, to develop craft to be sold to the government as target drones. The control systems of the time were large, heavy, and very crude compared to modern radio-control (RC) systems. Proportional control, the idea of providing incremental flight control displacement that matches what a pilot could do in the cockpit, was but a dream to the early RC pilots. But as with many dreams, persistence on the part of passionate pioneers, the development of faster and cheaper computer technologies, the creation of microelectric mechanical systems (MEMS), GPS navigation on a chip, miniature power plants, and advanced radio systems have created an environment conducive to a rapid transformation of a toy into a viable tool. It can be argued that the rapid advancements in the hobbyist radio-controlled aircraft and the development of miniature automated stability and navigation systems is creating the commercial revolution of a technology that is rivaled only by computers and the mobile phone. The revolution has happened so quickly that regulatory entities such as the United States' Federal Aviation Administration have had difficulty controlling its proliferation in many commercial industries. Some might say that, on one hand, the military industrial complex is pressing for the development of unmanned aircraft from one side while the hobbyists are pushing for the commercial use of unmanned aircraft from the other.

1.18 Will Unmanned Aircraft Systems Replace Manned Aircraft?

The band of unmanned aircraft control runs from a completely autonomous flight control system independent of any outside signals to one that employs a constant data link enabling a pilot to remotely fly the aircraft and, of course, variations in between. A fully autonomous aircraft could in theory fly without the effects of enemy signal jamming and carry out a variety of complex missions. The disadvantage is that a fully autonomous flight control system can be simulated in a computer, enabling the enemy to develop counters to the system much in the same way as video gamers do with autonomous opponents. Once the program flaws are identified, it becomes a simpler task to defeat the autonomous system. Additionally, fully autonomous systems would most likely not be allowed to employ lethal force, since the chain of responsibility is nonexistent. At the other end of the spectrum, an aircraft that depends on an outside signal, no matter how well it is encrypted, has the potential to be jammed, or worse, directed by the enemy through a false coded message. Even if true artificial intelligence is developed enabling an unmanned aircraft to act autonomously with the intuitiveness of a human being, the responsibility factor will prevent the UAS from fully replacing manned aircraft. This is even truer with civil applications of passenger travel where at least one "conductor" on board will be required to be held accountable for the actions of the aircraft and to exercise authority over the passengers.

DISCUSSION QUESTIONS

1.1 Sir George Caley, Otto Lilienthal, Samuel Pierpont Langley, and the Wright Brothers: All used unmanned aircraft in the development of their concepts of manned aircraft. What might their reasoning have been and what might some of the advantages have been in so doing?

1.2 Some have argued that the Wright Brothers' greatest accomplishment was figuring out how to provide a method to control an aircraft about the lateral axis. The Kettering Bug did not use wing warping or Curtiss' ailerons for such control. How might this have facilitated or hindered the early unmanned aircraft's success?

1.3 Replacing the human in the manned aircraft with a suitable mechanical system proved to be a daunting problem in the past, and even, in some ways, to this day. How did the efforts of Tesla and Sperry serve to pave the way to an automated system?

1.4 What was the most significant unmanned aircraft of WWII? What influenced your choice?

1.5 A lack of accuracy in automated navigation systems prevented nearly all militarized unmanned aircraft from being effective strategic weapons prior to the Gulf War. What technological advancements changed this situation?

1.6 Many can point to myriad reasons that allowed the successful development of unmanned aircraft systems. Which technologies played a major role in the development of commercial UASs? What technology advancements will provide the next great leap in the capabilities of UAS?

2

UAS Applications

Mark Blanks

CONTENTS

2.1 Introduction

Unmanned technology, and robotics in general, is a revolutionary technology that has applications in nearly every industry. There are new applications discovered on a daily basis for unmanned systems that will improve the efficiency or safety of countless tasks. This chapter will provide an overview of some of the most common applications for Unmanned Aircraft Systems (UAS) that are at various stages of maturity. Additionally, the basic technology that differentiates applications and operational considerations will be explored.

The applications for the UAS range from simple video capture to precise scientific measurement. The knowledge required by the UAS operator to conduct a specific mission can vary dramatically from little prerequisite knowledge to extensive specialized training in a scientific discipline. This range in expertise is a clear demonstrator that UASs are tools utilized to obtain a desired set of information for a particular application. The operation of an unmanned system is rarely conducted solely for the purpose of flying an aircraft outside of the flight training environment. Therefore, it is critical to the purpose of the mission and the data collection required prior to the actual launch of the UAS.

The applications discussed in this chapter all have their own set of challenges and variables. The best practices associated with each application could cover an entire book. This chapter provides only a broad cross section of the industries and missions.

2.2 Basic Technology

Before launching into a discussion of the many and varied uses for UASs, providing some foundational information on methods of vehicle control and stabilization, as well as sensor design, may prove beneficial, particularly to those having limited exposure to the technology. Consequently, the topics included in the next few subsections, preceding the main body of this chapter devoted to applications, will provide a fundamental understanding of basic platform control and sensor technology.

2.2.1 Control Methods

To begin the discussion, we must consider the various methods of aircraft control and how different missions are conducted. The operation of the aircraft ranges from full manual control, to stabilized or "remote control," to automated flight profiles without direct flight path control. The level of automation in the flight mission is dependent upon several factors, including, but not limited to, the amount of repetitious aircraft movements required, aircraft proximity to other objects, and the dynamic nature of the mission.

2.2.1.1 Manual Control

Under manual control the operator has direct, unassisted control of the aircraft's flight path. The control input is typically applied through a handheld console that allows the operator to make fine changes in aircraft pitch, roll, yaw, and throttle (see Figure 2.1). The console can be configured to provide exponential control depending on the degree of input applied so that fine inputs can be made with small inputs and large inputs will result in exponentially larger commands. The operator may also have direct control over other aircraft subsystems such as flaps, landing gear, and brakes.

Manual aircraft control provides a skilled operator with precise control over the aircraft's flight path and predictable outcomes to control inputs. However, manual control requires extensive operator training and experience to accomplish effectively and safely. Due to the difficulty of manually controlling an aircraft, many operators that are capable of full manual control have spent a lifetime flying remote control aircraft as a hobby.

2.2.1.2 Stabilized Control

Under stabilized control the operator has direct, assisted control of the aircraft's flight path. This type of aircraft control typically routes the operator's inputs from a handheld console through an autopilot onboard the aircraft that translates the direct inputs into desired outputs. Stabilized control allows the operator to maintain direct control of the

FIGURE 2.1
Example of an external pilot (EP) console that can be used to manually control the aircraft. (Courtesy of Kansas State University Salina.)

aircraft's position, but reduces the need for fine control to ensure a fixed-wing aircraft returns to wings level or a Vertical Takeoff and Landing (VTOL) aircraft returns to hover. Some VTOL aircraft are equipped with a magnetometer that maintains a single direction as "away" from the operator so that, regardless of the aircraft's orientation, away, left, right, and toward the operator remains constant. Stabilized control greatly reduces the operator skill level required to effectively and safely control the aircraft while still providing dynamic control of the flight path. The majority of VTOL systems are capable of stabilized control and this has resulted in significant growth of the VTOL market due to the ease of aircraft operation. However, stabilized control means that the operator must be able to see the aircraft clearly enough to determine the precise orientation of the aircraft in relation to the object(s) being observed. Applications that require repetitive, precise positioning of the aircraft over an area of interest, such as aerial mapping, are difficult to conduct from only the ground-level perspective of an operator.

2.2.1.3 Automated Control

Under automated control the operator has indirect, assisted control of the aircraft's flight path. This type of control is typically conducted through a graphical software interface that provides an overhead view of the aircraft's position overlaid on aerial or satellite imagery (see Figure 2.2). The operator can usually plan the mission in advance through the software's planning tools and also upload commands to the aircraft during flight to alter the flight path. The aircraft's autopilot determines the control surface and throttle inputs to position the aircraft on the desired flight path in a 3-D space, and the operator observes the behavior of the aircraft to ensure that mission is conducted as desired.

Automated control requires the least amount of direct operator skill for aircraft control; however, the multitude of software interfaces for UASs vary greatly in complexity. Some interfaces are designed to provide only basic functionality and may be custom tailored to a specific aircraft and therefore only need high-level inputs from the operator. Other interfaces may require operator input for every possible variable in the mission

FIGURE 2.2
The Mission Planner software is an example of a graphical interface used for automated missions. (Courtesy of Kansas State University Salina.)

and can take a significant time to learn. Regardless of the interface, automated control can greatly increase the efficiency and reduce the workload required for a particular mission. Repetitive flight paths, such as orbits and mapping missions, are particularly well suited to automated control.

2.2.2 Payloads

A detailed discussion of payloads is not within the scope of this chapter; however, understanding the basic types of payloads in the context of various applications is important to the overall comprehension of how different missions are conducted. In this chapter's context, payloads are divided into three categories: still imagers, full motion video, and other payloads. Applications tend to be most effectively completed using only one of the previous types, although some applications will use a combination of multiple payload types. Each type has advantages and disadvantages that must be carefully weighed prior to conducting the mission. Selection of the proper payload for the application is the first significant platform decision that must occur and this decision should be based on the desired information to be collected.

2.2.2.1 Still Imagers

Still imagers are available in a multitude of varieties ranging from simple off-the-shelf consumer cameras to specialized spectral imagers. The common baseline for still imagers is that they acquire a static image at a single point in time. This means that movement of a dynamic environment may not be captured through these payloads, and the scene may change before or after the image is taken. However, still imagers typically provide the highest resolution imagery and images can be taken within a short time period of each other, up to multiple times per second. Once still imagery acquisitions occur more than approximately 15–20 frames per second, the imaging process is regarded to be Full Motion Video. Imagers that can capture FMV typically suffer degradation in image quality and thus are treated here separately. Due to the high resolution or unique characteristics of still imagers, they are well suited to applications where FMV will not suffice. For example, most remote sensing applications are conducted with still imagers.

Many still imager payloads are capable of capturing data associated with the image that is helpful for data processing after collection. This data is called "metadata," which means that it describes other data. The metadata that can be collected with still imagery includes, but is not limited to, date and time, GPS location, camera orientation, focal length, shutter speed, aperture setting, ISO level, camera type, lens type, etc. Some applications require processing of the imagery after collection that cannot be conducted without certain metadata parameters; therefore, it is critical for the operator to know that all required metadata is collected at the time of acquisition.

2.2.2.2 Full Motion Video

Full Motion Video (FMV) payloads are imagers that collect continuous imagery allowing playback in real-time or after collection at a rate that captures "real world" motion (see Figure 2.3). For example, most cinematic movies are captured at a rate of 24 frames per second. FMV can capture dynamic motion of objects and/or persons that is used in a variety of applications, often in real-time, to understand the scenario that is unfolding within the imager's field of view. FMV is particularly well suited to active environments such as

FIGURE 2.3
The Sony Nex-7 can be used for still image capture and FMV. (Courtesy Kansas State University Salina.)

cinematography and law-enforcement applications. However, in order to capture imagery at a sufficient rate, most FMV imagers are of significantly lower resolution than still imagers. Even 1080p "high definition" video only contains about 2 million pixels versus still imagery that might contain over 30 million pixels in each image, meaning that 1080p high definition video has only 1/15th the resolution of high quality still imagery.

Similar to still imagery, FMV can be tagged with metadata describing the imagery. Certain software can "exploit" the imagery to its fullest extent by using this metadata. Typically, the metadata is used to graphically display the location of the camera, the location of the object where the camera is viewing, or to automatically adjust the imagery to remove distortion or other artifacts. The operator must be aware of the metadata requirements prior to the flight mission to ensure that the desired information is captured for the application.

2.2.2.3 Other Payloads

There are many other types of payloads that different applications may utilize beyond imagers. These can range from air sampling devices to communications relays to radars. Applications that rely on a specific type of payload, other than imagers, will be explained in greater depth in Section 2.3. However, the most important consideration in any application is to ensure the appropriate data are collected with the correct method. Prior to any flight operation, the operator should understand the desired information to be collected and the method by which the data must be acquired. The operator may not be a technical expert on the sensor used in the mission, but they should at least understand the parameters that must be met for the mission to be successful.

2.3 Applications

The remainder of the chapter is devoted to a discussion of the use of unmanned aircraft as sensor platforms to acquire data. Subsections will discuss the concept of remote sensing, metric and mapping applications, and imaging for a variety of applications that include inspection of structures and infrastructures, news gathering, cinematography, commercial promotion, law enforcement, emergency management, search and rescue, and reconnaissance, among others. Non-visual applications will also be covered and the chapter will conclude with a discussion of relevant factors critical to the successful acquisition of remotely sensed data.

2.3.1 Remote Sensing

Remote sensing is defined as "the science of gathering data on an object or area from a considerable distance, as with radar or infrared photography, to observe the earth or heavenly body" (Dictionary.com 2015). This definition is purposefully broad and it encompasses many common UAS applications. While it is not possible to list every remote-sensing application of UASs in the context of this chapter, it is worthwhile to note a few key examples. The common thread of these examples is that they involve the remote observation of the earth in order to measure some characteristics thereof. These measurements can range from plant health to the topography of a given area. Most remote-sensing applications require careful control of how the data are collected in order to make precise measurements, thus making remote sensing one of the more challenging applications to conduct accurately.

2.3.1.1 Photogrammetric Applications

Photogrammetry is defined as "the science of making reliable measurements by the use of photographs and especially aerial photographs" (Merriam-Webster.com 2015). The science of photogrammetry has existed for decades with many early methods being developed during World War II. Photogrammetry allows 3-D measurements to be made from 2-D images through the same technique as human eyes enable us to see in 3-D. By collecting overlapping images from different perspectives, the shape of an object can be mathematically determined with great accuracy. Photogrammetric principles lie at the core of numerous UAS applications, three of which are discussed here: aerial mapping, aerial surveying, and volumetric calculations.

2.3.1.1.1 Aerial Mapping

Aerial mapping is the process of building a map from aerial imagery. An aerial photograph is inherently deformed by camera lens distortion, angle of view, and topography of the area imaged and cannot be used as a map without correction. However, an aerial map is corrected to account for deformation through a process known as orthorectification, which allows the map to be used to measure distances and scales. Aerial mapping is the baseline process that is necessary for many UAS applications such as aerial surveying, precision agriculture, and natural resource management. These applications often start with a basic aerial map generated from any number of different sensors and then analyzed to interpret what the map data means for the particular application (see Figure 2.4).

The process of creating an aerial map with unmanned aircraft begins with the collection of aerial imagery covering the geographic area of interest. Simply joining consecutive images together can form a basic 2-D "mosaic" of the area, but the distortion from the camera and angle of view in the images prevents this mosaic from being accurate. To create an accurate aerial map, a photogrammetric process is utilized to correct the deformation in individual images and to understand the 3-D shape of the terrain. The photogrammetric process commonly used for UAS aerial mapping is based on "structure from motion"

FIGURE 2.4
A 3-D model of a corn field as generated from a point cloud. (Courtesy of Kansas State University Salina.)

principles that use multiple perspectives of a single object to determine the object's shape. When applied to aerial mapping, this means that the imagery must be collected so that each image overlaps another one in the forward direction and consecutive strips of images overlap other strips, typically by at least 2/3rds (or 66%), so that multiple perspectives of any single point are captured.

After the aerial imagery is collected, advanced software performs the structure from motion calculations to generate a 3-D model of the targeted area. The general process utilized in this software is

- Identify key points (or features) in each image
- Match key points from each image with similar key points from other images
- Develop a "cloud" of key points that were found in multiple images (the "point cloud")
- Scale the point cloud using ground control points or camera GPS locations
- Increase point cloud density by finding additional key points after the scale and model shape are generally known
- Connect the points in the cloud to create a solid surface, or "mesh"
- Overlay image texture onto the mesh to create a solid, textured 3-D model
- The resulting 3-D model can be exported in many formats, both 3-D and 2-D

The model that results from the Structure from Motion process is often exported to other software suites for further processing or analysis depending upon the type of application. For example, a precision agriculture application that utilizes near-infrared or multispectral imagery might use additional software to calculate a vegetation index assessing crop health. Regardless of the application, aerial mapping relies on proper collection of aerial imagery. Poorly collected imagery, such as blurry images or inadequate image overlap, will result in subsequently low quality maps that may contain significant inaccuracies or even holes in the data.

2.3.1.1.2 Aerial Surveying

Aerial surveying is often confused with aerial mapping and sometimes the terms are used interchangeably. However, the term "surveying" is differentiated from mapping by the reference to measurement of physical characteristics. Specifically, Esri (2015) defines surveying as "measuring physical or geometric characteristics of the earth. Surveys are often classified by the type of data studied or by the instruments or methods used. Examples include geodetic, geologic, topographic, hydrographic, land, geophysical, soil, mine, and engineering surveys." Many aerial surveys begin with an aerial map of an area of interest, but the process of surveying implies that characteristics are being measured beyond simple aerial imagery.

Topographical maps are a good example of an aerial map that is actually a type of survey. The process of creating a topographical (elevation) map from UAS data begins with the collection of imagery as described in Section 2.3.1.1.1, which is then processed into a 3-D model. This 3-D model can then be exported into Geographic Information System (GIS) software to create contour lines for elevation changes. The contour lines can either be overlaid on top of the imagery or they can form their own map, often with shading to indicate elevation or reliefs. A topographic map allows for rapid identification of the elevation of any point on the map.

2.3.1.1.3 Volumetrics

Measuring the volume of physical objects or empty spaces is critical to numerous industries, particularly those that rely on knowing the quantity of stockpiled materials or the amount of material removed from an area. The same 3-D model derived for aerial surveying applications can be used to measure the volume that lies below or above objects in the model. This is especially useful for the mining industry, which must know precisely how much material is removed from a mine for both regulatory compliance and productivity assessments. A coal power plant that must maintain a certain number of burn days in stockpiled coal must know exactly how much coal is stored on site. Topographical land surveys have traditionally verified the volume of stockpiled coal but a UAS can fly stockpiles and mines in a relatively short time and build a 3-D model of the area to estimate volume. The process of creating this volume calculation is the same as discussed earlier, although volumetric surveys are particularly sensitive to errors in the 3-D model. Even small errors can result in significant difference in the calculated volume.

2.3.1.2 Precision Agriculture

A March 2013 report from the Association for Unmanned Vehicle Systems International (AUVSI) stated that agriculture was expected to be the largest market application for UASs by a wide margin (AUVSI 2013). As the largest industry in the world, agriculture is present in nearly every country and employs millions of people around the globe. As the global population continues to rise, so must global food production. For this reason, new methods for increasing production efficiency and decreasing costs are essential. Many of these operations are turning to a new* technique known as "precision agriculture." Precision agriculture is a farm management system that utilizes information and technology to

* According to Oliver (2010, 4) the term, "precision agriculture," was first used in 1990, a quarter century ago, "... as the title of a workshop held in Great Falls, Montana." When viewed in the historical context of the evolution of the art, science, and practice of farming, which has been fundamental to the maintenance of civilization for millennia, the use of precision agriculture is, indeed, a very recent development.

enhance the production of the farm. Applications of UASs in precision agriculture are numerous and include the following:

- *Crop health assessment*: Every crop is capable of producing a certain yield when it is 100% healthy, but natural and man-made factors reduce the health and subsequent yield potential for all crops. Farmers must carefully manage their crop's health to ensure maximum yields and this includes supplying the crops with adequate nutrients and water while also limiting the harmful effects of pests and weeds. UASs can assist in the assessment of crop health by remotely sensing the photosynthetic activity of the plants. One way to determine this photosynthetic activity is by calculating a vegetation index, which is a relative index of plant "greenness," or health. There are a variety of vegetation indices that can be used to assess crop health, but the most common index is the Normalized Difference Vegetation Index (NDVI) (see Figure 2.5). The NDVI is calculated by comparing the difference between visible and near-infrared light that is reflected by the plant. Since plants absorb visible and near-infrared light at different rates based on their photosynthetic activity, NDVI can indicate the relative health of the plant. A farmer or an agronomist can view an NDVI map of a field and rapidly determine which parts of the field are more productive than others. This is very useful information that can inform farm management decisions.
- *Stand counts*: Most agricultural operations sow a desired number of plants per acre for the maximum yield that the nutrients and soil of the particular field can support (i.e., 25,000 plants per acre). However, many factors will affect how many of these plants will emerge and grow into healthy crops. Understanding early in the growing season the number of crops that actually emerged can assist the farmer in making decisions about whether or not to replant certain areas or assist in developing a reasonable expectation for the field's yield. A UAS can be flown during the early season before the plants' leaves start to overlap one another to provide accurate stand counts of the crops. This process involves the creation of an aerial map with sufficient resolution to see individual crops and then using a software algorithm to count individual plants.

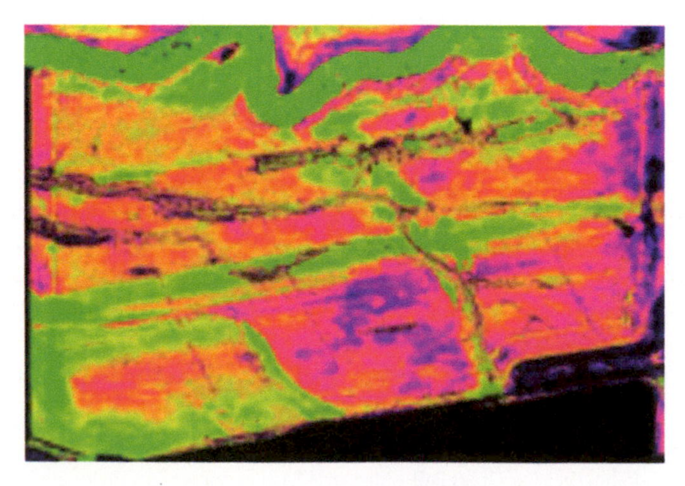

FIGURE 2.5
An example of an NDVI analysis done on a corn field. (Courtesy of Kansas State University Salina.)

- *Crop damage assessments*: Natural events, such as hailstorms or droughts, may cause significant loss of large quantities of crops. Many farmers purchase crop insurance that protects them in the event of such a loss. Insurance companies will reimburse the farmer for lost yield based on the amount of actual versus typical yield for a field that has suffered significant damage. A UAS can verify the extent of widespread crop damage so that insurance companies can reimburse the farmer for the proper amount of yield loss, especially for claims that are filed as 100% losses.

2.3.1.3 Natural Resource Management

The protection of the world's natural resources is vital to the sustainability of the planet. In the United States, agencies such as the Bureau of Land Management (BLM), U.S. Geological Survey (USGS), and U.S. Fish and Wildlife Service (USFWS) are tasked with managing the nation's natural resources. These agencies, among others, must continuously monitor the health of our natural resources and make management decisions ensuring that these resources thrive. For several years, the USGS and BLM have been flying surplus military UASs to assess a variety of natural resources ranging from the effect of dam removals to bird population counts. Natural resource applications for a UAS range from aerial surveys to wildlife monitoring and utilize a host of different technologies. A couple of examples include

- *Impact of the Elwha dam removal*: In the summer of 2012, the USGS collaborated with the Bureau of Reclamation and the National Park Service to assess the impact of the Elwha dam removal in the state of Washington. An AeroVironment Raven UAS was flown over the site during and after the removal of the dam. The Department of Interior uses the subsequent 3-D models for evaluation of sediment distribution throughout the river basin and the effects it may have on wildlife and the environment (USGS 2012).
- *Census of ground-nesting pelicans*: The USFWS and USGS tested UASs as a means for collecting population data of pelicans nesting in south-central North Dakota over the summer of 2014. Manned aircraft had been used previously for this application, but they often delivered sub-par image resolution and were flown at potentially hazardous low altitudes. The test showed that small UASs could deliver accurate bird population counts without disturbing the animals and provide documentation for assessing population trends over time (USGS 2014).

2.3.2 Industrial Inspection

Industrial inspection is a relatively new application for unmanned systems and is becoming one of the most prevalent uses for small UASs. Industrial inspection includes numerous applications within different industries that share a common goal of inspecting equipment, infrastructure, or hardware for defect identification. Items that might be inspected range from highway bridges to flare stacks and each application has its own set of challenges.

2.3.2.1 Civil Infrastructure

Civil infrastructure is composed of the fundamental facilities and structures, such as roads, bridges, tunnels, sewers, and the like, that enable our everyday lives. This infrastructure wears over time, be it from vehicle traffic, weather, or other forces, and eventually the structures will degrade and become unsafe or unusable. Routine inspection of civil infrastructure is crucial for defect identification before loss of infrastructure integrity.

The techniques for infrastructure inspection range greatly, depending on the type of structure being inspected. UASs are primarily capable of remotely sensing defects and have little ability to interact with the object being inspected. This limits UAS to visual, infrared, or other imagers that can sense defects such as cracks or bending of structural components. A few examples of civil infrastructure inspection include

- *Bridge inspection*: Small Vertical Takeoff and Landing (VTOL) UASs are utilized underneath and beside bridge structures to look for cracks in structural members, excessive weathering, loose hardware, or other defects. Most bridge inspections use FMV or a combination of FMV and still imagery. UASs are advantageous for bridge inspections because they reduce human risk by not requiring a person to be hoisted or suspended while inspecting. Additionally, UASs can often inspect more quickly and cost effectively than other mechanisms currently available.
- *Road condition monitoring*: UASs are effective for identifying deteriorating road conditions for both paved and unpaved surfaces (Zhang 2011). Road deterioration in the form of cracks, cupping, or potholes lead to further surface failures and hazards for drivers. Unpaved roads also deteriorate, although often more dynamically than paved roads, and can lead to hazardous conditions. Fixed-wing and VTOL aircraft can be used to map road condition using visual or LiDAR sensors. Typically, VTOL aircraft are used in applications where a short stretch of road needs an extremely detailed inspection and fixed-wings are used for longer road lengths that do not require as much resolution. One challenge with road inspection is the potential safety hazard of flying a UAS over a road being traveled by motorists. Current U.S. aviation regulations require that no traffic be on the roads during an inspection, which necessitates the shutdown of the road and thus limits some of the utility of a UAS for this application.
- *Levee and dam inspection*: As the levee and dam system ages across much of the globe, there is an increasing need to monitor the deterioration of these structures. UASs are particularly effective in mapping levees for erosion (USACE 2015) and inspecting dam faces for cracking. Levee erosion detection requires significant precision in terrain modeling and is difficult to obtain from a manned aircraft flying at higher altitudes than most UASs. Additionally, dam inspection that would be performed by a human rappelling down the dam face can be accomplished with less risk by a VTOL aircraft flying close to the face inspecting for cracks.

2.3.2.2 Electric Power Industry

The electric power industry provides a valuable resource to homes and businesses and is also subject to a host of adverse conditions, both natural and man-made. Inspecting and monitoring electric power infrastructure is also inherently dangerous. The combination of adversity and danger caused the electric power industry to be a focus of many early start-ups in the commercial UAS sector as they attempted to reduce the cost and risk associated with maintaining this infrastructure.

There are many applications within the electric power industry, including, but not limited to

- *Detailed structure inspection*: Small UASs are capable of delivering extremely high quality close-up imagery of hardware components of electric power structures such as transmission poles, transformers, and insulators. Loose hardware, damaged insulators, and reduced structural integrity can all be detected by a small UAS flying either a FMV or still imagery payload. A UAS has the unique ability to get a "top down" view of transmission components on lines that are energized without having a lineman climb the pole or utilizing a bucket truck. This means that a detailed inspection can be completed faster and with less human risk than traditional inspection methods.
- *Long distance transmission line inspection*: Many UASs, including small gas-powered aircraft, are capable of extraordinarily long endurance. This long endurance makes the UAS ideal for flying over transmission lines for long distances to quickly reveal any major damage to structures or lines. Some UASs can fly for 8+ hours and inspect hundreds of miles of line in a single flight. However, the current U.S. regulatory environment prohibits most Beyond Visual Line-of-Sight (BVLOS) operations, which prevents this application from being fully exploited. Streaming video or imagery from a low-flying aircraft that is many miles from the ground control station is a challenge due to terrain obstruction of the radio link.
- *Right-of-Way encroachment and management*: Electric power companies must continuously monitor the right-of-way (ROW) around their transmission lines for encroachment of either vegetation or man-made structure. Any objects that fall within the ROW of a transmission line have potential to cause damage to the lines if they fall near or on the lines or poles. Private landowners will often build structures on a ROW that may hinder access for the electric power company maintaining the lines. The UAS can rapidly fly down a ROW and map any encroachments and assist the electric power company in managing the ROW, while also providing documentation of ROW encroachments that may become contentious issues with landowners.
- *Corona inspections*: Electrical coronas can waste large amounts of transmission grid energy and have been known to ignite devastating fires. Special cameras that are sensitive to the ultraviolet (UV) light spectrum are able to detect the "corona discharge" that occurs when there is a significant ionization of the air around an arcing electric power component. A corona camera sees a corona discharge as flashes of light that are often overlaid on visual imagery. Recent developments in sensor technology have reduced the size of corona cameras to the point where they can be mounted on a small UAS and used for routine inspections.

2.3.2.3 ***Wind Turbine Inspection***

As the demand for "clean energy" grows, wind turbines are an increasingly common sight across the continental United States. Wind turbines can stretch as high as 400 feet above the ground posing a significant challenge for maintaining these complex mechanical devices. Wind turbines are exposed to harsh environmental forces such as lightning strikes, airborne particles, and bird strikes. Over time, these forces will erode or damage the turbine blades making routine inspection of the blades, hub, and tower of wind turbines necessary to keep them in continual operation.

Small UAS have proven to be very effective in detecting blade erosion and damage using both FMV and still imagery. Instead of having a human rappelling from the top of a wind turbine tower, a UAS operator can stand on the ground and fly a VTOL aircraft along stationary turbine blades and look for defects. Often a damaged blade will "whistle" as it travels through the air which is a strong indicator that the turbine needs immediate inspection. Unfortunately, there is currently no technology proven to remotely identify subsurface blade defects such as delamination, and this type of inspection still requires physical contact from a human inspector.

2.3.2.4 Tower/Antenna Inspection

In much the same manner as electric power and wind turbine inspections, radio, cell phone, and other types of towers can be rapidly inspected for damage or loose hardware (see Figure 2.6). Any time that a human has to climb to great heights there is a risk of falling, even with appropriate protective equipment. VTOL UASs are able to keep inspectors on the ground for many of these inspections reducing risk and increasing efficiency. However, high power transmitters can cause communication link failure between the aircraft and ground control station. The UAS operator must be aware of what the potential effects are from transmitters prior to inspecting a tower and take appropriate measures to ensure that control of the aircraft can be maintained.

2.3.2.5 Oil and Gas Inspection

The oil and gas industry is one of the largest industries in the world, with over 30 billion barrels of oil consumed globally each year (CIA 2013). The infrastructure required to support this industry is massive, and since uncontrolled oil leakage can cause tremendous environmental damage, it is critical that this infrastructure is properly maintained. UASs have been used for multiple applications in the oil and gas industry, including

- *Pipeline patrol and inspection*: Similar to electric power transmission line inspection, pipelines can be patrolled by a UAS to identify leaks or damage to the pipeline.

FIGURE 2.6
Radio towers are one example of infrastructure benefiting from UAS inspection methods. (Courtesy of Kansas State University Salina.)

Pipeline inspection is still primarily conducted using low-flying manned aircraft, but there is increasing pressure to perform these patrols with unmanned aircraft. A low-flying UAS can detect leaks or damage by looking for the effects of a leak on the surrounding vegetation. The visible browning of nearby vegetation will often be the first indication of a leak in a buried gas pipeline. Oil leaking from an above-ground pipeline will have similar detrimental effects on vegetation. Therefore, inspection of pipelines is often conducted using either FMV or still imagery and can simply be streamed back to an operator or formed into an aerial map.

- *Flare stack inspection*: Flare stacks are used to burn off the excessive gas that may accumulate as the result of oil extraction or refinement processes. These stacks are often mounted on tall towers to reduce the risk associated with an open flame burning close to the ground. Historically, inspection of flare stacks for deterioration and damage has required the stack and entire extraction/refinement process be shut down while a human climbed up to visually inspect the hardware. Unmanned aircraft can inspect flare stacks without shutting them down and without risking human life to accomplish the inspection. Thus, the operation of a UAS as a sensor platform increases efficiency and safety of the operation.
- *Oil and gas exploration*: The location of petroleum deposits across the globe is an ongoing effort that requires many types of data. UASs give geologists and geophysicists a new tool to use in their exploration efforts. There are characteristics that indicate the presence of subsurface oil and gas deposits that can be detected from an airborne platform. UASs can use aerial mapping and surveying techniques to identify these characteristics and indicate areas where ground-based crews should perform further analysis. They can also collect data from ground-based seismic sensors remotely when seismic tests are conducted to map the subsurface geology of an area.

2.3.3 Aerial Filming and Photography

For the purposes of this chapter, "aerial filming" applications refer to operations that primarily utilize FMV for the sole purpose of providing a moving picture of some scene (see Figure 2.7). Aerial photography refers to still images that are acquired of a scene from an airborne perspective. These terms are used to capture applications that may not be as scientific as other applications such as aerial mapping and remote sensing.

2.3.3.1 Filmmaking

The filmmaking industry captured headlines in the latter part of 2014 as the first approved commercial application for small UAS under the FAA's exemption process contained in Section 333 of the FAA Modernization and Reform Act of 2012. Filmmakers have long used aerial video from manned helicopters and airplanes to capture unique perspectives for a movie scene. UASs enable aerial shots to be taken from lower altitudes than ever before, even to the extent of flying a camera into or out of a building. They can produce stable, "movie quality" video without the complications of cranes and jibs or the risk of low altitude helicopter flight. Most filmmakers have chosen to fly large multirotor aircraft with professional cameras mounted on stabilized gimbals. These aerial shots are often conducted with the UAS operator flying the aircraft in a stabilized control mode while a camera operator controls the camera through a live video feed to obtain the shot desired.

FIGURE 2.7
Aerial views of scenes can now be captured using low-cost, relatively easy-to-use UAS. (Courtesy of Kansas State University Salina.)

2.3.3.2 Real Estate

In an increasingly competitive real estate industry, realtors are constantly looking for new ways to show off properties. The advent of the small UAS has made it possible for the average realtor to obtain high-quality aerial images of both residential and commercial properties so that prospective buyers can easily see the layout of the property that is for sale. There has been a surge of realtors operating low-cost small UASs to obtain this aerial perspective in almost every major real estate market both urban and rural. The urban real estate industry is a challenging environment for UAS operations. The UAS operator must protect the safety of people who may not be aware an aircraft is operating over their heads. The risk of losing sight of the aircraft behind urban structures is very real. Realtors must be cautious when operating a UAS, especially in urban areas, to ensure that they are following the FAA's guidance and policies for these operations.

2.3.3.3 Marketing

UASs are now being used to market everything from cars to homes to concerts. The advertising and marketing industry has quickly learned that an aerial perspective can be a powerful instrument in the persuasion of consumers toward their products. Marketing with a UAS can range from aerial filming to banner towing. In June 2014, a startup company flew a 3-foot by 12-foot banner using a multirotor aircraft on the Las Vegas strip to advertise for the new company (Velotta 2014). There is potential that future banner towing that is currently accomplished utilizing manned aircraft will someday be conducted almost entirely by UASs.

2.3.3.4 News Reporting

Major news outlets across the globe are interested in using UASs to provide live video reporting of events as they unfold, ranging from public demonstrations to traffic reporting. News stations in large cities often have their own manned helicopters that they use to report on live events, but UASs bring this same capability closer to earth and at a cost that almost every news station could afford. By their very nature, most events on which news stations report routinely have many people in the scene. This poses a problem with UAS flight operations over crowds and the FAA has yet to establish the protocol for operation of unmanned systems over large groups of people. There is a substantial risk to persons on the ground if even a small UAS were to fall out of the sky into a crowd and therefore most of these operations are not currently permitted.

2.3.4 Intelligence, Surveillance, and Reconnaissance and Emergency Response

Intelligence, surveillance, and reconnaissance (ISR) missions have historically referred to military operations where an enemy is located and monitored to assist future or ongoing combat missions. However, this same term can be applied to other types of applications in the civil and commercial markets that are not related to observation of an enemy. Instead, ISR can be used to refer to the collection of information about something or someone.

2.3.4.1 Law Enforcement

The military origin of UASs used to collect intelligence about enemy combatants creates a natural connection to law-enforcement applications of unmanned systems. The ability of a small UAS to collect real time, on-demand video in a covert manner is viewed by many as the primary application of "drones" in law enforcement. However, there are many other uses beyond covert surveillance that can benefit law enforcement. Two of these applications include

- *Accident and crime scene reconstruction*: The same photogrammetric process that is used in aerial surveying can be applied to accident and crime scenes to create accurate 3-D models of the scene for later analysis. Determining the causal factors behind an automotive accident can be extremely challenging. The aerial perspective of a UAS image and/or a 3-D model of the scene can help investigators identify where the accident first began and who may have been responsible. Additionally, there has been some initial work performed in using 3-D models derived from UAS data to reconstruct crime scenes and rule out various theories of what occurred during the crime (Miller 2013).
- *Tactical operations support*: Conducting a tactical operation is both stressful and dangerous for law-enforcement officers (see Figure 2.8). Circumstances such as an active shooter in a shopping mall or a hostage taker require police to assess the situation rapidly and precisely. A UAS can provide another tool to enhance situational awareness during these events that could potentially save lives by identifying dynamics in the scene that cannot be identified from a ground-based perspective. Tactical operations support applications typically require live-streaming FMV that is then distributed to those on the scene that need to make rapid decisions to support the mission.

2.3.4.2 Search and Rescue

There is perhaps no other application of UAS that has more potential to save human lives than search and rescue. Depending on the situation, minutes can make the difference between saving and losing a life. UASs, particularly small UASs, can be launched rapidly and cover more ground than rescuers on foot. A thermal infrared sensor can locate victims

FIGURE 2.8
Obtaining an aerial perspective provides a strategic advantage. (Courtesy of Kansas State University Salina.)

quickly by sensing the difference in their body heat from the surroundings. Manned aircraft can also cover ground rapidly during a search and rescue mission, but they often take significant time to be dispatched from a local airport and fly to the scene. Small UASs can be deployed directly by those on the scene, thus saving precious time that may make the difference in rescuing a victim or recovering a body.

The Royal Canadian Mounted Police demonstrated a vivid example of the lifesaving capability a UAS provides when they utilized a small UAS in May 2013 to save the life of a driver who was injured in an automotive accident and then stranded in the cold. In this instance, the officers responding to the accident were unable to locate the driver, although they were able to communicate with him via a cell phone. The officers were able to obtain a GPS location from the driver's phone, but still could not locate him. They ultimately launched a small multirotor aircraft and were able to identify the driver's heat signature within minutes and found him unresponsive but alive at the base of a tree (RCMP 2013).

2.3.4.3 Communications Relay

In the aftermath of a major disaster, one of the major challenges faced by first responders is communicating with each other and potential victims. Often cell phone towers are either overwhelmed with call volume or destroyed and radios cannot reach far enough to ensure consistent communications. As an example, the aftermath of Hurricane Katrina in 2005 left many first responders and families with no way of communicating with victims or each other (AP 2005). One potential application for ensuring that communications can continue in such situations is to use a UAS as a communications relay for either cell phones or radio networks. In this application, an unmanned aircraft is equipped with a payload that can serve as a temporary cell phone or radio repeater and be positioned in a long-term orbit over the affected area. The U.S. military has been using UAS as communications relays for several years, primarily in the Middle East theaters with significant success (Carr 2009). However, operating an unmanned aircraft over a civilian disaster area poses some challenges to air traffic control when there may be numerous rescue aircraft flying in the same airspace and therefore requires substantial coordination with the governing aviation authority before flight.

2.3.4.4 Signals Intelligence

Signals intelligence, often abbreviated as SIGINT by the military, refers to the collection of intelligence through electronic and communications signals. In a military context, SIGINT is used to help determine the location of an enemy and potentially their intentions. In some cases, all that is needed to locate someone of interest is to discover the location of a signal source. In other instances, it may be necessary to intercept the signal in a manner that allows the content of the signal to be interpreted. There is strong potential for SIGINT to be used in civilian applications in addition to military operations, primarily for search and rescue purposes. It may not be possible for a lost person to directly contact rescuers, but SIGINT can be used to identify signals radiating from the lost person's electronic devices to triangulate the individual's position and locate the person quickly. The use of SIGINT for non-military purposes must be conducted with caution to ensure that there is no intrusion on a person's reasonable right to privacy.

2.3.5 Atmospheric Information Collection

Atmospheric sampling is one of the few applications for UAS that does not relate directly to imaging or interacting with ground-based objects. Instead, atmospheric sampling involves the sensing or collection of airborne particulates or gases to identify the characteristics of the atmosphere. Atmospheric sampling has been performed for many years using manned aircraft and balloons, but UASs bring a new capability to sample air more effectively and in regions that were previously more challenging to sample, such as extreme low and high altitudes. UASs can also assist in understanding weather patterns and forecasting by providing information about the temperature, wind speeds, humidity, and other variables at multiple altitudes.

2.3.5.1 Meteorology

In 2010, NASA conducted the first unmanned flights over a tropical cyclone to research new ways to predict the path and strength of tropical storms (NASA 2010). This is a prime example of how UASs are being used to further the understanding of weather, including dangerous weather conditions. The information collected from NASA's Global Hawk UAS provided much higher resolution data than is obtainable from satellite-based sensors. Other meteorological applications capture data variables such as temperatures, wind speeds, and ozone content to better understand how weather patterns may be changing. The primary benefit of a UAS versus manned aircraft and balloons is its ability to be launched quickly, operate for long durations, and maintain precise control of the sensors' positioning. A UAS can also fly into weather phenomena that are too dangerous to operate other aircraft in, such as tornados. For example, the University of Colorado Boulder's Research and Engineering Center for Unmanned Vehicles (RECUV) has performed significant research into the use of UASs for helping understand tornado development and prediction (Elston 2011).

2.3.5.2 Hazardous Material Detection

Another type of atmospheric sampling is the detection of airborne hazardous materials. Detecting toxic substances in the air is critical in identifying events that may be hazardous to humans or the environment. Gas leaks are likely the most common hazardous substance that may become airborne, and locating these leaks can be challenging because the gas may be odorless and colorless, while still being hazardous. UASs are ideal for this application since no human is onboard an unmanned aircraft, and the aircraft can be flown through areas with known hazards to identify and quantify the substance and potentially locate the source. Gas leaks from chemical plants, petroleum pipelines, or other sources can be found using this method, and early research indicates that multirotor aircraft can even enhance currently available sensors due to the air movement of the rotors increasing the flow over the sensors (Gerhardt et al. 2014).

2.3.5.3 Radioactive Material Detection

Nuclear radiation can be detected from an unmanned aircraft in much the same way as other hazardous materials. Sensors are installed on the aircraft that can sense radioactive particles and then flown through an area where radioactive material is suspected. The quantity and location of the radioactive particles can be observed to assist responders in dealing with the material (Pöllänen et al. 2009). The primary advantage of using an unmanned aircraft for this application is that it reduces the exposure of humans to potential harmful levels of radiation. On March 11, 2011, a tsunami caused the meltdown of the Fukushima nuclear power plant in Japan. The area remains too radioactive for humans to enter even years after the disaster, but robots and unmanned aircraft have been able to successfully assess the damage and levels of radioactivity (Siminski 2014). Additionally, it is possible that a UAS could be used to detect and prevent potential nuclear terroristic activities by locating radioactive sources before they are detonated.

2.3.6 Applications Requiring Physical Interaction with Substances, Materials, or Objects

All of the applications previously discussed in this chapter have involved remotely sensing something, collecting samples of airborne gases and particulates, or intercepting some type of signal. There are several UAS applications in development that would require some type of physical interaction with objects. Physical interaction is very challenging for unmanned aircraft since they can normally operate in an unconstrained 3-D environment, but physical interaction places limits on how the aircraft may be operated. A few examples of applications that interact with objects are listed here.

2.3.6.1 Aerial Chemical Application

Commonly called "crop dusting," aerial application of chemicals, primarily for agricultural purposes, has long been conducted using manned aircraft. In largely agrarian regions, such as the U.S. Midwest, low-level air traffic consists of a considerable number of aerial applicators that fly extremely close to the ground at high speeds to apply fertilizer, pesticides, and herbicides to vast agricultural operations. These flight operations are inherently dangerous due to the proximity of the aircraft to ground objects and their high speeds. Recent developments, especially in Japan, have shown that UASs can be effectively used for aerial application of chemicals. In fact, the Yamaha RMAX sprays much of Japan's rice crop. One Yamaha business planner notes that "in Japan more than 2,500 RMAX helicopters are being used to spray 40 percent of the fields planted to rice—that country's number one crop" (UC Davis 2013). These aircraft are much larger than the current "small UAS" category for which the FAA has promulgated regulations in the United States. Larger

UASs will likely require type certification under regulations similar to those currently applicable to manned aircraft. Without additional regulations, integration of large UASs into the National Airspace is likely not viable in the United States.

2.3.6.2 Water Sampling

Water quality and availability is increasingly becoming a major public concern with recent droughts and reductions in aquifer volumes. UASs are already being used to improve water management at multiple universities. Research underway at the University of Nebraska-Lincoln (UNL) has focused on the collection of water samples from bodies of water to determine water quality or to identify harmful algal blooms (UNL 2015). UNL created a multirotor UAS system that can pump water through a tube from a lake or stream up to collection bottles to assist rapid collection of samples. Additionally, Kansas State University researchers have successfully shown that small UASs can identify and characterize harmful algal blooms (Van der Merwe and Price 2015), which is a natural precursor to the actual sampling of water with active algal blooms. As this technology progresses out of the university research environment, it is likely that water quality management will be greatly improved through both remote sensing and physical sampling.

2.3.6.3 Cargo/Package Delivery

In December 2013, Internet commerce giant Amazon.com announced that it would begin delivering packages using small UASs within the next five years (60 Minutes 2013). This announcement, made by a large reputable company, created a media sensation around the possibilities of delivering packages via UAS. It was far from the first time that this application had been considered. In late 2011, a Kaman KMAX K-2000 became the first unmanned helicopter to deliver cargo in an active combat environment to U.S. Marines operating in Afghanistan (NAVAIR 2012). The KMAX operation in Afghanistan proves that it is possible to deliver packages and cargo via unmanned aircraft. The Amazon.com business model would require many small aircraft flying over highly populated urban areas to deliver small packages. There are many regulatory and technical challenges that must be overcome for this to occur, but testing of small UAS package delivery has already begun across the globe.

2.4 Additional Considerations

A few brief, though relevant, comments regarding mission planning and data processing and analysis, factors critical to the successful completion of any mission regardless of application, comprise the final section in this chapter.

2.4.1 Mission Planning

The methods for conducting the previous applications vary dramatically in the means of controlling the aircraft, the types of payloads that are utilized, and the type of information that is collected. Some of the applications, such as wind turbine inspection, must be conducted almost entirely under manual or stabilized control, while others may be fully automated. The payloads may vary from simple off-the-shelf consumer cameras to specialized radioactive particulate samplers. Prior to conducting a UAS mission for any application, the following items must be considered:

1. *What type of data is to be collected?*
   The type of data that is going to be collected on a UAS mission must be carefully defined before taking any further steps. This may be as simple as determining whether basic FMV will be adequate to determine the location of a missing person or as complicated as defining the ground sampling distance needed from aerial imagery to determine the difference between a gull and a pelican. Flight missions that are conducted without understanding the resulting data requirements often result in failed completion of the mission's goals.
2. *What type of sensor/payload is needed?*
   Once the type of data is known, the sensor/payload that can be flown on the aircraft can be selected. There is often a disparity between the data desired versus the reasonable payloads that can either be carried by the aircraft or fit within the mission's budget parameters. This is the stage where compromises must be made, and the viability of the mission determined.
3. *What type of aircraft must be used?*
   If possible, the aircraft should not be selected until the payload is known. Additionally, the data requirements may drive the type of aircraft to either a VTOL or fixed-wing platform. Selection of the aircraft before the sensor and data parameters are known can result in unusable end products.
4. *Will the flight environment enable the desired results to be achieved?*
   The natural, airspace, and regulatory flight environment may hinder the ability of the mission to achieve the desired results. Careful consideration must be given to all of these factors before conducting the operation.

2.4.2 Data Processing and Analysis

Many applications require data collected from the UAS to be processed into usable information and analyzed or interpreted. Simple forms of data processing may only locate where the data came from geographically, while complex data processing is needed to derive accurate products such as volumetric surveys. As new applications arise for unmanned systems, the methods for processing and analyzing the data that supports the application will continue to be an area of significant development for the UAS industry. In some cases, new methods of data analysis will drive entirely new applications that have not even been considered.

2.5 Conclusion

The applications presented in this chapter are only a small sampling of the multitude of uses for UAS. New applications are developing every day that are demonstrating the potential for UAS to affect almost every industry around the globe. Many applications will evolve to being commonplace events while there are certainly others that will prove to be less useful in the long run. The rapid development of advanced aircraft systems, improved sensors, and favorable regulations are likely to only increase the value of UAS to everyday life.

DISCUSSION QUESTIONS

2.1 Describe the three basic methods of control discussed in this chapter and discuss how the mission, application, and type of data to be acquired may determine which is used.

2.2 Describe each of the various sensors discussed in this chapter and list the various applications for which these would be used.

2.3 Discuss in detail each of the applications covered in this section. What do you believe the best platform design and payload package would be for each? Support your answer. Operations known as 3-D missions are discussed elsewhere in this book. Give examples of applications that would be considered 3-D operations. Provide examples of applications requiring physical interaction with substances, materials, or objects.

2.4 List and describe in detail those items which must be considered prior to conducting any UAS mission. Mention why you believe each is significant. Reflecting back on the types of applications discussed, which do you believe would most likely rely on extensive data processing? Which would likely be the least data dependent?

References

60 Minutes. 2013. Amazon's Jeff Bezos Looks to the Future. CBS News. December 1. Accessed April 26, 2015. http://www.cbsnews.com/news/amazons-jeff-bezos-looks-to-the-future/.

AP (Associated Press). 2005. Katrina outages reveal phone system quirks. NBCNews.com. Last modified August 31, 2005. http://www.nbcnews.com/id/9120503/ns/technology_and_science-tech_and_gadgets/t/katrina-outages-reveal-phone-system-quirks/#.VT1D6842ybE.

AUVSI (Association for Unmanned Vehicle Systems International). 2013. *The Economic Impact of Unmanned Aircraft Systems Integration in the United States.* Arlington, VA.

Carr, D.F. 2009. Communications Relay Grows with Expansion of UAV Missions. *Defense Systems.* April 13. http://defensesystems.com/articles/2009/07/29/c4isr-1-uav-relay.aspx.

CIA (Central Intelligence Agency). 2013. *The World Factbook 2013–2014.* Washington, DC: Central Intelligence Agency. https://www.cia.gov/library/publications/the-world-factbook/.

Dictionary.com. 2015. Remote Sensing. Dictionary.com Unabridged. Random House, Inc. Accessed April 26. http://dictionary.reference.com/browse/remote%20sensing.

Elston, J.S. 2011. Semi-Autonomous Small Unmanned Aircraft Systems for Sampling Tornadic Supercell Thunderstorms. PhD dissertation. University of Colorado. http://tornadochaser.colorado.edu/data/publications/2011-[elston]-dissertation.pdf.

Esri. 2015. Surveying—GIS Dictionary. Esri GIS Dictionary. Accessed April 26. http://support.esri.com/en/knowledgebase/GISDictionary/term/surveying.

Gerhardt, Nathan, Reece Clothier, Graham Wild. 2014. Investigating the Practicality of Hazardous Material Detection Using Unmanned Aerial Systems. *Metrology for Aerospace (MetroAeroSpace)*, 2014 IEEE: 133–137. doi: 10.1109/MetroAeroSpace.2014.6865908.

Merriam-Webster.com. 2015. Photogrammetry. Merriam-Webster, Inc. Accessed April 26. http://www.merriam-webster.com/dictionary/photogrammetry.

Miller, B. 2013. *The Future of Drones in America: Law Enforcement and Privacy Considerations. Before the Committee on the Judiciary.* United States Senate. (Written testimony of Benjamin Miller, Unmanned Aircraft Program Manager, Mesa County Sheriff's Office and Representative of the Airborne Law Enforcement Association).

NASA (National Aeronautics and Space Administration). 2010. NASA's Global Hawk Drone Aircraft Flies Over Frank on the GRIP Hurricane Mission. Last modified September 1, 2010. Accessed April 26, 2015. http://www.nasa.gov/mission_pages/hurricanes/missions/grip/news/frank-flyover.html.

NAVAIR (Naval Air Systems Command). 2012. Marines Find First Deployed Cargo Unmanned Aerial System "Reliable." NAVAIR News. July 23. Accessed April 26, 2015. http://www.navair.navy.mil/index.cfm?fuseaction = home.NAVAIRNewsStory&id = 5073.

Oliver, M.A. 2010. An Overview of Geostatistics in Precision Agriculture. In *Geostatistical Applications for Precision Agriculture,* ed. M.A. Oliver, 1–34. New York: Springer.

Pöllänen, Roy, Harri Toivonen, Kari Peräjävi, Tero Karhunen, Tarja Ilander, Jukka Lehtinen, Kimmo Rintala, Tuure Katajainen, Jarkko Niemelä, Marko Juusela. 2009. Radiation surveillance using an unmanned aerial vehicle. *Applied Radiation and Isotopes* 67, iss. 2: 340–344. doi:10.1016/j.apradiso.2008.10.008.

RCMP (Royal Canadian Mounted Police). 2013. Single Vehicle Rollover—Saskatoon RCMP Search for Injured Driver with Unmanned Aerial Vehicle. *RCMP in Saskatchewan,* May 9. http://www.rcmp-grc.gc.ca/sk/news-nouvelle/video-gallery/video-pages/search-rescue-eng.htm.

Siminski, J. 2014. Fukushima Plant's Radiation Levels Monitored with an UAV. *The Aviationist.* January 29. http://theaviationist.com/2014/01/29/fukushima-japan-uav/.

UC Davis (University of California, Davis). 2013. Remote-Controlled Helicopter Tested for Use in Vineyard Applications. June 5. http://news.ucdavis.edu/search/news_detail.lasso?id = 10623.

UNL (University of Nebraska-Lincoln). 2015. Water-Slurping Drones Have Broad Potential. Accessed April 26. http://research.unl.edu/annualreport/2014/water-slurping-drones-have-broad-potential/.

USACE (US Army Corps of Engineers). 2015. Unmanned Aerial Vehicle. US Army Corps of Engineers Jacksonville District. Accessed April 26. http://www.saj.usace.army.mil/Missions/UnmannedAerialVehicle.aspx.

USGS (United States Geological Survey). 2012. Monitoring River Impacts During Removal of Elwha and Glines Dams. National Unmanned Aircraft Systems (UAS) Project Office. Last modified March 18, 2015. Accessed April 26, 2015. http://rmgsc.cr.usgs.gov/UAS/WA_BORRiverSedimentMonitoring.shtml.

USGS (United Geological Survey). 2014. Census of Ground-nesting Pelicans. National Unmanned Aircraft Systems (UAS) Project Office. Last modified March 13, 2015. Accessed April 26, 2015. http://rmgsc.cr.usgs.gov/UAS/ND_ChaseLakeNWRPelicans.shtml.

Van der Merwe, Deon, Kevin P. Price. 2015. Harmful Algal Bloom Characterization at Ultra-High Spatial and Temporal Resolution Using Small Unmanned Aircraft Systems. *Toxins,* 7, no. 4: 1065–1078.

Velotta, R.N. 2014. DroneCast introduces advertising by Octocopter. *Las Vegas Review-Journal,* July 9. http://www.reviewjournal.com/news/las-vegas/dronecast-introduces-advertising-octocopter.

Zhang, C. 2011. *Monitoring the condition of unpaved roads with remote sensing and other technology.* Final Report for US DOT DTPH56-06-BAA-002. South Dakota State University.

3

The "System" in UAS

Joshua Brungardt with Richard Kurt Barnhart

CONTENTS

3.1 Introduction

3.1.1 What Makes Up an Unmanned Aircraft System

In this chapter we will briefly discuss the elements that combine to create an Unmanned Aircraft System (UAS). Most civilian unmanned systems consist of an unmanned or remotely piloted aircraft, the human element, payload, control elements, and data link communication architecture. A military UAS may also include other elements such as a weapons system platform. Figure 3.1 illustrates a common UAS and how the various elements are combined to create the system.

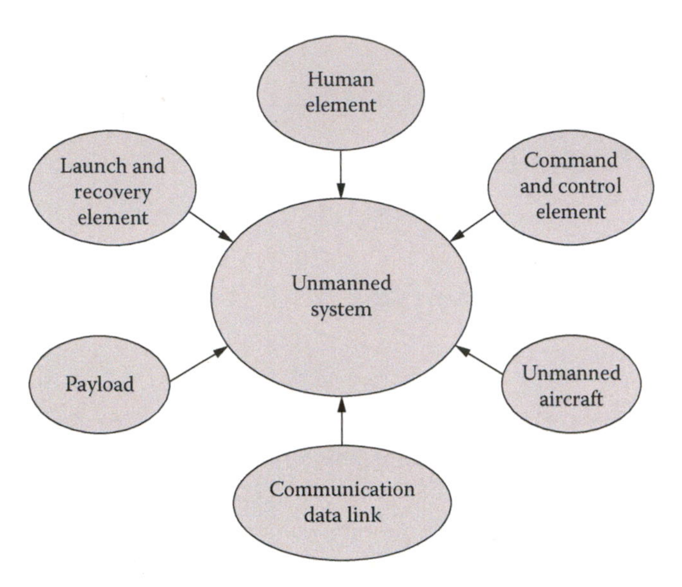

FIGURE 3.1
Elements of the unmanned aircraft system (UAS).

3.2 UAS/RPA

The term "unmanned aircraft" means an aircraft that is operated without the possibility of direct human intervention from within or on the aircraft. In more recent years in some sectors, particularly some branches of the military, there has been a push to change the term *unmanned aircraft* (UA) to remotely piloted aircraft (RPA) or remotely piloted vehicle (RPV). Unmanned aircraft is really a misnomer considering how much human involvement is crucial to the operation of the system. UASs are categorized into five groups by the U.S. Department of Defense as seen in Table 3.1. In general terms, the FAA has traditionally differentiated "sUAS" or small unmanned aircraft systems (under 55 lbs) from other larger UASs. In the future, the FAA will likely classify a UAS into risk-based classifications depending on their potential impact on public safety. sUAS are the first to be integrated into the National Airspace System (NAS), a process which is underway now.

TABLE 3.1
U.S. Department of Defense UAS Classification System

UAS Category	Max Gross Takeoff Weight	Normal Operating Altitude (ft)	Airspeed
Group 1	<20 pounds	<1200 above ground level (AGL)	<100 knots
Group 2	21–55 pounds	<3500 AGL	<250 knots
Group 3	<1320 pounds	<18,000 mean sea level (MSL)	
Group 4	>1320 pounds		Any airspeed
Group 5		>18,000 MSL	

Note: If a UAS has even one characteristic of the next higher level, it is classified in that level.

3.2.1 Fixed-Wing

A fixed-wing UAS has many missions including intelligence gathering, surveillance, and reconnaissance, or ISR. Some military fixed-wing UASs have adapted a joint mission combining ISR and weapons delivery, such as the General Atomics Predator series of aircraft. The Predator™ was originally designed for an ISR mission with an aircraft designation of RQ-1. In the military aircraft classification system the *R* stands for reconnaissance and the *Q* classifies it as an unmanned aerial system. In recent years, however, the Predator's designation has been changed to MQ-1, the *M* standing for multirole, having recently been used to deliver hellfire missiles.

Fixed-wing UAS platforms have the advantage of offering operators long flight duration for either maximizing time on station or maximizing range. Northrop Grumman's RQ-4 Global Hawk™ has completed flights of more than 30 hours covering more than 8200 nautical miles. Fixed-wing platforms also offer the ability to conduct flights at much higher altitudes where the vehicle is not visible with the naked eye. Fixed-wing aircraft are also more capable of carrying substantial payloads for an extended period of time as compared to vertical takeoff and landing aircraft (VTOL).

The disadvantages of fixed-wing UAS platforms are that the logistics required for launch and recovery (L&R) can be very substantial (known as a large logistical "footprint"). Some may require runways to land and takeoff, whereas others may require catapults to reach flying speed for takeoff and then recover with a net or capture cable. Some small fixed-wing platforms such as AeroVironment's Raven™ are hand launched and recovered by stalling the aircraft over the intended landing spot or by deploying a parachute. In recent years there has also been a move toward small, very simple fixed-wing UASs utilizing simple polystyrene foam or 3-D printed parts. These aircraft often offer relatively high capability and durability and have proven very popular with many users for their ease of deployment.

3.2.2 Vertical Takeoff and Landing

As with all UASs, vertical takeoff and landing UAS platforms have numerous applications. A VTOL platform can be in the form of a helicopter, a fixed-wing aircraft that can hover, or even a tilt-rotor. Some examples of a VTOL UAS would be the Northrop Grumman MQ-8 Fire Scout™ or the Bell Eagle Eye™ tilt-rotor. Other notable examples include the Aerovel Flexrotor, the DJI Inspire, the Aeryon SkyRanger, and the Aibotix X6. These UAS platforms have the advantage of small L&R footprints. This means that most do not need runways or roads to takeoff or land. Most also do not require any type of equipment such as catapults or nets for the L&R. Unlike fixed-wing platforms, the VTOL UAS can monitor from a fixed position requiring only a small space to operate. An added advantage of VTOL aircraft is their inherent ability to observe objects in very close proximity such as for inspection or low-altitude mapping missions.

Smaller electric helicopters, radio-control size, have advantages of very rapid deployment times making them ideal for search and rescue, disaster relief, or crime fighting. Simple helicopter systems can be stored in a first responder's vehicle and launched within minutes. These small helicopters also offer the advantage of being somewhat covert when in operation at low altitudes. With no gasoline engine, the electric motor is quiet enough to allow it to operate at altitudes without creating a nuisance. The disadvantages of small electric helicopters are that battery technology to date has not enabled long endurance to be achieved beyond 30 to 60 minutes.

3.3 Command and Control Element

3.3.1 Autopilot

The concept behind UAS automation is the ability for an unmanned system to execute its mission following a set of preprogrammed instructions without operator intervention. A fully autonomous UAS is able to fly without operator intervention from takeoff to touchdown. The amount of autonomy in a UAS varies widely from none to full autonomy. On one end of the spectrum the aircraft is operated completely by remote control with constant operator involvement (an external pilot). The aircraft's flight characteristics are stabilized by its autopilot system; however if the pilot were to be removed from the controls, the aircraft would eventually crash. These concepts will be covered in greater detail in a subsequent chapter.

On the other end of the spectrum the vehicle's onboard autopilot controls everything from takeoff to landing, requiring no pilot intervention. The pilot-in-command can intervene in case of emergencies, overriding the autopilot if necessary to change the flight path or to avoid a hazard. The autopilots for these vehicles are used to guide the vehicle along a designated path via predetermined waypoints.

Many commercial autopilot systems have become available in recent years for small UASs (sUASs). These small autopilot systems can be integrated to existing radio-controlled (hobby) aircraft or onto custom-built sUAS platforms. Commercial autopilot systems (often referred to as COTS for commercial-off-the-shelf systems; COTS is a widely used acronym for many different technologies) for sUAS have become smaller and lighter in recent years. They offer many of the same operational advantages that large UAS autopilots offer and are far less expensive. For example, the Cloud Cap Technology's Piccolo series of autopilots (see Figure 3.2) offers multivehicle control, fully autonomous takeoff and landing, VTOL and fixed-wing support, and waypoint navigation. Increasingly, open-source autopilots are being used for many applications and the utility and functionality of this technology is quite good for many sUAS operations. A brief Internet search will yield numerous current and emerging examples of open-source programmable automated controllers for a variety of unmanned vehicles, not limited to air.

FIGURE 3.2
Piccolo™ SL autopilot unit. (Copyright Cloud Cap Technology, a UTC Aerospace Systems Company.)

Autopilot systems for UASs are programmed with redundant technology. As a safety feature of most UAS autopilots, the system can perform a "lost-link" procedure if communication becomes severed between the ground control station and the air vehicle. There are many different ways that these systems execute this procedure. Most of these procedures involve creating a lost-link profile where the mission flight profiles (altitudes, flight path, and speeds) are loaded into the memory of the system prior to aircraft launch. Once the aircraft is launched, the autopilot will fly the mission profile as long as it remains in radio contact with the ground control station. The mission or lost-link profile can be modified when necessary if connectivity remains during flight. If contact with the ground station is lost in flight, the autopilot will execute its preprogrammed lost-link profile.

Other examples of lost-link procedures include having the vehicle:

1. Proceed to a waypoint where signal strength is certain in order to reacquire connectivity.
2. Return to first waypoint and loiter or hover for a predetermined time in an attempt to reacquire the signal and then returning to landing waypoint to land if this is unsuccessful.
3. Remain on current heading for a predetermined amount of time. During this time, any secondary means of communication can be attempted with the aircraft.
4. Climb to reacquire link.

Most commercially available technology by default require the aircraft to return and land immediately if the link is not reestablished within a given timeframe.

3.3.2 Ground Control Station

A ground control station or GCS is a land- or sea-based control center that provides the facilities for human control of unmanned vehicles in the air or in space (Figure 3.3). GCSs vary in physical size and can be as small as a handheld transmitter (Figure 3.4) or as large

FIGURE 3.3
General Atomics MQ-1 Predator GCS. (Copyright General Atomics.)

(a)

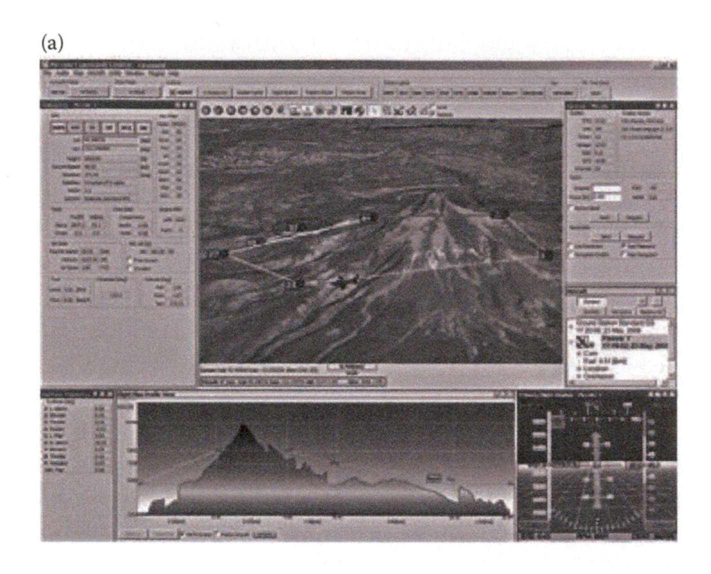

(b)

FIGURE 3.4
(a) Piccolo™ Command Center GCS. (Copyright Cloud Cap Technology, a UTC Aerospace Systems Company.) (b) AeroVironment handheld GCS. (Copyright AeroVironment.)

as a self-contained facility with multiple workstations. Larger military UASs require a GCS with multiple personnel to operate separate aircraft systems. One of the foremost goals for future UAS operation will be the capability for one crew to operate multiple aircraft from one GCS, however many challenges remain before this concept can be safely integrated into the NAS.

For larger military UASs, a GCS usually consists of at least a pilot station and a sensor station. The pilot station is for just that: the pilot-in-command who operates the aircraft and its systems. The sensor station is for the operation of the sensor payload and radio communications. There can be many more operations, depending on the complexity of the UAS, which may require more workstations. For smaller less complex UASs, these workstations may be combined requiring only one operator.

3.4 Communication Data Link

Data link is the term used to describe how the UAS command and control information is sent and received both to and from the GCS and autopilot. There are also often separate data links for some payload systems. With respect to radio frequency transmission, UAS operations can be divided into two categories: radio frequency line-of-sight (LOS) and beyond line-of-sight (BLOS). This is different from beyond-visual-line-of-sight or BVLOS, which merely refers to that distance beyond which the naked eye cannot detect an aircraft, which is a much shorter distance than radio frequency BLOS.

3.4.1 Line-of-Sight

Line-of-sight (LOS) operations refer to operating the UAS via direct radio waves. In the United States, civilian LOS operations are usually conducted on the 915 MHz, 2.45 GHz, or the 5.8 GHz radio frequencies. These frequencies are unlicensed industrial, scientific, and medical (ISM) frequencies that are governed by Part 18 of the Federal Communications Commission (FCC) regulations. Other frequencies such as 310–390 MHz, 405–425 MHz, and 1350–1390 MHz are discrete LOS frequencies requiring a license on which to operate. Depending on the strength of the transmitter and receiver, and the obstacles in between, these communications can travel several miles. Signal strength can also be improved utilizing a directional tracking antenna. The directional antenna uses the location of the UAS to continuously adjust the direction in which it is pointed in order to always direct its signal at the UAS. Some larger systems have directional receiving antennas onboard the aircraft thereby improving signal strength even further.

ISM frequency bands are widely used making them susceptible to frequency congestion, which can cause the UAS to lose communication with the ground station due to signal interference. Rapid frequency hopping has emerged as a technology that minimizes this problem. Frequency hopping is a basic signal modulation technique used to spread the signal across the frequency spectrum. It is this repeated switching of frequencies during radio transmission that minimizes the effectiveness of unauthorized interception or jamming. With this technology, the transmitter operates in synchronization with a receiver, which remains tuned to the same frequency as the transmitter. During frequency hopping a short burst of data is transmitted on a narrowband, then the transmitter tunes to another frequency and transmits again, a process that repeats. The *hopping* pattern can be from several *times* per *second* to several thousand *times* per *second*. The FCC has allowed frequency hopping on the 2.45 GHz unlicensed band.

3.4.2 Beyond Line-of-Sight

Beyond line-of-sight (BLOS) operations refer to operating the UAS via satellite communications or using a relay vehicle, usually another aircraft. Civilian operators have access to BLOS via the Iridium satellite system, which is owned and operated by Iridium LLC. Most sUASs do not have the need or ability to operate BLOS since their missions are conducted within line-of-sight range. Military BLOS operations are conducted via satellite on an encrypted Ku band in the 12–18 GHz range. Some UASs in the market operate almost continuously through Ku band. Their launch phase is usually conducted using LOS and then transferred to BLOS data link. They are often then transferred back to LOS for its recovery. One drawback of BLOS operations is that there can be several seconds of delay time once a command is sent to the aircraft, for it to respond to that command. This delay is caused by the many relays and systems it must travel through. With technological improvements over the past several years, it is possible to conduct launch and recovery of the aircraft through BLOS data link.

3.5 Payload

Outside of research and development, most UASs are aloft to accomplish a mission and the mission usually requires an onboard payload. The payload can be related to surveillance, weapons delivery, communications, aerial sensing, cargo or many other applications. UASs are often designed around the intended payload they will employ. As we have discussed, some UASs have multiple payloads. The size and weight of payloads are two of the largest considerations when designing a UAS. Most commercial application sUAS platforms require a relatively small payload, generally less than 5 lbs. Manufacturers of some sUAS have elected to accommodate interchangeable payloads that can be quickly removed and replaced.

In reference to the missions of surveillance and aerial sensing, sensor payloads come in many different forms for different missions. Examples of sensors can include electro-optical (EO) cameras, thermal infrared (IR) cameras, spectral sensors, synthetic aperture radars (SAR), or laser range finder/designators. Optical sensor packages (cameras) can be either installed by permanently mounting them to the UAS aircraft giving the sensor operator a fixed view only, or they can employ a mounted system called a gimbal or turret (Figure 3.5). A gimbal or turret mounting system gives the sensor a predetermined range of motion usually in three axes. The gimbal or turret receives input either through the autopilot system or through a separate receiver. Some gimbals are also equipped with vibration isolation, which reduces the amount of aircraft vibration that is transmitted to the camera thus requiring less electronic image stabilization to produce a clear image or video. Vibration isolation can be performed by either an elastic/rubber mounting or using an electronic gyrostabilization system.

3.5.1 Electro-Optical

Electro-optical cameras are so named because they use electronics to pivot, zoom, and focus the image. These cameras operate in the visible light spectrum. The imagery they yield can be in the form of full motion video, still pictures, or even blended still and video images. Most sUAS payload EO cameras use narrow to mid field of view (FOV) lenses. Larger UAS camera payloads can also be equipped with wide or ultrawide FOV (WFOV) sensors. An EO sensor can be used for many missions and combined with different types of sensors to create blended images. They are most frequently operated during daylight hours for optimal video quality.

3.5.2 Thermal Infrared

Thermal Infrared cameras operate in the infrared range of the electromagnetic spectrum (~700 nm–1 mm). IR, or sometimes called FLIR for forward-looking infrared, sensors form an image using IR or heat radiation. Two types of IR cameras used for UAS payloads are

(a)

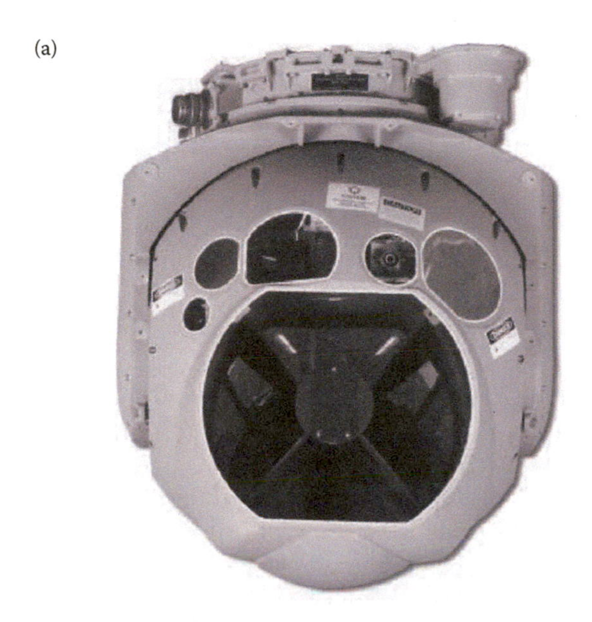

(b)

FIGURE 3.5
(a) Multi-spectral Targeting System-B (MTS-B) sensor pod. (Copyright Raytheon Co.) (b) General Atomics MQ-9 Reaper (formerly called "Predator B"). (Copyright General Atomics.)

cooled and non-cooled. Cooled cameras are often more expensive and heavier than non-cooled cameras. Modern cooled cameras are cooled by a cryo-cooler, which lowers the sensor temperature to cryogenic temperature (below 150°C). These systems can be manufactured to produce images in the mid-wave infrared (MWIR) band of the spectrum where the thermal contrast is high. These types of cameras can also be designed to work in the longwave infrared (LWIR) band. A cooled IR camera's detectors are typically located in a vacuum sealed case and require extra power to cool. In general, cooled cameras produce a higher quality image than uncooled cameras.

Non-cooled cameras use sensors that are at or just below ambient temperature and work through the change of resistance, voltage, or current created when heated by the infrared radiation it detects. Non-cooled sensors are designed to work in the LWIR band from 7 to 14 μm in wavelength, where terrestrial temperature targets emit most of their infrared energy.

This discussion of payloads merely scratches the surface; payload systems are discussed in greater depth in Chapter 4.

3.5.3 Spectral

One class of payloads has become particularly useful when it comes to missions that are related to plant growth. A myriad of multi- and hyper-spectral imaging payloads have, and are being developed, which can detect energy wavelengths that exist outside of normally visible light as detected by the naked eye (visible light—400–799 nanometers [nm], which is related to wavelength on the electromagnetic spectrum). The energy most often sought for analysis in plant-based applications is either in the blue/green/red band (450–690 nm) or the infrared band (IR, 700 nm–1 mm). The infrared band can be further divided into near infrared (NIR, 800 nm–2.5 μm), short-wave infrared (SWIR, .9–1.7 μm), medium-wavelength infrared (MWIR, 3–8 μm), long-range infrared (LWIR, 8–15 μm), and far-infrared (FIR, 15–1000 μm). There is also ultra-violet light (UV, 100–400 nm), which is sometimes useful in these types of applications but there are many other bands that can be studied. Scientists involved in agricultural and plant-based applications often study the amount of this energy that is either absorbed or reflected by plant vegetation and this information is then analyzed in order to make plant health/state determinations depending on the specific type information needed. One product these scientists use is the NDVI or normalized-difference vegetation index as denoted by the formula: NDVI = (NIR-RED)/(NIR + RED). Knowing the NDVI, scientists can make determinations with respect to the need for crop chemical application and overall plant health including drought/disease stress etc.

3.5.4 Laser

A laser range finder uses a laser beam to determine the distance to an object. A laser designator uses a laser beam to designate a target. The laser designator sends a series of invisible coded pulses that reflect back from the target and are detected by the receiver. There are, however, drawbacks to using a laser designator on an intended target. The laser may not be accurate if atmospheric conditions are not clear, such as rain, clouds, blowing dust, or smoke. The laser can also be absorbed by special paints or reflect incorrectly or not at all such as when aimed at glass.

3.6 Launch and Recovery

The launch and recovery element (LRE) of the UAS is often one of the most labor-intensive aspects of the UAS operation. Some UASs have very elaborate LRE procedures, whereas others have virtually none. Larger systems have procedures and dedicated personnel that prepare, launch, and recover the UAS. Runway lengths of up to 10,000 feet and support equipment such as ground tugs, fuel trucks, and ground power units or GPUs are needed for these large UASs. Small VTOL UASs tend to have the least complex procedures and

FIGURE 3.6
Soldier hand-launching a Raven RQ-11. (Copyright AeroVironment.)

equipment when it comes to LRE, most of which consists of only a suitable takeoff and landing area. Other UASs, such as the Raven manufactured by AeroVironment Inc., have very small LRE since they can be hand launched and recovered by an onboard parachute (Figure 3.6).

There are many ways to conduct launch and recovery operations for current UASs. Some of the most common involves using a catapult system to get the aircraft to flight speed in a very short distance. The ScanEagle™ manufactured by Insitu, a Boeing company, utilizes a catapult for takeoff and an arresting cable Insitu calls the SkyHook™ for its recovery (Figure 3.7). In this system, the vehicle is equipped with a hook on the end of its wing tips as well as employing a very precise dual global positioning system to fly into the suspended cable for recovery.

The Aerosonde Mark 4.7 manufactured by Aerosonde, a Textron Inc. company, has optional LRE equipment depending on the variant model. It can be launched using a car top launcher whereby a ground vehicle is used to enable the UAS to reach flying speed (Figure 3.8). A catapult system is also available for launching the Aerosonde. For the landing phase, it can "belly land" on grass or hard surfaces, or it can recover into a moving net.

3.7 Human Element

Although covered in much greater depth in a subsequent chapter, the most important element of the UAS is the human element. At this point the human element is required for the operation of the UAS. This element consists of a pilot, a sensor, and supporting ground crew. As previously mentioned, some of these positions can be combined into one depending on the complexity of the system. In the future, the human element will likely get smaller as technological capability increases. As with commercial airliners of the past, automation will require less human interaction. The UAS pilot in command is responsible for the safe operation of the aircraft. This element is described in greater detail in Chapter 11.

(a)

(b)

FIGURE 3.7
(a) Insitu ScanEagle™ catapult launcher. (b) Insitu ScanEagle™ arresting cable the "SkyHook™." (Copyright Insitu Inc. a Boeing Company.)

FIGURE 3.8
AAI Aerosonde vehicle-top launch cradle. (Copyright AAI Aerosonde a Textron Systems Company.)

DISCUSSION QUESTIONS

3.1 Why have some elements of the military gravitated away from the term "unmanned?"

3.2 What are some advantages and disadvantages of VTOL UAS?

3.3 What are some advantages and disadvantages of the fixed-wing UAS?

3.4 Research commonly used open-source UAS autopilot technology and discuss their relative strengths and weaknesses.

3.5 Reference the difference discussed in this chapter between the terms BLOS and BVLOS. In what scenario might the term BVLOS be important? BLOS?

4

UAS Sensing: Theory and Practice

Gabriel B. Ladd

CONTENTS

4.1 Introduction

In this age of computer-centric systems what we do is driven by the concept of data. These pieces of information that represent something else have come to have huge importance in both our endeavors and our daily lives. It is for the creation of these pieces of data that we fly unmanned aircraft systems (UAS). Almost exclusively we send our robotic aircraft into the skies to collect data.

While bouncing and jouncing around in the skies, these vehicles collect all sorts of information about a wide range of topics using an equally wide range of sensors. The data collected can be broken down into two general categories: *in situ* and remote sensing. These two methods are used to collect vast amounts of information about our world and what is going on in it in the hope that it will help improve our understanding of our environment.

4.1.1 *In Situ* Sensing

The phrase *in situ* comes from the Latin for "in place." Merriam Webster's Dictionary defines it as "in the natural or original position or place" (Staff 2015b). In the case of *in situ* sensing with UAS, the aircraft is brought into the location where the measurements are to be made. There are two general purposes for doing this; either to force the air vehicle to respond to some sort of environmental parameters or to measure some specific environmental attribute at a location of interest. If the former, this can run the gambit from testing the air vehicle's response to control input as measured by the onboard systems or forcing the air vehicle through adverse environmental conditions such as storms to measure the vehicle's response, assuming it survives the encounter. In either of these cases, the point is to measure the air vehicle's performance and its responses to the environment. The other major form of *in situ* sensing is to measure attributes about the environment or medium through which the air vehicle is moving. These attributes range from temperature to gas composition and type. In many applications, both forms of *in situ* measurements are used in concert to achieve the overall goal of the flight. For example, aircraft dynamics may be used in conjunction with the gas analysis to allow researchers to account for the vehicle's motion in their analysis.

4.1.2 Remote Sensing

Remote sensing is the process of measuring an object or phenomenon of interest from a distance. This is done by detecting and measuring the effects of said object or phenomenon on the physical universe, usually in the form of emitted or reflected particles and/or waves. Remote sensing is by definition independent of platform and application. In general, there are three broad categories of remote sensing: terrestrial, airborne, and space based. What is covered in this chapter is not all of remote sensing, but an overview of the subset "airborne remote sensing from UAS."

Airborne remote sensing as a discipline has a large range of sensor options from large multi arrays, to single sensor pickup systems. In general, there are four classes of remote-sensing sensors: framing, pushbroom, scanner, and receiver systems. These sensors work on most of the electromagnetic (EM) spectrum. As a rule of thumb it is easiest to think of the UAS subset of airborne remote sensing as basically the same as manned aircraft remote sensing, but just constrained to smaller platforms. Not all UASs are small, and in the cases where the airframe is large enough, there is no need to distinguish between the sensors used in UAS remote sensing and the manned aircraft remote sensing. On the other hand, one can argue that if all of the hobby and toy "drones" in the world now are counted, the majority of UASs are small indeed. The array of choices for different UAS platforms can be found in Figure 4.1.

The emitted or reflected particles and/or waves most often associated with airborne remote sensing is sunlight, or what is called on the EM spectrum Visible (VIS) light. For the purposes of this discussion we are working within a paradigm where its behavior is well known, so we will leave the debate on the nature of light as a particle or wave to the physicists.

To illustrate how light works in remote sensing, consider what happens when you take a photograph outside on a sunny day. Perhaps you use your phone or a point and

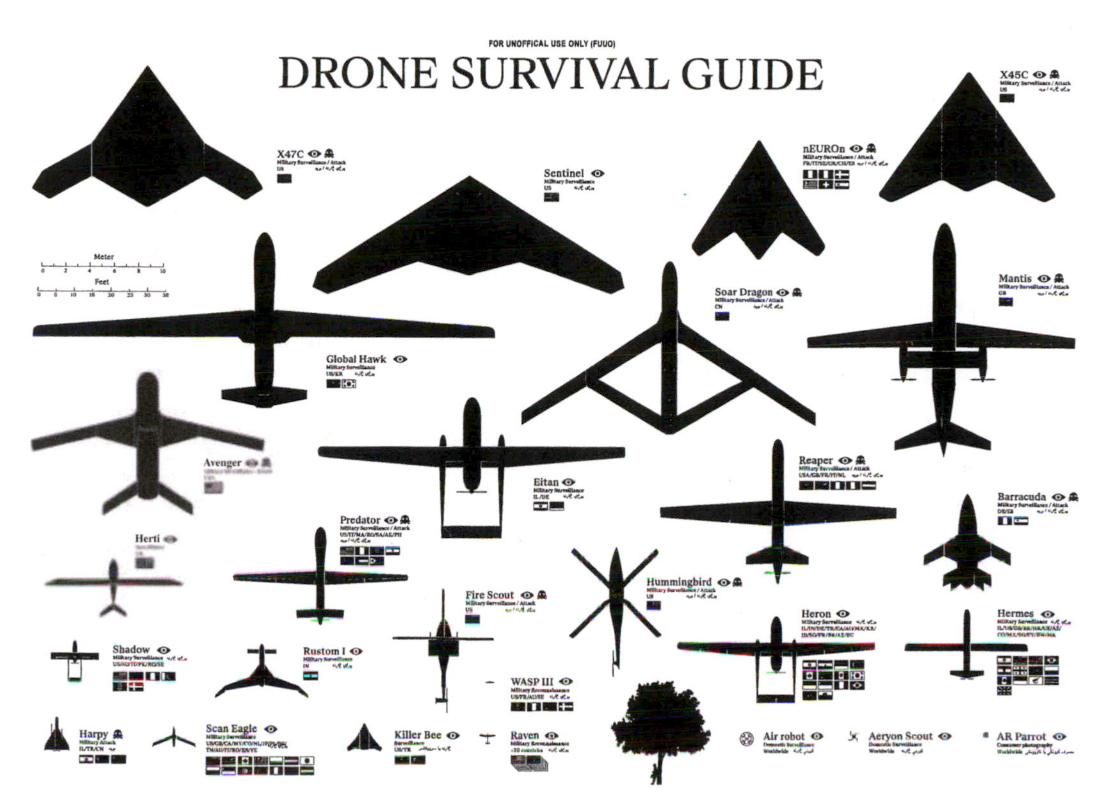

FIGURE 4.1
UAS (drone) silhouette chart for scale comparison. (Ruben Pater. http://www.dronesurvivalguide.org/DSG.pdf)

click camera or even an "old fashioned" film camera; in any case, you are capturing and recording sunlight. In the case of electrooptical (EO), or digital cameras, as the consumer market calls them, the light comes through the lens and strikes a sensor. In the case of film camera, light strikes a thin piece of sensitive material commonly referred to as film. For most applications, this material is a thin piece of plastic with a light-sensitive coating (Wolf and Dewitt 2000). To return to our example, by pointing the camera at the individual you are in fact turning the light receiving lens toward the individual and collecting the sunlight they are reflecting. The fact that you are capturing and measuring the reflected light for the purpose of understanding what is going on with the distant object that is reflecting it is the critical difference between remote sensing and the *in situ* measurement methods.

In the example of a digital picture of a person taken outside, the light entering the camera is measured by an array of light-sensitive sensors. These sensors react to the light, and that reaction is measured and stored as the fundamental building block of the image. The digital image you are used to seeing on your phone or computer screen is constructed by the image sensor based upon the array's reaction. Not enough reaction and the image is too dark; too much reaction and the image is washed out. The sensor reaction can be moderated and adjusted in many ways, which can be broken down into two general categories: physical and software. The physical modification methods are what most people would think of as traditional photographic methods such as aperture, focus, and ISO. The software modifications include what would have been done in the darkroom when developing film, but in digital image manipulation can go much further than that. It should be noted here that the UAS remote-sensing community has almost exclusively embraced digital imaging and thus film-based remote sensing will not be explored further here.

4.1.3 Platform Considerations

The combination of high efficacy electric motors, powerful mini computers, and battery technology that has allowed for the consumer explosion of UASs enabled a new age of airframe experimentation. The proliferation of unique designs leveraging on these technologies has brought about what is arguably the fourth great age in airframe experimentation. In this period, where the lines between traditional airframes are becoming blurred, it is important to understand the basic pros and cons of different configurations when evaluating an aircraft as a sensor platform.

For the purposes of this discussion, UASs are going to be divided into three general size categories: small, medium, and large, by their maximum takeoff weight. The small UASs are those under 55 pounds. The medium UAS will be considered those between 55 pounds and 400 pounds. Large UASs are then considered anything over 400 pounds.

The aircraft as a sensor platform means that the user has to play to the strengths and weaknesses of the air vehicle configuration. Almost everyone is used to doing this when choosing a form of ground transportation. We don't expect a sports car to be able to do the things a dump truck can, nor would we expect someone to use a racing motorcycle for a multi-day or multi-week winter trek. Likewise, aircraft should be used in accord with their strengths and weaknesses. For the purposes of this discussion, the field of aircraft will be broken down into five morphological groupings: tethered, buoyant, fixed wing, rotor craft, and hybrid.

The tethered aircraft group is made up of such aircraft as kites and tethered balloons. These systems, while aircraft in the strictest sense of the word in that they can move up and down in the air column, are not usually considered candidates for the remote and autopilot systems that are the defining features of a UAS. They are mentioned here because they can be very good sensing platforms, and from a technical perspective there is fundamentally nothing preventing the use of remote pilot and autopilot systems to provide a degree of autonomy to these systems.

The buoyant group of aircraft uses buoyancy in some form to help counteract gravity and achieve flight about the earth's surface. Traditionally, we think of this group being represented by descendants of the form pioneered by Count Ferdinand von Zeppelin in the late 1800s. Modern day blimps, while often similar in exterior shape, lack the rigid metal super structure of the Zeppelin's design and rely instead upon a combination of envelope material strength and lightweight bracing material. Both the Zeppelin and the blimp are characterized by a central gas vessel, which provides the lift, and motors arranged around the vessel to propel it, with yaw and pitch controlled by tail fins. Because of this configuration, these aircraft are often referred to as "lighter-than-air" vehicles. Because of the nature of gas and buoyancy, these vehicles must contain a large volume of lighter-than-air gas to lift both the airframe and the payload. This also tends to limit how much propulsion power they can carry. Which in turn means they are not capable of controlled operations in high wind speed. How much wind and how it affects the aircraft depends very much on the individual aircraft design. But as a rule of thumb these vehicles tend to be lightly built and move at low speeds.

Unlike fully buoyant aircraft, the small category of semi-buoyant aircraft are not fully dependent on buoyancy to produce lift but instead use airflow to generate lift and buoyancy to offset a large portion of their weight. Which means they use smaller and often more streamlined gas vessels instead of the large gas vessel so central to blimps and Zeppelins. This configuration makes these aircraft less susceptible to high winds, but they are still lightly built and highly dependent upon buoyancy to generate lift, which tends to make them slower moving and more sensitive to winds than traditional aircraft. As a group the buoyant aircraft generally operate in lower winds for takeoff and landing and require large takeoff and landing areas. But conversely, they have the potential for very long duration operations at a wide range of altitudes.

Fixed-wing aircraft are vehicles where the lift is created by induced airflow over wings that are affixed to fuselage of the aircraft. The air flows over the wings as the vehicle is propelled by an engine. This need for horizontal motion to generate lift means that space is required for takeoff and landing. In the case of a UAS, the engine usually drives a propeller, or less often a jet turbine accelerates the air, causing the craft to move. There are some less common examples of rocket-based propulsion for fixed-wing aircraft, but these are generally very expensive and used to test airframes in extreme environments.

The characteristics and operating environments of fixed wing aircraft are driven largely by their wing and engine designs. Aspect Ratio (AR), the ratio of the wing span to its area, dictates how a fixed-wing aircraft performs (Anderson 2001). As a rule, longer and thinner wings generate more efficient lift, but as they get longer they are harder to roll. The shorter and thicker wings are less efficient but roll very easily. This is why jet airliners, designed in part for fuel economy, have long thin wings, whereas fighter jets, primarily concerned with maneuverability, have short thick wings. The long thin-winged aircraft have more endurance and thus more range, and they are also more aerodynamically stable and thus tend to make better camera platforms. The shorter, thicker winged aircraft are more maneuverable and less stable, and they also tend to have less endurance and shorter range. With the advent of autopilot technology, control of unstable systems has become much easier. The ability of the autopilot to keep the shorter wing aircraft in level flight with much more precision than can a human remote pilot has enabled them to serve as camera platforms. This stability, coupled with better maneuverability, allows some fixed-wing aircraft to takeoff and land in smaller and more challenging spaces, consequently making them much more viable as aerial imaging platforms.

Rotorcraft, or rotor wing aircraft, use spinning wings as their primary source of lift. These take the form of propellers, similar to the ones used to generate motion in the fixed-wing and buoyant aircraft, the chief difference being these are designed to lift the aircraft's entire weight and control it in flight. Because the spinning blades are used to generate lift, the aircraft is capable of vertical takeoff and landing (VTOL). There are two general types of rotorcraft, single rotor and multi-rotor. The single rotor, what we grew up calling helicopters, use a single main lift rotor to both lift and control the vehicle. The single rotor lift system is marked by complicated mechanical linkages that allow for the adjustment of blade pitch in both the cyclic and collective senses, which allows the vehicle to pitch and roll while varying the overall amount of lift generated. To counteract the single lift rotor's torque, they also utilize a much smaller tail rotor, the speed of which is coupled with the lift rotor speed and enables the aircraft to yaw on command.

Multi-rotor systems predominantly use electric motors. In general, they use between three and eight individual lifting motors directly connected to propellers. These systems are mechanically much simpler than single rotor systems because the lift, pitch, roll, and yaw are controlled by varying motor revolutions per minute (RPMs) in concert. In either single or multi-rotor configuration, rotorcraft have the primary advantage of being capable of VTOL. They, unlike buoyant systems, are heavier than air, so depending on the configuration they are more tolerant to surrounding wind speed than buoyant and fixed-wing aircraft. As sensor platforms rotorcraft have benefits in their ability to operate out of small, tight spaces, their major drawbacks being their vibrational environment, their rotor wash, shorter endurance times, and correspondingly shorter ranges compared to the other aircraft types. Despite the potential downsides, the ease of use, mechanical simplicity, and ability to operate out of tight spaces has made the electric multi-rotor one of the most popular consumer grade UAS platforms.

In this age of computer-aided flight, airframe experimentation on hybrid aircraft design, particularly in the small and midsized UAS realm, has exploded. This has produced a range of very interesting hybrid aircraft designs, which are intended to combine the advantages of the VTOL takeoff with the long range and endurance of fixed-wing aircraft. The initial attempts at this configuration resulted in the "tilt rotor" aircraft design. These aircraft utilized complex and heavy mechanisms to tilt the engines and in some cases the entire fixed wing into a position to allow for vertical flight. More recent designs effectively use two separate propulsion systems: A multi-rotor system consisting of four or more lift rotors, and a separate propulsion system for forward flight. In either case these hybrid systems attempt to strike a balance between the VTOL operational flexibility and the endurance of a fixed-wing aircraft. As a result, they tend to have a reduced payload capacity and endurance of comparable AR fixed wing, while being more susceptible to the winds on takeoff and landing than a comparable multi-rotor aircraft.

All the aircraft types discussed earlier in brief have different advantages and disadvantages. These pros and cons must be weighed and examined when considering an aircraft as a sensor platform for a specific data collection purpose. As one goes about designing a UAS system, the task quickly becomes one of pairing the best sensor to collect the data with an airframe to give your sensor the wings it needs to answer the overarching question for which you are flying.

4.2 Remote Sensing

4.2.1 Overview

The systems used for remote sensing can be divided into two general categories: active and passive. Active sensors are those that emit some form of electromagnetic (EM) radiation that is directed at a target and then measures the reflected signal. Passive sensors do not emit EM radiation, instead measuring what is emitted by other sources after it is reflected or as the target emits it. In the technique of remote sensing, especially the subset of UAS remote sensing, the external EM source is almost always the sun.

Active sensors provide an EM source. Analogously, what most people have experienced goes back to their digital camera. Most digital cameras have a built-in flash, which emits light just prior to the picture being taken. This EM spectrum radiation leaves the camera and goes out into the space in front of your camera and bounces into anything in front of you and all over the scene you are trying to capture before being bounced back at the camera. The goal of this is to create enough light reflecting back at the camera that what light your target reflects will create an image of the target. Active sensors do this with very selective sections of the EM spectrum.

The electromagnetic spectrum is broken down into bands. EM Bands are segments of energy grouped together by a common property and defined by a specific range of

TABLE 4.1

EM Spectrum and What Is Measured

Spectrum	Wavelength (nm)	Radiation Source	Property Measured
Visible (VIS)	400–700	Sun	Surface reflection
Near infrared (NIR)	700–1100	Sun	Surface reflection
Short-wave infrared (SWIR)	1100–2500	Sun	Surface reflection
Mid-wave infrared (MWIR)	2500–5000	Sun, thermal irradiance	Surface reflection, temperature emitted
Long-wave infrared (LWIR) or thermal infrared (TIR)	8000–14,000	Thermal irradiance	Temperature emitted

Note: For more details see Table 1.3 in Schowengerdt (2007).

vibrational wavelengths. In remote sensing, the bands most commonly dealt with are outlined in Table 4.1. As can be seen from Table 4.1, different spectrums have different sources and different surface properties of interest that are measured by the wavelength. These surface properties can seem a bit abstract until you think about what it means from a fundamental physics perspective. Think about any digital picture you have taken and you will see the surface properties manifest in the color of the object. As we discussed before, this color as we think of it day to day is actually the color of white light reflected by the object. The color white would be the reflectance of all colors and the color black would be absorbance of all colors. Hence the reason black objects get hotter in the sun more quickly than white objects.

The imaging sensor responds to specific wavelength of incoming EM radiation. This response can be measured and calibrated. This calibration process is called radiometric calibration. Once a sensor is calibrated for a particular subset of the EM spectrum, the sensor response can be tied directly to a very small, if not individual, component wavelength of the EM spectrum being reflected by the object. The ability to determine the spectral reflectance of an object can be very powerful, but depending on your application it may not be required. That is to say, knowing the range of the sensor's EM sensitivity may suffice to answer the question without pinning down specific wavelengths.

One of the most powerful aspects of remote sensing is spatial continuity, which is the ability to know what is going on at an individual location and every other location around for a substantial distance. The spatially contiguous data enables all sorts of analyses not possible without the perspective provided by continuous coverage. Going back to the example of a picture you have taken, it is possible, even without the camera being radiometrically calibrated, to tell that the person in the picture is wearing a red knit winter hat and a windbreaker. The nature of both of these items you can guess at from the texture of the object and the fact that it is being worn as an article of clothing. This is done by observing the sublet variation, or lack thereof, in the material texture, along with the overall context derived from having the spatial continuity of the image. Thus the image can provide a lot of information, in some cases more than a thousand words of information. To be fair, there is an underlying assumption here, one of resolution. The assumption is that the person in the picture is close enough to have the sensor record the level of detail needed to express the texture. If the image were of too low a resolution or the person is too far away you might be able to tell they are wearing a red cap of some sort, but you may not have enough textual information to say more.

Another important advantage to spatial data is the complete or nearly complete sampling. This is illustrated very clearly in crop monitoring. When an agronomist goes into a corn field to pick corn ears for yield prediction he or she will look at between 10 and 100 individual ears from a small number of individual locations across a field. This is a very small fraction of the ears of corn growing in the field. When you use remote sensing to produce a map, data will be collected on almost every plant in the field, if not all of them. The down side to this is that by collecting literally every plant in the area of interest (AOI), you have to develop ways of excluding things you are not interested in measuring. It also means the analyst has to use a different form of analysis. Rather than doing analysis from a position of having measured a small subset and making assumptions of population normalcy, one has to approach the analysis with an understanding of what is representative and what isn't, which leads very quickly into spatial statistical analysis.

4.2.2 Sensor Types

There are two general types of remote-sensing sensors: Those that build images and those that do not. Imaging sensors are the predominant sensor type in the remote-sensing world, especially the UAS subset, but the spot sensors can be very useful tools.

4.2.2.1 Spot Sensors

Spot sensors are sensors that measure single locations and do not attempt to create imagery. They forgo the spatial continuity, instead opting for simplicity. These systems can still provide useful data and are often used in conjunction with imaging sensors to help inform the imagery. NASA's Microspectrometer Instrument Suite system flew over the Arctic Ocean in 2009 on the NASA Sierra UAS as a prototype system that used spot sensors to inform visible light spectrum (VIS) imagery. The Microspectrometer Instrument Suite was designed to identify surface melt water on sea ice using point measurement from spectrometers. The 2009 mission was to validate the technique by comparing the spectrometer data to the VIS spectrum images collected (Ladd et al. 2011).

4.2.2.2 Imaging Sensors

There are several basic methods of generating images from remotely sensed data: Line scanners, pushbrooms, and array sensors. Line scanners are sensors that move a single sensing element, or a very small number of sensor elements, back and forth rapidly to build up a picture of what is below the aircraft. What this means is that the width of the image is dictated by how far the element can be pivoted and how quickly the sensor can collect a reading from the moving platform. In many ways, this is effectively a spot sensor being swung back and forth across the aircraft path fast enough to generate a picture. Pushbroom sensors are sensors that have a fixed array of sensors, which are the width of the final image product across the aircraft's path, but only one element tall. This means they are continually sampled and the height of each pixel along the aircraft's direction of travel is dictated by the aircraft's speed and how long it takes for the sensor array to collect a sample. Array sensors are made up of a two dimensional array of sensors. These sensors are used to collect an image simultaneously. Array sensors are very similar, if not identical to what you are used to using in consumer digital cameras, in that when they are triggered the sensor's value is recorded for all individual sensors in the array at once. Thus the data that makes up an individual image is collected all in one exposure. There usually is some lag time associated with collecting and saving all the data (Schowengerdt 2007).

4.2.3 Common Sensors

The most common remote-sensing sensor for UAS is what is conventionally thought of as a camera. A camera is defined by Merriam and Webster's Dictionary as "a device that consists of a lightproof chamber with an aperture fitted with a lens and a shutter through which the image of an object is projected onto a surface for recording" (Staff 2015a). The current trend in modern digital cameras is to move away from mechanical shutters. This is in part to simplify the system. Because of this trend, anything with a lens and a sensor in a radiation proof chamber is considered a "camera." To that end there are four general types of cameras in widespread use in UAS remote-sensing community: Visible Spectrum, Near Infrared, Infrared, and Hyperspectral.

4.2.3.1 Visible Spectrum Cameras and Near-Infrared Cameras

The visible light spectrum (VIS) and the near-infrared (NIR) cameras operate on essentially identical principles, the only difference being that the NIR camera sensor is sensitive to the NIR wavelengths of the EM spectrum (see Table 4.1). A VIS image generally consists of three different primary colors combined to make one full color image. In digital parlance, this is referred to as a three band image, or an RGB after the three primary colors red, green, blue from which the full color image is created.

Because of the proximity of the NIR portion of the spectrum to the VIS spectrum, film used during World War II was able to capture green, red, and NIR reflectance. To address the inability of the human eye to see NIR, the image was printed with the NIR spectrum shifted to another color visible to the human eye. Each band had to take a primary color so NIR became red, red became green, and green became blue (Wolf and Dewitt 2000). This false color standard became known as color infrared (CIR) and is widely used in vegetation applications (Schowengerdt 2007).

4.2.3.2 Long-Wave Infrared Cameras

Infrared sensors beyond 1800 nm of the EM spectrum range can no longer use the same sensor material as visible light cameras. Instead they have to rely on different materials with different properties. Fundamentally, the concept of the influx of radiation into the 2-D array of sensor elements is the same. The difference stems from the fundamental physical and material properties involved in reacting to the specific wavelengths of the EM spectrum. There are several methods for creating thermographic images. Some involve cryogenic cooling of the camera sensors, while others can be done at room temperature. Not surprisingly, the room temperature focal plan array systems tend to be more popular for their simplicity and lower cost. The room temperature sensors work by having individual elements in the 2-D array respond to the incoming radiation. In the case of long-wave infrared (LWIR) cameras, which are also known as thermographic cameras, the individual elements heat up and that temperature change causes an electrical signal to be generated (Rogalski 2002). The array dictates the portion and precision of the camera's response to the thermal section of the EM spectrum. The thermographic image is effectively the wide spectrum monochromatic image representing the thermal emission intensity of what the camera is seeing. This is manifested in either black and white imagery, or false color imagery where color is assigned to by pixel value.

These sensors effectively respond to the intensity of the heat striking the sensors array, which means a very hot object can wash out much of the image. A similar photographic analogy would be the appearance of reflective material like that on traffic cones or motorcycle apparel worn by someone when you are taking a picture of them at night with a flash. In both cases a bright spot can wash out much if not all of the surrounding area.

4.2.3.3 Hyperspectral Imagers

Hyperspectral imagers utilize several data collection techniques, which are very similar to the cameras we have already discussed. The primary difference is that they collect co-registered pixels in many spectrums. Instead of one three channel color image (R, G, B), you end up with a many channel image. According to the *ASPRS Manual of Photogrammetry*, Fifth Edition, to be a truly hyperspectral imager a system must have 100 or more spectral channels per image. This results in a highly multiband image, also called a hyper cube that can be thought of as many pictures all taken at the same time. This achievement is what differentiates a hyperspectral imaging system from a normal 1–3 band VIS spectrum camera or even a many band multispectral camera (McGlone et al. 2004). The requirements for spectral resolution tend to drive the weight and price of the hyperspectral imagers higher than cameras with less spectral bands, even the multispectral ones.

The power of hyperspectral imaging systems lies in their spectral, not spatial resolution. A hyperspectral data cube can be analyzed in the 2-D image-based analyses that all cameras are capable of creating. Color images or VI can be created depending on what spectra are recorded. It can also be analyzed for per pixel spectral signature. This spectral signature is unique to materials and thus can be used to identify materials. Depending upon the spectral resolution, hyperspectral image analysis can differentiate between real and fake vegetation or go so far as to identify specific materials and/or soil types (Exelis 2015, Landgrebe 2002). The problem with hyperspectral cameras on a UAS is the required size and vibrational sensitivity. Because of the need to address so many different spectral bands, the camera has to have additional components, which add to size and weight. To reduce the unit size, manufacturers either keep the sensor array size small or they use a pushbroom design (Resonon 2015, Rikola 2015). Either option is a viable solution but they affect how the system is operated to collect data and how much it costs to collect that data. The small resolution of an array affects how much spatial resolution each image has, which in turn means that the camera system and thus the aircraft needs to be flown at very different altitudes depending on desired resolution. A pushbroom hyperspectral camera can potentially have the same resolution issues coupled with the additional need of very high precision position and orientation system (POS) data to align the individual single pixel tall image slices. Past work in this area has revealed that autopilot orientation and positioning data was not good enough for seamless image creation (Hruska et al. 2012). The need for high precision POS data means a substantially increased cost, not to mention additional weight, power consumption, and cost, which on the medium and large aircraft can be justified. It will be some time until that level of complex hardware is operated on the entry-level quadcopter.

4.2.3.4 LiDAR

Light detection and ranging (LiDAR) has changed the face of remote sensing since its wide acceptance in the 1990s. LiDAR uses a laser beam and a receiver to measure the distance to the ground. A simple way to think of LiDAR is to think about what happens when you take a laser pointer into a dark room and rapidly shine it back and forth in front of you. You can quickly build up an idea of where all the furniture is and how far away it is. This is a simple but effective way to visualize the conceptual underpinnings of LiDAR. The actual sensor systems are able to discriminate between multiple return reflections of a single light pulse. There are several ways of doing this. The early systems used a line scanner method with optically coupled laser and receiver (Wehr and Lohr 1999). Later systems used separate scanners and emitters, and leading products now are trending toward a single instant point source emission variant called flash LiDAR (Gelbart et al. 2002). Also on the threshold of commercial availability are systems called Geiger Mode LiDAR that are able to measure the return of a single photon. Both the Geiger mode LiDAR and the flash LiDAR are close to being operational and are poised to change how LiDAR data is collected and processed.

As one might expect, all these systems require a very accurate measurement of time, sensor-pointing angles, and sensor-spatial location. LiDAR sensor ground resolution is, as a general rule, dependent on attitude and point density. As a rule the higher a system is flown, the further apart the individual pulses are spaced on the ground, which reduces the data resolution. One major difference with LiDAR and the passive camera systems is that there is an upper limit on operational range. As a rule of thumb the lower power the LiDAR system, the shorter the range. Power is a limiting factor for UAS systems. The midsized and large UAS systems running combustion or turbine engines can produce their own power, so they are less limited than the small systems using electric motors, which are dependent upon their battery packs.

The manned aircraft systems can operate at altitudes of 10,000 m above ground level. The smaller LiDAR units used for the midsized UAV systems over the past few years have a useful range of a few hundred meters or less. For example, the RIEGL VUX-1UAV specifies an operational ceiling of 1000 ft or about 333 m (RIEGL 2015). The smallest systems are able to be deployed on small multi-rotor systems (Wallace et al. 2012) and have a still shorter range, usually on the order of 100 m (Velodyne 2015).

4.2.3.5 Synthetic Aperture Radar

Synthetic aperture radar (SAR) systems are radar systems that map by using a band of the EM spectrum in the range of 1 m–1 mm (Schowengerdt 2007). They send out a radar pulse and use the reflected radar signal and the receiver's motion to construct a 2-D or 3-D image of the returning echoes. The resolution is dependent on the wavelength of the EM spectrum being used, but in airborne SAR the ground resolution is usually less than a meter and more often in the centimeter range (McGlone et al. 2004).

A simple way to think of SAR is to imagine standing at the bottom of a narrow canyon yelling loudly and hearing the yell echoed back several times, each time a little softer and perhaps a little more distorted. Every time you hear your voice repeated back to you it is the sound bouncing off some surface in the distance. Now imagine you were to yell over and over again while walking across the canyon and each yell was different. Eventually, if your ears were good enough and your voice held out you would probably get a pretty good idea of where the first and loudest echo had been returned. SAR works in a very similar way. The major difference being that SAR works on every echo and determines, from the myriad of echoes it picks up, where in the distance each object is located relative to the moving receiver. This is very computationally intensive and requires a high degree of precision in timing and positional orientation information.

One of the major advantages to SAR is that it utilizes long wavelengths that can pass through most cloud layers, smoke, and dust in the atmosphere. It is also able to achieve high spatial resolution and pick up on surface features such as waves and other textures.

FIGURE 4.2
NANO SAR Circa 2008. (From Barnard Microsystems. 2015. Synthetic Aperture RADAR. Barnard Microsystems Accessed 3/30/15. http://www.barnardmicrosystems.com/UAV/features/synthetic_aperture_radar.html.)

This capability enables some very interesting scientific and commercial applications (Jackson and Apel 2004).

There have been relatively few commercial applications of small SAR technology on a UAS since the widely publicized flights in 2008 (see Figure 4.2) on the Boeing/Insitu ScanEagle™ system (Barnard Microsystems 2015). General Atomics has created a SAR system that operates on predator-sized aircraft, but that system, which is specified as having a range of 80 km and weighing between 80 and 120 pounds, depending on which variant is used, is only viable on large UAS systems (General Atomics 2015).

The area of SAR remote sensing from a UAS is unique in UAS remote sensing at the moment because technology exists, but there is no published work about it being used in the academic or commercial fields. Assuming that the systems are being employed by the world's militaries as advertised by their vendors, it is unclear why the research community and/or commercial sectors are not using the technology more. Still, it remains a powerful data collection technique, and as UAS utilization increases, one can only assume the use of such systems will likewise grow.

4.2.4 Common Applications

There are many applications to which UAS remotely sensed data can be put. The usability of the data is highly dependent first on the airframe and second on how the sensors are configured. It is generally true that anything that is built is designed for a primary purpose, and if it is used for some other purpose it doesn't work as well as was intended. This seems to be doubly true for small UASs, which are more often than not sensors with wings as much as they are aircraft. So in designing and utilizing a UAS sensor system, it is imperative to keep in mind what it is meant to do, why it is flying. This will make operations and data processing much easier.

What follows is an overview of UAS applications, largely derived from academic research and the writer's personal experience. It is not all-inclusive, nor is it meant to be. As a trend, what we are seeing from UAS remote sensing is an expansion of the discipline into new niches, rather than a radical change in what the technology is capable of. This means, as a rule, if you think an operation can be done conventionally, odds are there is a way to collect your remote sensing data with a UAS. The only remaining question is, given current technology and regulations, is it economical?

4.2.4.1 Live Video

What most people envision when they think of drones is live video streamed somewhere so a government agent or military commander can see what is going on in real time. These applications are usually accomplished using a gimbaled system, which allows an operator to point a camera in virtually any direction. This allows for continuous observation of an object or target on the ground. The limitations of these systems tend to come into play when the observer attempts to look at the target through objects on the ground like trees and/or parts of the aircraft. Trying to look through the aircraft is less than informative and is usually avoidable with a well-thought-out integration of the system into the airframe and good flight planning.

Many of the UAS gimbal systems on the market are designed for military applications (see Figures 4.3 through 4.5). This means they are designed with intelligence, surveillance, and reconnaissance (ISR) purposes in mind. To that end they are built with a range of sensors, which tend to be less useful outside of ISR applications.

There are increasing numbers of commercial gimbals being built for the small UAS market, specifically targeted at the entry-level hobby market through the cinematography sector. These gimbals, unlike the military application gimbals, tend to be designed to mount a wide range of commercial VIS spectrum cameras. They are often not capable of as large

FIGURE 4.3
TASE 500 Gimbal with VIS, MWIR, and LWIR cameras. (From CloudCap. 2015. TASE 500 Gimbal. http://www.cloudcaptech.com/gimbal_tase500.shtm.)

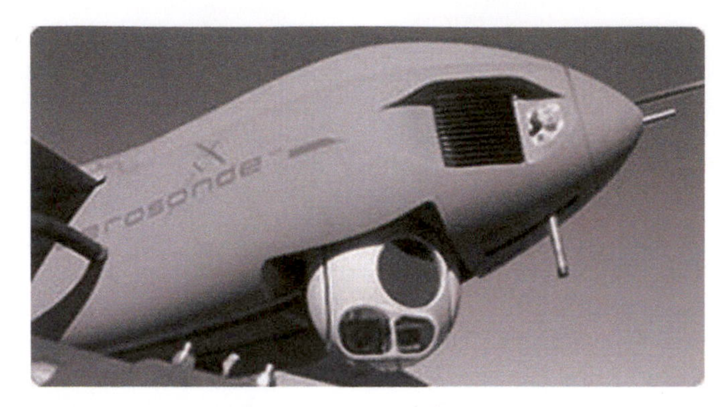

FIGURE 4.4
TASE 400 Gimbal with VIS, LWIR, and slot for laser range finder on an Aerosonde. (From Keller, J. 2011. Stabilized camera gimbal for day and night surveillance for UAVs introduced by Goodrich Cloud Cap Technology. *Military & Aerospace Electronics*, October, 2011.)

FIGURE 4.5
Small UAS Gimble. (From Staff. 2012. New Sensor Payload for Raven UAV Unveiled. *Unmanned Ground, Aerial, Sea and Space Systems.*)

a range of motion and tend to rely more on positioning the aircraft to enable the shot. They also tend to be lighter weight and more fragile than the military grade systems. An example of a GoPro gimbal on an entry-level quadcopter can be found in Figure 4.6.

4.2.4.2 Emergency Response

Emergency response is an interesting type of operation for UASs in the commercial sector. It is an application with many of the same ISR requirements pioneered by the military, so the technology is available; the real question remaining to be fully tested at this point is commercial viability. Can the UAS be integrated into the existing structures safely and in a way that expands the existing systems and methods economically? That question has not been answered conclusively and probably will never be fully answered until the systems are fully integrated and it becomes a moot point. Several attempts have been made to integrate sensing system beyond the initial ISR applications (Choi and Lee 2011), but few have made into the commercial sector.

The area of emergency response in general has not been as heavily investigated by academia as of the writing of this chapter, despite it being called out in the AUVSI forecast

FIGURE 4.6
Entry level quadcopter with gimbal. (From Staff. 2015c. IRIS+. 3D Robotics Accessed 4/22/15. https://store.3drobotics.com/products/IRIS.)

(Jenkins and Vasigh 2013). The lack of scholarly articles may be due to time lag in funding and publishing, though new articles and tradeshow conversations clearly show that emergency response is a topic of interest for manufacturers and the commercial sector (Bodeen 2014, iRevolution 2014).

Some companies have been conducting search and rescue operations, which can be thought of as a specific subset of emergency response, for years now. These companies operate visible and multispectral camera systems to search for missing persons. How these systems will fit into the search and rescue system of the future is not yet fully clear. The one thing that does seem to be clear about UAS search and rescue is that it can make a difference, which is being noticed (Mortimer 2014).

4.2.4.3 Background Imagery

One of the most interesting areas for the UAS in the commercial sector is the creation of imagery for mapping. The demand for high-resolution current imagery data for maps has increased radically with the advent of geographic information systems (GIS), especially with their introduction into daily life through applications like Bing Maps™ and Google Earth™. The use of small format camera systems employed in small UASs and for area mapping is something that has been enabled by the remote sensing and computer vision communities through the development of automated systems that combine the many images required to cover an area into one seamless orthorectified mosaic. These mosaicked images are a critical element in making imagery of all spectrums from small format cameras on UAS viable data products for GIS analysis (Liu et al. 2006, Grenzdörffer et al. 2008).

Mosaicked images are created by the combination of many frames into a single large image. This image can then be used for multiple applications. The single-most common use is as an updated background map to inform the user about what is going on in the target area. This allows the GIS operator/analyst to quickly and easily determine what has changed since the last set of images.

4.2.4.4 3-D Point Cloud/Modeling

The creation and exploitation of the 3-D point clouds started as a LiDAR source, but with the advent of the photogrammetric point cloud as part of the data-processing work flow for so many UAS data-processing programs, it has become a desired data product of many orthomosaic work flows. The 3-D point cloud can be used for a wide range of applications regardless of its source. The 3-D point cloud's power lies in the ability to work in the three dimensions. It may appear strangely self-evident to the uninitiated in the GIS fields, but the vast majority of the GIS data is planometric, so the advent of 3-D data in abundance is changing how most GIS professionals set up and plan analysis.

The UAS not only allows for 3-D, but very high-resolution 3-D data. This enables detailed analysis of small areas; for example, being able to accurately calculate the volume of small buildings allows for analyses of heated volumes and estimation of what upgrades will do to such facilities. It is now possible to calculate with some level of precision the volume of small dirt piles at construction sites, and to estimate how much of the reserves remain on site or how much will be spent in dirt removal costs. The high-resolution nature of the UAS imagery also allows for much more detailed creation of surface models from imagery (Hugenholtz et al. 2013). Another advantage of detailed 3-D models derived from UAS data is the ability to generate more detailed calculations of biomass volumes (Zarco-Tejada et al. 2014).

4.2.4.5 Vegetation Health Measurements

4.2.4.5.1 Vegetation Index an Overview

Biomass, crop health, crop yield, and other plant characteristics can be estimated from remotely sensed data using a Vegetation Index (VI) (Jang et al. 2006). VIs are created by using different bands of the EM spectrum captured by remote-sensing systems (Stark et al. 2000). The most common bands used in vegetation analysis applications are the NIR, the red, the green, and the blue (Pinter et al. 2003). The most common VI used to measure plant stress is the Normalized Difference Vegetative Index (NDVI). The NDVI is created by manipulating the red and the NIR bands as shown in Equation 4.1. The second most common VI used in plant health analysis is the Green Normalized Vegetation Index (GNDVI). This is made by manipulating the green and the NIR bands as described in Equation 4.2 (Vygodskaya et al. 1989).

$$\mathrm{NDVI} = \frac{(\mathrm{NIR} - \mathrm{Red})}{(\mathrm{NIR} + \mathrm{Red})} \tag{4.1}$$

Equation 4.1: NDVI equation

$$\mathrm{GNDVI} = \frac{(\mathrm{NIR} - \mathrm{Green})}{(\mathrm{NIR} + \mathrm{Green})} \tag{4.2}$$

Equation 4.2: GNDVI equation

Once created, the VI maps can be analyzed and used to develop a variety of data products. The NDVI and GNDVI are health analogs. As such they can be used to inform many types of analyses. In agriculture, they can be used to inform prescription maps for fertilizing and seeding crops, as well as crop yield estimates so farmers can maximize their profits (Ladd 2007, Prasad et al. 2006). In forestry they are used for, among other things, change detection and biomass estimation (Sader and Winne 1992, Spruce et al. 2011, Vogelmann 1990).

4.2.4.5.2 UAS in Agriculture-Vegetation Indices

In a 2013 report the Association for Unmanned Vehicles International (AUVSI) estimated that agriculture was likely to be the biggest growth sector for UASs through 2025

(Jenkins and Vasigh 2013). The concept of Precision Agriculture (PA) is one in which previously unused or unavailable data is brought into play to better understand and manage crops in increasing detail.

UASs have the potential to change agriculture operations. The concept of farmers being able to fly, collect, and analyze their own data has been around for a long time (Nagchaudhuri et al. 2005, 2006). The technology for flying UASs and collecting data has arrived. The current challenge for famers is that the skillset required to operate and maintain an aircraft system is not one most farmers already possess. This is forcing the UAS industry to develop highly intuitive, highly automated, and easy-to-maintain aircraft systems. The UAS industry is still working on some aspects of aircraft operations and collection of data for larger areas or highly complicated terrain. As Zhang and Kovacs pointed out in their 2012 survey article, the largest stumbling block at the present time is the processing of the large amount of individual frames captured into a meaningful analysis (Zhang and Kovacs 2012). The companies who create the image-processing software are working hard to reduce the technical skill level required to run this software, but there still exists a large gap between the technology-savvy research scientists, or their students, and the average farmer. As a consequence, there is a proliferation of features that create Vegetation Indices (VI) from the imagery collected automatically. These VIs are often mistaken for a final result which in agriculture or any plant assessment is usually not the case.

The concept of VI health assessment has been used in many different crop types. The general premise is to capture CIR imagery or multiband imagery, which can be utilized to create the VI. Based upon the spectral bands, either NDVI or GNDVI can be used as health predictors; both are good general indicators, though they perform differently depending on the crop (Moges et al. 2005). Once the VI has been calculated, it can be used for many applications, including vegetation identification and general health mapping, as shown in Figures 4.7 through 4.9.

One fact that is often overlooked by those not in the agriculture community is most crop analysis requires more than just aerial imagery and its derivative products such as an

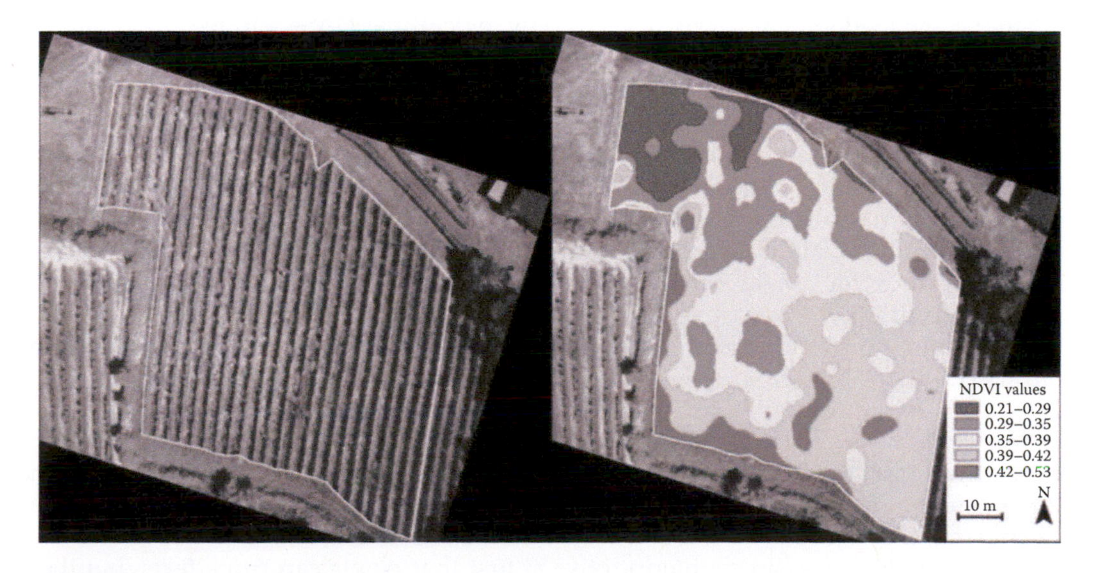

FIGURE 4.7
Multispectral images of the Monteboro vineyard (left) false color image at 5 cm of spatial. (From Primicerio, J. et al. 2012. *Precision Agriculture* 13 (4):517–523.)

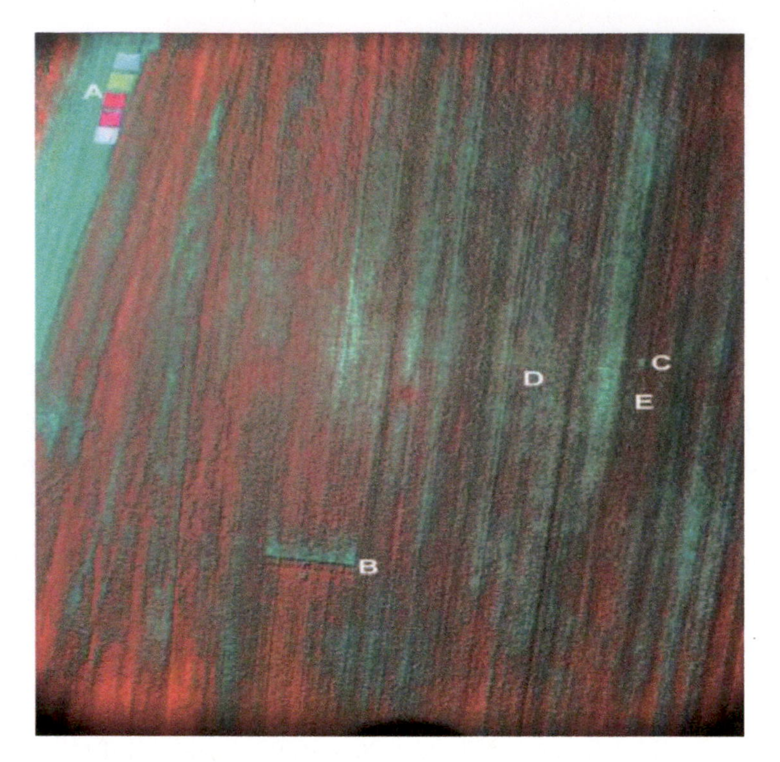

FIGURE 4.8
CIR image made from bands NIR, green, blue. (From Hunt, E.R. et al. 2010. *Remote Sensing* 2 (1):290–305.)

NDVI. The data product created from a VI is by definition relative, giving relative health of the crop area at the time collected. Crop modeling and prediction takes input from other *in situ* sources such as soil type, soil chemistry, soil moisture, and daily temperature, to name a few (Sadler et al. 2000). It seems only logical that UASs will feed data into the PA systems to help inform the predictive models, and this has been a matter of academic study for some time (Ladd et al. 2006, Nagchaudhuri et al. 2006), but the large-scale adoption probably will not happen until the major agricultural equipment manufacturers start making turnkey systems that incorporate data from multiple sources. Even then odds are there will be a human in the loop for some time to adjust and modify the model recommendations as many agronomists do now.

It is clear from the AUVSI report (Jenkins and Vasigh 2013) and the proliferation of companies targeting this market sector and the wide range of applications across almost all vegetation types (Lehmann et al. 2015, Pinter et al. 2003, Sullivan 2005), that this industry will rapidly move toward the automation that enables widespread use beyond the university.

4.2.4.5.3 UAS in Agriculture-Thermal Mapping

A growing subset of UASs for PA is the use of LWIR cameras to identify plant stress. The fundamental biological concept being exploited by thermal imaging is that if the plants are stressed, they will not uptake water from the soil as much as healthy plants and thus will be hotter. It may seem counterintuitive at first that plants can run a fever, but there has been an increasing amount of study of this phenomenon with improvement of small commercial thermal cameras (Chaerle et al. 2004). This has logically moved into the realm

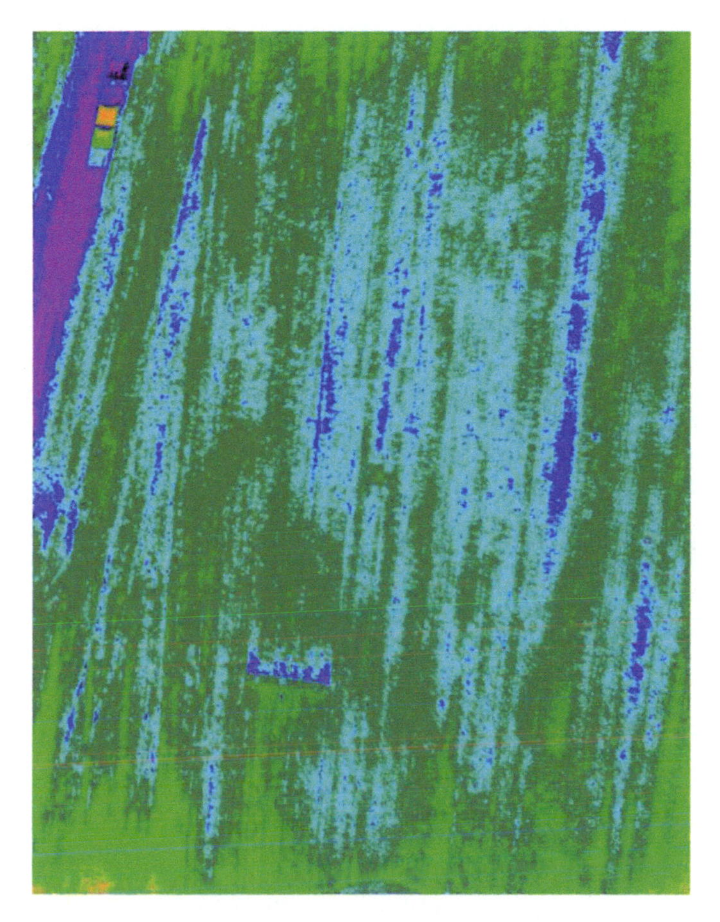

FIGURE 4.9
GNDVI created from NIR, green, blue image. (From Hunt, E.R. et al. 2010. *Remote Sensing* 2 (1):290–305.)

of small UASs, where the ability to operate the aircraft at low altitudes for increasing time periods has enabled the creation and evaluation of systems to make thermal maps of crops (Bendig et al. 2012).

4.2.4.5.4 Broader Vegetation Management

Beyond precision agriculture, the use of UASs to create VI for vegetation management and monitoring has been increasing over the years. Today, commercial agriculture operations are seeing applications of UAS technology in a wide range of areas across disciplines and subject matters. A good example is the use of UASs in forestry, particularly the management of deciduous forests where research is actively ongoing with regard to pest infestation (Grenzdörffer et al. 2008, Lehmann et al. 2015).

Not all vegetation management utilizes VIs. Andrea Laliberte and her team in New Mexico had very good luck using object detection algorithms to identify individual plants in the arid rangeland of New Mexico (Laliberte et al. 2007). While this approach probably wouldn't work with high-density vegetation, there are some logical extensions into research where object differentiation can be useful, such as orchard management.

Spectral signature analysis using hyperspectral image analysis is an established discipline within the remote-sensing field. Its use can be seen with both satellite and airborne platforms. Hyperspectral imagers are complicated and difficult to integrate into aircraft and still maintain precision. The majority of them are "pushbroom" sensors, which in turn means they lack the overlap between image slices that has enabled the multi-frame mosaic boom in UAS imagery discussed in Section 2.4.4. Consequently, they require much more sophisticated POS systems, which currently cannot be integrated on the smaller foam, or multi-rotor systems. There is potential in the largest size multi-rotor systems. Work with these systems is now being done using larger fixed-wing aircraft with 4–5 m wingspans. Both the teams at the Idaho National Lab (Mitchell et al. 2012) and the Instituto de Agricultura in Córdoba, Spain (Calderón et al. 2013) have had good results using different hyperspectral sensors, leveraging the spectral discrimination of hyperspectral sensors with image classification and object classification, respectively. While both studies were confined to relatively small research areas, there is clearly potential for much larger areas and wider applications.

In the United States, one of the research areas where LiDAR sensors carried on UAS have been utilized is in forestry. The ability to accurately measure canopy height and properties is of great importance to forestry management. Canopy data such as height and crown diameter are analogs or enablers for several other measurements or indicators, such as diameter at breast height, biomass, age, and health (Dubayah and Drake 2000, Popescu et al. 2003). The use of LiDAR on commercial UAS system has been slower to adopt, particularly for the small UAS systems where the higher-end POS systems cannot be carried. Larger UAS platforms are capable of carrying bigger systems such as the Army's Buckeye Project (Fischer et al. 2008). Smaller UASs are showing good results in forestry applications such as calculating tree height and crown size (Wallace et al. 2012). An advantage to the small UA systems is their ability to fly close to the target to achieve high point density. The point density affects the sensor's ability to discriminate target height and details, as illustrated in Figure 4.10.

One interesting thing to note is that the smaller airframes such as the multi-rotor systems used by Wallace and his team at the University of Tasmania are range limited and

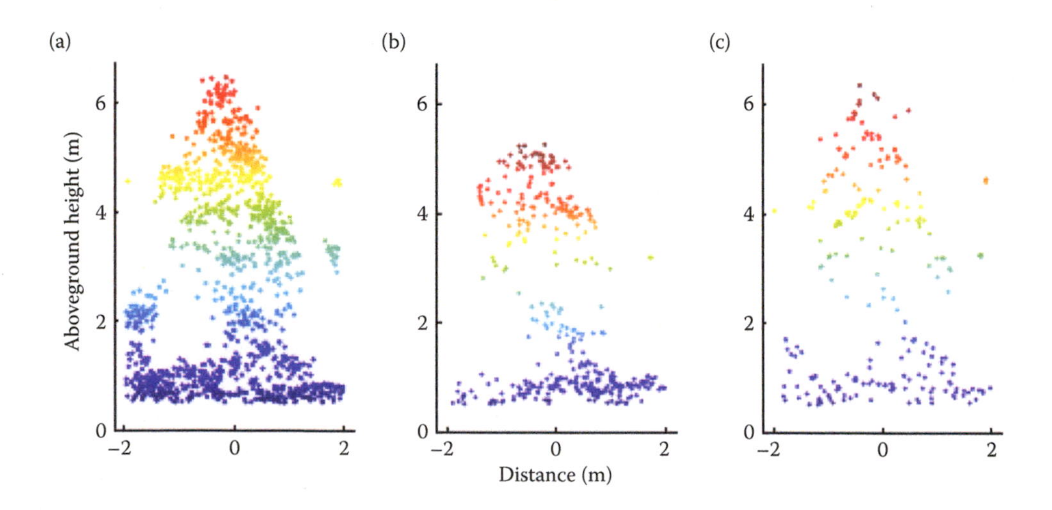

FIGURE 4.10
Decreasing point density affects tree height measurements. Point density changes from 70 points per meter to 10 points per meter in c. (From Wallace, L.O., A. Lucieer, C.S. Watson. 2012. Assessing the feasibility of UAV-BASED LiDAR for high resolution forest change detection. In *International Archives of the Photogrammetry, Remote Sensing and Spatial Information Sciences*. Melbourne, Australia.)

need to be within about 50 meters of the ground (Wallace et al. 2014) due to return signal degradation, while the larger systems such as the BuckEye or VUX-1 are not so limited. If history is any guide, the smaller systems will improve with time.

4.2.4.5.5 Airframes for Vegetation Applications

The literature describing the earlier work in UASs for precision agriculture and vegetation mapping reveals that the work was done with fixed wing platforms and has since progressed to multi-rotor systems. This is attributable to the recent proliferation of commercially available multi-rotor systems with their autopilot derived stability and camera-pointing accuracy. The downside of the multi-rotor system is short endurance. It is foreseeable that small area applications, for certain farm sizes and crop types, along with R&D efforts, will continue to utilize the multi-rotor system. Large area monitoring will most likely require the use of fixed-wing platforms, which are either on the large end of the small category, or small end of the medium-sized UAS category. The endurance provided by such platforms will help reduce operating costs.

4.3 Conclusions

UAS sensing as a subset of airborne remote sensing is still in its adolescence. UAS remote sensing is an unusual discipline because of the range of choices of operational airframes is so large (see Figure 4.1). On the large end of the spectrum the technology used in manned aircraft can virtually be directly integrated into unmanned airframes, while on the smallest end, systems can carry only the smallest of sensors. What is more, the large airframes can carry a payload nearly to the edge of space, while the smallest scale quadcopters fly out of range after a few hundred yards.

Since the turn of the century we have seen an explosion of digital technology that has been the foundation for the growth of the UAS market as a whole and of the remote-sensing component in particular. The advent of the compact high-resolution digital camera has enabled small UAS remote sensing. What we are seeing now is a confluence of capabilities in sensors, software, and airframes. The large airframes can cover the spectrum of passive and active sensors, whereas the midsized airframes have only been able to fly the complete spectrum in the past few years. The small airframes can fly almost all the passive sensors and active sensors are just starting to come online. The one capability exception in the category of small UAS sensors is the availability of small hyperspectral imager systems (more than 100 bands). These imagers can be found on the largest of the small systems or the smallest of the medium systems, depending on which size classification is used, but they have not been operated on the multi-rotor systems yet, due to their size and need for very good POS to integrate the pushbroom data. One can only imagine how quickly they would be adopted if an array sensor were to be developed that could utilize the structure from motion techniques critical to the small UAS visible camera orthomosaic processes.

The field of UAS remote sensing is in its adolescence and still exploring its possibilities. As fast as new sensors become available they are being used for new applications and, as often as not, unimagined or designed for applications. What is still being worked out by the industry is how to make these systems commercially viable. We know they work for the military and research scientists. They are fun and useful platforms for hobbyists. The realtors and the movie industry have shown great interest in them. What has not been definitively proven is how and where they fit into the economics of commercial remote sensing and data collection. We know they have opened new niches and are competing with existing technologies in others, but they simply have not been in commercial use long enough for any final evaluation. What can be said is that the next 10–20 years will be a very interesting time to be involved with the UAS in general and remote sensing and applications in particular.

Glossary

In situ—Latin, meaning in place, this is used to describe the way measurements are made. Anything measured in the medium or location of the phenomena is call ***in situ***.

Light detection and ranging (LiDAR)—"An instrument capable of measuring distance and direction to an object by emitting timed pulses of light in a measured direction and converting to equivalent distance the measured interval of time between when a pulse is was emitted and when the echo was received (ASCE, ACSM, and ASPRS 1994)."

Orthophotomap—"A map made by assembling a number of orthophotographs into a single, composite picture. Also called an orthophotomosaic (ASCE, ACSM, and ASPRS 1994)."

Orthorectified—Imagery that has been processed to remove distortions due to terrain or objects on the surface being mapped. This applies to orthophotograph, an orthophotomap, or an orthomosaic.

Planometric—"A map that displays only the x,y locations of features and represents only horizontal distances (ESRI 2015)."

REMOTE SENSING QUESTIONS

1. Does the following definition apply to LiDAR, SAR, both, or neither? Please explain your reasoning.
   "An instrument capable of measuring distance and direction to an object by emitting timed pulses of electromagnetic energy in a measured direction and converting to equivalent distance the measured interval of time between when a pulse was emitted and when the echo was received."
2. Unmanned or manned, the aircraft has proven its viability as a platform for remote sensing time and time again. Given the myriad issues one faces when attempting to utilize unmanned aircraft in non-segregated airspace, discuss reasons why a research team would choose an unmanned aircraft platform over its manned counterpart?
3. When an agronomist evaluates images of a particular section of land (one square mile), noting the reflectance of a particular wavelength of EM can yield a measure of emergence of a desired crop. Explain how just using the percent presence of reflected EM might lead to errors in an accurate percentage of emergence of a desired crop.
4. Near Infrared and Infrared EM are both outside the visible band of the EM spectrum. Though they sound similar in name, they are very different with respect to

the nature of the data they collect. Discuss how they are both similar and different while differentiating between the useful natures of the collected data of each.

5. Emissivity is the measure of the amount of long-wave radiation given off by a particular material at a specific temperature. How might an object's emissivity make it more difficult to see even if it is hot?
6. In order for radar systems to distinguish between two objects in close proximity to each other, the wavelength of the transmitted signal needs to be relatively short. SAR, however, can use very long wavelengths and get remarkable "photo-like" images. In simple terms, try to explain how this capability of SAR might be possible.
7. Still imagery and video—both are common forms of data collection used on small unmanned aircraft. Compare and contrast both collection methodologies and describe situations where each might be more appropriate than the other.
8. A gimbal camera mount usually allows a camera to be moved in the vertical direction while stabilizers maintain the camera level with the ground. Explain why it might not be necessary to provide a pan function on a gimbal mount when affixed to a quadrotor.
9. Prior to the use of unmanned aircraft to gather aerial imagery, satellite imagery was used by agronomists to help determine plant health. What might some of the benefits be in using unmanned aircraft systems to gather this imagery versus satellites?
10. Explain the surface-level differences between the "multi-frame mosaic" systems and the "pushbroom" systems concentrating which type of system is most likely to be found on small UAS and why.

Bibliography

Anderson, J.D. 2001. *Fundamentals of Aerodynamics*. Third Edition. Boston: McGraw-Hill.

ASCE, ACSM, and ASPRS. 1994. "Orthophotomap" page 370 and "lidar" page 299, In *Glossary of the Mapping Sciences*. Edited by Stephen R. DeLoach. Bethesda, MD 20814: American Society for Photogrammetry and Remote Sensing; American Congress on Surveying and Mapping; American Society of Civil Engineers.

Barnard Microsystems. 2015. Synthetic Aperture RADAR. Barnard Microsystems Accessed 3/30/15. http://www.barnardmicrosystems.com/UAV/features/synthetic_aperture_radar.html.

Bendig, J., A. Bolten, and G. Bareth. 2012. Introducing a low-cost mini-UAV for thermal- and multispectral-imaging. *The International Archives of the Photogrammetry, Remote Sensing and Spatial Information Sciences* 39:345–349.

Bodeen, C. 2014. Volunteers step up in China's response to quake. Associated Press Accessed 4/15/15. https://news.yahoo.com/volunteers-step-chinas-response-quake-125026145.html.

Calderón, R., J.A. Navas-Cortés, C. Lucena, and P.J. Zarco-Tejada. 2013. High-resolution airborne hyperspectral and thermal imagery for early detection of *Verticillium* wilt of olive using fluorescence, temperature and narrow-band spectral indices. *Remote Sensing of Environment* 139:231–245.

Chaerle, L., Dik Hagenbeek, Erik De Bruyne, R. Valcke, and D. Van Der Straeten. 2004. Thermal and chlorophyll-fluorescence imaging distinguish plant-pathogen interactions at an early stage. *Plant and Cell Physiology* 45 (7):887–896.

Choi, K. and I. Lee. 2011. A UAV-based close-range rapid aerial monitoring system for emergency responses. *The International Archives of the Photogrammetry, Remote Sensing and Spatial Information Sciences* 38:247–252.

CloudCap. 2015. TASE 500 Gimbal. http://www.cloudcaptech.com/gimbal_tase500.shtm.

Dubayah, R.O. and J.B. Drake. 2000. Lidar remote sensing for forestry. *Journal of Forestry* 98 (6):44–46.

ESRI. 2015. Planimetric. ESRI Accessed 4/22/15. http://support.esri.com/en/knowledgebase/GISDictionary/term/planimetric%20map.

Exelis. 2015. Basic Hyperspectral Analysis Tutorial. Accessed 4/7/15. http://www.exelisvis.com/docs/HyperspectralAnalysisTutorial.html.

Fischer, R.L, B.G. Kennedy, M. Jones, J. Walker, D. Muresan, G. Baxter, M. Flood, B. Follmer, X. Sun, and W. Chen. 2008. Development, integration, testing, and evaluation of the US Army Buckeye System to the NAVAIR Arrow UAV. SPIE Defense and Security Symposium. Orlando World Center Marriott Resort and Convention Center, Orlando, FL.

Gelbart, A., B.C. Redman, R.S. Light, C.A. Schwartzlow, and A.J. Griffis. 2002. Flash LiDAR based on multiple-slit streak tube imaging LiDAR VII. *Laser Radar Technology and Applications* 4723, pp. 9–18.

General Atomics. 2015. *Lynx Multi-Mode Radar.* General Atomics Aeronautical Systems, Inc. http://www.ga-asi.com/Websites/gaasi/images/products/sensor_systems/pdf/LynxSAR021915.pdf

Grenzdörffer, G.J., A. Engel, and B. Teichert. 2008. The photogrammetric potential of low-cost UAVs in forestry and agriculture. *The International Archives of the Photogrammetry, Remote Sensing and Spatial Information Sciences* 31 (B3):1207–1214.

Griffiths, S. 2013. Drone Survival Guide. *Daily Mail*, 12/24/2013. http://www.dailymail.co.uk/sciencetech/article-2528902/21st-century-bird-watching-Drone-guide-lets-sky-gazers-spot-flying-military-robots-silhouettes.html.

Hruska, R., J. Mitchell, M. Anderson, and N.F. Glenn. 2012. Radiometric and geometric analysis of hyperspectral imagery acquired from an unmanned aerial vehicle. *Remote Sensing* 4 (12):2736–2752. doi: 10.3390/rs4092736.

Hugenholtz, C.H., K. Whitehead, O.W. Brown, T.E. Barchyn, B.J. Moorman, A. LeClair, K. Riddell, and T. Hamilton. 2013. Geomorphological mapping with a small unmanned aircraft system (sUAS): Feature detection and accuracy assessment of a photogrammetrically-derived digital terrain model. *Geomorphology* 194:16–24.

Hunt, E.R., W.D. Hively, S.J. Fujikawa, D. S. Linden, Craig ST Daughtry, and G.W. McCarty. 2010. Acquisition of NIR-green-blue digital photographs from unmanned aircraft for crop monitoring. *Remote Sensing* 2 (1):290–305.

iRevolution. 2014. Humanitarian UAV Missions During Balkan Floods. irevolution.net Accessed 4/15/15. http://irevolution.net/2014/07/07/humanitarian-uav-missions-during-balkan-floods/.

Jackson, C.R. and J.R. Apel. 2004. Principles of synthetic aperture radar, Chapter 1. In *Synthetic Aperture Radar Marine User's Manual.* edited by U.S. Department of Commerce. Washington DC: National Oceanic and Atmospheric Administration.

Jang, G-S., K.A. Sudduth, S.Y. Hong, N.R. Kitchen, and H.L. Palm. 2006. Relating hyperspectral image bands and vegetation indices to corn and soybean yield. *Korean Journal of Remote Sensing* 22 (3):153–197.

Jenkins, D. and B. Vasigh. 2013. *The Economic Impact of Unmanned Aircraft Systems Integration in the United States*: Association for Unmanned Vehicle Systems International (AUVSI).

Keller, J. 2011. Stabilized camera gimbal for day and night surveillance for UAVs introduced by Goodrich Cloud Cap Technology. *Military & Aerospace Electronics*, October, 2011.

Ladd, G.B., A. Nagchaudhuri, T.J. Earl, M. Mitra, and G.L. Bland. 2006. Rectification, georeferencing, and mosaicking of images acquired with remotely operated aerial platforms. *Proceedings of the American Society for Photogrammetry and Remote Sensing Annual Conference ASPRS*. Reno, Nevada.

Ladd, G.B. 2007. Assessing Soil pH Impacts on Corn Yield through Aerial Imaging and Ground Measurements. Master of Science Research, Marine Estuarine Environmental Sciences Graduate Program, University Of Maryland, Eastern Shore Campus (24835668).

Ladd, G., G. Bland, M. Fladeland, T. Miles, J. Yungel, and M. Linkswiler. 2011. Microspectrometer Instrument Suite Results from NASA SIERRA Flights in July 2009. Milwaukee 2011 ASPRS Annual Conference, Milwaukee.

Laliberte, A.S., A. Rango, J.E. Herrick, Ed L. Fredrickson, and L. Burkett. 2007. An object-based image analysis approach for determining fractional cover of senescent and green vegetation with digital plot photography. *Journal of Arid Environments* 69 (1):1–14.

Landgrebe, D. 2002. Hyperspectral image data analysis as a high dimensional signal processing problem. *IEEE Signal Processing Magazine* 19 (1):17–28.

Lehmann, J.R.K., F. Nieberding, T. Prinz, and C. Knoth. 2015. Analysis of unmanned aerial system-based CIR images in forestry—A new perspective to monitor pest infestation levels. *Forests* 6 (3):594–612.

Liu, L., I. Stamos, G. Yu, G. Wolberg, and S. Zokai. 2006. Multiview geometry for texture mapping 2d images onto 3d range data. Computer Vision and Pattern Recognition, 2006 IEEE Computer Society Conference on.

McGlone, J.C., E.M. Mikhail, J.S. Bethel, R. Mullen, and American Society for Photogrammetry and Remote Sensing. 2004. *Manual of Photogrammetry*. 5th ed. Bethesda, Md.: American Society for Photogrammetry and Remote Sensing.

Mitchell, J.J., N.F. Glenn, M.O. Anderson, R.C. Hruska, A. Halford, C. Baun, and N. Nydegger. 2012. Unmanned Aerial Vehicle (UAV) Hyperspectral Remote Sensing for Dryland Vegetation Monitoring. edited by US Department of Energy: Idaho National Labratory.

Moges, S.M., W.R. Raun, R.W. Mullen, K.W. Freeman, G.V. Johnson, and J.B. Solie. 2005. Evaluation of green, red, and near infrared bands for predicting winter wheat biomass, nitrogen uptake, and final grain yield. *Journal of Plant Nutrition* 27 (8):1431–1441.

Mortimer, G. 2014. sUAS News has 9840 followers on Twitter Gene Robinson of RP Search Services Wins First Annual Spectra Award at SUSB Expo *sUAS News*, 5/12/14. http://www.suasnews.com/2014/05/29116/gene-robinson-of-rp-search-services-wins-first-annual-spectra-award-at-susb-expo/.

Nagchaudhuri, A., M. Mitra, C. Brooks, T.J. Earl, G. Ladd, and G. Bland. 2005. Initiating Environmentally Conscious Precision Agriculture at UMES. *Proceedings of 2005 Annual Conference of American Society of Engineering Education*, Oregon Convention Center, Portland, Oregon.

Nagchaudhuri, A., M. Mitra, C. Brooks, T.J. Earl, G. Ladd, and G. Bland. 2006. Integration of Mechatronics, Geospatial Information Technology, and, Remote Sensing in Agriculture and Environmental Stewardship. 2006 ASME International Mechanical Engineering Congress and Exposition, Chicago, IL., USA, November 5–10.

Pinter, P.J., J.L. Hatfield, J.S. Schepers, E.M. Barnes, M.S. Moran, C.S. Daughtry, and D.R. Upchurch. 2003. Remote sensing for crop management. *Photogrammetric Engineering & Remote Sensing* 69 (6):647–664.

Popescu, S.C., R.H. Wynne, and R.F. Nelson. 2003. Measuring individual tree crown diameter with lidar and assessing its influence on estimating forest volume and biomass. *Canadian Journal of Remote Sensing* 29 (5):564–577.

Prasad, A.K., L. Chai, R.P. Singh, and M. Kafatos. 2006. Crop yield estimation model for Iowa using remote sensing and surface parameters. *International Journal of Applied Earth Observation and Geoinformation* 8 (1):26–33.

Primicerio, J., S.F. Di Gennaro, E. Fiorillo, L. Genesio, E. Lugato, A. Matese, and F.P. Vaccari. 2012. A flexible unmanned aerial vehicle for precision agriculture. *Precision Agriculture* 13 (4):517–523.

Resonon. 2015. Hyperspectral Imaging Cameras. http://www.resonon.com/data-sheets/Resonon-HyperspectralCameras.Datasheet.pdf.

RIEGL. 2015. RIEGL VUX-1UAV Specifications. http://www.riegl.com/products/uasuav-scanning/riegl-vux-1uav/.

Rikola. 2015. Snapshot Hypersepctral Camera is Airborne. http://www.rikola.fi/Hyperspectral_camera.pdf.

Rogalski, A. 2002. Infrared detectors: An overview. *Infrared Physics & Technology* 43:187–210.

Sader, S.A. and J.C. Winne. 1992. RGB-NDVI colour composites for visualizing forest change dynamics. *International Journal of Remote Sensing* 13 (16):3055–3067.

Sadler, E.J., B.K. Gerwig, D.E. Evans, W.J. Busscher, and P.J. Bauer. 2000. Site-specific modeling of corn yield in the SE coastal plain. *Agricultural Systems* 64 (3):189–207.

Schowengerdt, R.A. 2007. *Remote Sensing: Models and Methods for Image Processing.* Third edition. Amsterdam: Academic Press Publications.

Spruce, J.P., S. Sader, R.E. Ryan, J. Smoot, P. Kuper, K. Ross, D. Prados, J. Russell, G. Gasser, and R. McKellip. 2011. Assessment of MODIS NDVI time series data products for detecting forest defoliation by gypsy moth outbreaks. *Remote Sensing of Environment* 115 (2):427–437.

Staff. 2012. New Sensor Payload for Raven UAV Unveiled. *Unmanned Ground, Aerial, Sea and Space Systems.*

Staff. 2015a. Camera. In *Merriam Webster's Online Dictionary.* http://www.merriam-webster.com/dictionary/camera: Merriam and Webster.

Staff. 2015b. *In Situ.* In *Merriam Webster's Online Dictionary.* http://www.merriam-webster.com/dictionary/in%20situ: Merriam and Webster.

Staff. 2015c. IRIS+. 3D Robotics Accessed 4/22/15. https://store.3drobotics.com/products/IRIS.

Stark, R., A.A. Gitelson, U. Grits, D. Rundquist, and Y. Kaufman. 2000. New technique for remote estimation of vegetation fraction: Principles, algorithms, and validation. *Aspects of Applied Biology* 60 (Remote Sensing of Agriculture):241–247.

Sullivan, J. 2005. Small UAV's for Agricultural Applications. *Unmanned Science: The Science behind Unmanned Systems.*

Velodyne. 2015. High Definition LiDAR HDL-32E . http://velodynelidar.com/lidar/hdldownloads/HDL-32E_datasheet.pdf.

Vogelmann, J.E. 1990. Comparison between two vegetation indices for measuring different types of forest damage in the north-eastern United States. *Remote Sensing* 11 (12):2281–2297.

Vygodskaya, N.N., I.I. Gorshkova, and Ye. V. Fadeyava. 1989. Theoretical estimates of sensitivity in some vegetation indices to variation in the canopy conditions. *International Journal of Remote Sensing* 10:1857–1872.

Wallace, L.O., A. Lucieer, and C.S. Watson. 2012. Assessing the feasibility of UAV-BASED LiDAR for high resolution forest change detection. In *International Archives of the Photogrammetry, Remote Sensing and Spatial Information Sciences.* Melbourne, Australia. XXXIX-B7, pp. 499–504. http://www.int-arch-photogramm-remote-sens-spatial-inf-sci.net/XXXIX-B7/

Wallace, L., A. Lucieer, and C.S. Watson. 2014. Evaluating tree detection and segmentation routines on very high resolution UAV LiDAR Data. *IEEE Transactions on Geoscience and Remote Sensing* 52 (12):7.

Wehr, A. and U. Lohr. 1999. Airborne laser scanning—An introduction and overview. *ISPRS Journal of Photogrammetry & Remote Sensing* 54:68–82.

Wolf, P.R. and B.A. Dewitt. 2000. *Elements of Photogrammetry: With Applications in GIS.* 1 vols. Boston: McGraw-Hill.

Zarco-Tejada, P.J., R. Diaz-Varela, V. Angileri, and P. Loudjani. 2014. Tree height quantification using very high resolution imagery acquired from an unmanned aerial vehicle (UAV) and automatic 3D photo-reconstruction methods. *European Journal of Agronomy* 55:89–99.

Zhang, C. and J.M. Kovacs. 2012. The application of small unmanned aerial systems for precision agriculture: A review. *Precision Agriculture* 13 (6):693–712.

5

U.S. Aviation Regulatory System

Douglas M. Marshall

CONTENTS

5.1 U.S. Aviation Regulatory System

5.1.1 Introduction

Aviation regulations in the United States have existed nearly as long as the technology that is being regulated. All levels of government in civilized countries impose various regulations on their citizens and their activities.

Regulations in any technical environment such as aviation are typically driven by original equipment manufacturers and operators. As users experience incidents, problems, or anomalies, those events are properly reported to the Federal Aviation Administration (FAA). Should the number of events reach a certain critical mass or the outcome is sufficiently severe (fatalities, injuries, or property damage), the data generated may provoke a review of the relevant regulation, if any.

The introduction of a new technology or procedure into the National Airspace System (NAS) requires a comprehensive safety analysis before the FAA can allow the change. The safety analysis includes a review of the relevant regulation and supporting advisory circulars (AC) or special Federal Aviation Regulations (SFARs) to determine whether the proposed technology or procedure can comply with current regulation. The FAA may grant exceptions in particular cases after performing the safety review as a way to manage unique, perhaps nonrecurring circumstances, or when the event that led to the review is determined to be unlikely to recur.

These circumstances may lead to the rulemaking process, which provides the mechanism for the FAA to fulfill its statutory mandate to ensure the safety of the aviation environment. This chapter describes the history of the U.S. federal as well as international aviation regulations; the structure of those statutes and regulations; the purpose and intent of the rules; how rules are made, changed, and enforced; how this system affects the development of unmanned aircraft system (UAS) technologies and operations; and concludes with a look into the future of UAS regulations.

5.1.2 History of U.S. Aviation Regulation

Aviation regulations in the United States have enjoyed a long and colorful history, beginning with the commencement of airmail operations by the U.S. Post Office in 1918, only 15 years after the first manned powered flight. Three years before that President Wilson signed a bill that created a National Advisory Committee on Aeronautics, which was intended to oversee a scientific study of the "problems" of flight. No fewer than six federal statutes enacted to regulate some aspects of aviation followed these early efforts. Most were directed toward safety concerns and the perceived need to bring some order to the commercial aspects of aviation. The issues that generated the greatest concern were the number of crashes, the need for a regulated civil airport network, the lack of a harmonized or common system of air navigation, and demand for a civil aviation infrastructure that would support growth and stability of the industry, both for military and nonmilitary applications.

5.1.3 Federal Aviation Administration

The Federal Aviation Act of 1958 created the Federal Aviation Administration.[*] The statute was enacted in response to a series of fatal accidents and midair collisions involving commercial passenger aircraft. The FAA is part of the Department of Transportation and derives its rulemaking and regulatory power from Title 49 of the United States Code, Section 106. The Commerce Clause of the U.S. Constitution (Article I, Section 8) grants Congress broad authority to "regulate commerce with foreign nations, and among the several states." The U.S. government therefore has exclusive power to regulate the airspace of the United States.[†] A citizen of the United States has a public right of transit through the navigable airspace.[‡] Among other powers the statute confers upon the administrator of the FAA is the mandate to develop plans and policy for the use of the navigable airspace and assign by regulation or order the use of the airspace necessary to ensure the safety of aircraft and the efficient use of airspace.[§] The administrator may modify or revoke a regulation, order, or guidance document when required in the public interest. The administrator shall prescribe air traffic regulations on the flight of aircraft (including regulations on safe altitudes) for navigating, protecting, and identifying aircraft; protecting individuals and property on the ground; using the navigable airspace efficiently; and preventing collision between aircraft, between aircraft and land or water vehicles, and between aircraft and airborne objects.[¶]

[*] Public Law 85-726, 85th Cong., 2nd Sess.; 72 Stat. 731; 49 U.S.C § 1301, as amended.
[†] 49 U.S.C. § 40103 (a)(1).
[‡] 49 U.S.C. § 40103 (a)(2).
[§] 49 U.S.C. § 40103 (b)(1).
[¶] 49 U.S.C. § 40103 (b)(2).

Pursuant to its rulemaking authority, the FAA has set forth the standards for the operation of aircraft in the sovereign airspace of the United States.[*] Commonly known as the Federal Aviation Regulations (FARs), these regulations are the "rules of the road" for certification of all civil aircraft,[†] airmen,[‡] and airspace;[§] certification and operations for air carriers and operators for compensation or hire;[¶] air traffic and general operating rules;[**] and schools and other certificated agencies,[††] airports,[‡‡] and navigational facilities.[§§]

The first section of 14 CFR, Part 1.1, lists the definitions and abbreviations to be observed in the ensuing parts and subparts of the FARs. Of more than passing interest to the unmanned aircraft community is the fact that the terms *UAV* or *UAS* or *unmanned system* or *unmanned* aircraft or any other term referring to remotely piloted aircraft were excluded from the FARs and, for that matter, any other federal regulation or statute until passage of the FAA Modernization and Reform Act of 2012.[¶¶] The term *aircraft* is defined as "a device that is used or intended to be used for flight in the air."[***] Similarly, "*airplane* means an engine-driven fixed-wing aircraft heavier than air, that is supported in flight by the dynamic reaction of the air against its wings."[†††] "*Air traffic* means aircraft operating in the air or on an airport surface, exclusive of loading ramps and parking areas."[‡‡‡]

The FAA regulates aircraft, airmen, certain categories of employees of airlines and commercial or common carrier operations, airports, and the national airspace. The FAA's "toolbox" is the system of regulations, rulemaking processes, certifications, ACs, special authorizations, and directives that the agency uses to carry out its regulatory functions of rulemaking, surveillance, compliance, and enforcement.

Three of the tools that the FAA uses to administer the FARs are ACs, airworthiness directives (AD), and policy statements. An AC or AD may be issued in response to a safety-related event or system anomaly, or a technical standards order (TSO) could be developed to remediate a technical problem. An AC provides guidance to owners or operators of aircraft or systems to facilitate compliance with the applicable regulations. An AD is a notification to owners and operators of certified aircraft that a known safety deficiency with a particular model of aircraft, engine, avionics, or other system exists and must be corrected. A TSO is a minimum performance standard for specified materials, parts, and appliances used on civil aircraft. When authorized to manufacture a material, part, or appliance to a TSO standard, this is referred to as TSO authorization. Issuance of a TSO authorization constitutes both design and production approval. However, issuance of a TSO authorization is not an approval to install and use the article in the aircraft. It simply means that the article meets the specific TSO and the applicant is authorized to manufacture it.

ACs are utilized to advise the aviation community on issues pertaining to the regulations, but are not binding upon the public. The exception would be when an AC is

[*] 14 CFR Part 1.1 et seq.
[†] 14 CFR Parts 21–49.
[‡] 14 CFR Parts 61–67.
[§] 14 CFR Parts 71–77.
[¶] 14 CFR Parts 119–135.
[**] 14 CFR Parts 91–105.
[††] 14 CFR Parts 141–147.
[‡‡] 14 CFR Parts 150–161.
[§§] 14 CFR Parts 170–171.
[¶¶] Public Law 112-95, 112th Congress, 126 Stat. 11.
[***] 14 CFR 1.1.
[†††] 14 CFR 1.1.
[‡‡‡] 14 CFR 1.1.

specifically referenced in a regulation.[*] The ACs are issued in a numbered-subject system corresponding to the subject areas of the FARs.[†] The AC that has created the most controversy in the unmanned aviation world is AC 91-57, which will be discussed in more detail later. That circular references 14 CFR Part 91 (Air Traffic and General Operating Rules), which contains the airspace regulations.

Another advisory tool is the policy statement. Administrative implementation (as announced or documented by a published policy statement) of a particular statutory provision shall be accorded deference by the courts when it appears that Congress delegated authority to the agency generally to make rules carrying the force of law and that the agency interpretation claiming deference was promulgated in the exercise of that authority. Delegation of such authority may be shown in a variety of ways, as by an agency's power to engage in adjudication or notice-and-comment rulemaking, or by some other indication of a comparable congressional intent.[‡] The FAA has issued a number of policy statements pertaining to unmanned aircraft, including AFS-400 UAS Policy Statement 05-01; a clarification published in the *Federal Register* February 6, 2007, titled "Unmanned Aircraft Operations in the National Airspace System"; and "Interim Operational Approval Guidance 08-01," which likewise references 14 CFR Part 91.[§¶] In addition, the FAA has published policies regarding Inspection and Maintenance Program requirements for Airworthiness Certification of Unmanned Aircraft Operating Under 55 Pounds, Aviation-Related Videos or Other Electronic Media on the Internet,[**] UAS Temporary Flight Restrictions for Sporting Events,[††] Education, Compliance, and Enforcement of Unauthorized Unmanned Aircraft Systems Operators,[‡‡] as well as no fewer than seven Orders, two additional ACs,[§§] three Guidance documents, four Legal Interpretations, and one special Rules Interpretation since 2007. These documents are all readily accessed on the FAA's website.[¶¶]

5.1.4 Enforcement and Sanctions

No system of rules regulations can be effective without a means to enforce them. The FARs are no exception. The FAA's mandate from Congress is to conduct surveillance of aviation activities, inspect aviation systems, investigate violations of the aviation regulations, and take appropriate measures to enforce the regulations in the event of a violation. The agency's investigative power extends to all provisions of the Federal Aviation Act of 1958; the Hazardous Materials Transportation Act; the Airport and Airway Development Act of 1970; the Airport and Airway Improvement Act of 1982; the Airport and Airway Safety and Capacity Expansion Act of 1987; and any rule, order, or regulation issued by the FAA. The FAA's central mission, pursuant to its own Order 2150.3A (Compliance and Enforcement Program), is to promote adherence to safety standards, but the agency recognizes that, due to the nature of aviation itself, it must largely depend upon voluntary compliance with the regulatory standards. The Fifth and Fourteenth Amendments to the

[*] Advisory Circular 00-2.11 (1997).
[†] Advisory Circular 00-2.11 (1997).
[‡] *United States v. Mead Corp.*, 533 U.S. 218; 121 S. Ct. 2164; 150 L. Ed. 2d 292.
[§] 72 FR 6689, Vol. 72, No. 29, February 13, 2007 (Interim Operational Approval Guidance 08-01).
[¶] Notice 8900.291.
[**] Notice 8900.292.
[††] FDC NOTAM 4/3621.
[‡‡] Notice 8900.268.
[§§] AC 21-12 and AC 45-2D.
[¶¶] www.faa.gov.

U.S. Constitution require that the FAA enforcement process provide "due process" in the procedures for ensuring compliance with the regulations. This means that no one shall be deprived of "life, liberty, or property without due process of law."* Thus, the FAA may not act arbitrarily or inconsistently in its efforts to enforce the regulations.

The enforcement process established by the FAA is designed to be fair, reasonable, and perceived to be fair by those who are subject to the regulations. It is a complicated process that provides a number of decision points that allow the FAA and the party being investigated to arrive at an informal resolution rather than taking the matter to a fully litigated trial. The range of possible outcomes varies from a case being abandoned by the FAA after it investigates an alleged violation to a trial, an outcome, and an appeal to the U.S. Court of Appeals or even to the U.S. Supreme Court (a rare event indeed). Trials are conducted like any other civil trial, and the FAA generally has the burden of proof in establishing a violation. Short of a civil penalty or certificate revocation or suspension, the FAA may issue warning letters or letters of correction, which are intended to bring the alleged offender into compliance with the regulations for violations that are not deemed sufficiently serious to warrant more severe sanctions. A constructive attitude of cooperation by the certificate holder often goes a long way toward resolving inadvertent or non-flagrant violations by first-time offenders. Civil penalties of up to $50,000 per violation are available in cases where there is no certificate to suspend or revoke, where revocation would impose an undue hardship, where qualification is not at issue, or where the violation is too serious to be handled administratively by use of remedial action. It is important to remember that the lack of an airman's certificate or other FAA-issued license does not immunize a person or entity from an FAA enforcement action backed up by imposition of a civil penalty.

The manner in which the FAA has chosen to enforce the FARs when dealing with unmanned aircraft will be dealt with below. As of this publication, the authors are not aware of any formal enforcement action that the FAA has taken against any UAS/remotely piloted aircraft operator, pilot, owner, manufacturer, or service.

5.2 International Aviation Regulations

As early as 1919, an international agreement (the Convention for the Regulation of Aerial Navigation, created by the Aeronautical Commission of the Peace Conference of 1919, otherwise known as the Versailles Treaty) recognized that the air above the high seas was not as "free" as the water of those seas. In that convention the contracting states acknowledged exclusive jurisdiction in the airspace above the land territory and territorial waters of the states, but agreed to allow, in times of peace, innocent passage of the civil aircraft of other states, so long as the other provisions of the convention were observed. States still retained the right to create prohibited areas in the interests of military needs or national security. During the global hostilities of the 1940s the United States initiated studies and later consulted with its major allies regarding further harmonization of the rules of international airspace, building upon the 1919 convention. The U.S. government eventually extended an invitation to 55 states or authorities to attend a meeting to discuss these issues, and in November 1944, an International Civil Aviation Conference was held in

* U.S. Constitution, Amendment 5 and Amendment 14.

Chicago. Fifty-four states attended this conference, at the end of which the Convention on International Civil Aviation was signed by 52 of those states. The convention created the permanent International Civil Aviation Organization (ICAO) as a means to secure international cooperation and the highest possible degree of uniformity in regulations and standards, procedures, and organization regarding civil aviation matters. The Chicago Conference laid the foundation for a set of rules and regulations regarding air navigation as a whole, which was intended to enhance safety in flying and construct the groundwork for the application of a common air navigation system throughout the world.

The constitution of the ICAO is the Convention on International Civil Aviation that was drawn up by the Chicago conference, and to which each ICAO contracting state is a party. According to the terms of the convention, the organization is made up of an assembly, a council of limited membership with various subordinate bodies, and a secretariat. The chief officers are the president of the council and the secretary general. ICAO works in close cooperation with other members of the United Nations family such as the World Meteorological Organization, the International Telecommunication Union, the Universal Postal Union, the World Health Organization, and the International Maritime Organization. Nongovernmental organizations that also participate in ICAO's work include the International Air Transport Association, the Airports Council International, the International Federation of Airline Pilots' Associations, and the International Council of Aircraft Owner and Pilot Associations.*

ICAO's objectives are many and are set forth in the 96 articles of the Chicago Convention and the 18 annexes thereto. Additional standards and guidelines are found in numerous supplements (Standards and Recommended Practices, or SARPS) and Procedures for Air Navigation Services, which are under continuing review and revision. Contracting states are free to take exceptions to any element of the annexes, and those exceptions are also published. Contracting States are also responsible for developing their own aeronautical information publication (AIP), which provides information to ICAO and other states about air traffic, airspace, airports, navaids (navigational aids), special use airspace, and weather and other relevant data for use by air crews transiting into or through the state's airspace. The AIPs will also contain information about a state's exceptions to the annexes and any significant differences between the rules and regulations of the state and ICAO's set of rules.

The annexes cover rules of the air, meteorological services for international air navigation, aeronautical charts, measurement units used in air and ground operations, operation of aircraft, aircraft nationality and registration marks, airworthiness of aircraft, facilitation (of border crossing), aeronautical communications, air traffic services, search and rescue, aircraft accident investigation, aerodromes, aeronautical information services, environmental protection, security-safeguarding international civil aviation against acts of unlawful interference, and the safe transportation of dangerous goods by air. The only reference in any ICAO document to unmanned aircraft is found in Article 8 of the convention, which states that:

> No aircraft capable of being flown without a pilot shall be flown without a pilot over the territory of a contracting State without special authorization by that State and in accordance with the terms of such authorization. Each contracting State undertakes to insure that the flight of such aircraft without a pilot in regions open to civil aircraft shall be so controlled as to obviate danger to civil aircraft.

* International Civil Aviation Organization: http://www.icao.int/.

ICAO's rules apply to international airspace, which is typically defined as the airspace over the high seas more than 12 miles from the sovereign territory of a state (country) as well as some domestic airspace by virtue of incorporation into a contracting state's own regulatory scheme. The rules apply to all contracting states (there are 188 of them), so any nation that elects not to become an ICAO member is not entitled to the protection of ICAO's rules. However, ICAO is a voluntary organization, and there are no provisions for enforcement of the regulations or standards such as those found in the FARs. As a founding member of ICAO and as a nation that has a substantial interest in preserving harmony in international commercial aviation, the United States enforces ICAO's rules against U.S. operators to the extent that the ICAO rule has been incorporated into the FARs and does not conflict with domestic regulations.

Additional international aviation organizations located in Europe that exercise some level of regulatory powers include EUROCONTROL (European Organization for the Safety of Air Navigation), EASA (European Aviation Safety Agency), and EUROCAE (European Organization for Civil Aviation Equipment). EUROCONTROL is an intergovernmental organization that acts as the core element of air traffic control services across Europe and is dedicated to harmonizing and integrating air navigations services in Europe and creating a uniform air traffic management system for civil and military users. The agency accomplishes this by coordinating efforts of air traffic controllers and air navigation providers to improve overall performance and safety. The organization is headquartered in Brussels and has 38 member states. The European Commission created a Single European Sky ATM Research (SESAR) initiative in 2001 and has delegated portions of the underlying regulatory responsibility to EUROCONTROL. EASA was established as an agency of the European Union in 2003 and has regulatory responsibility in the realm of civilian aviation safety, assuming the functions formerly performed by the Joint Aviation Authorities (JAA). Contrary to JAA's role, EASA has legal regulatory authority, which includes enforcement power. EASA has responsibility for airworthiness and environmental certification of aeronautical products manufactured, maintained, or used by persons under the regulatory oversight of European Union member states. EUROCAE reports to EASA, although it was created many years before EASA was formed, and deals exclusively with aviation standardization (with reference to airborne and ground systems and equipment). Its membership is made up of equipment and airframe manufacturers, regulators, European and international civil aviation authorities, air navigation service providers, airlines, airports, and other users. EUROCAE's Working Group 73 is devoted to the development of products intended to help assure the safe, efficient, and compatible operation of UASs with other vehicles operating within non-segregated airspace. WG-73 makes recommendations to EUROCAE with the expectation that those recommendations will be passed on to EASA.

More recently, EUROCAE WG-93, "Light Remotely Piloted Aircraft Systems Operations" has been tasked with developing standards and recommendations for guidance material for the safe operation of light RPAS (with a maximum takeoff weight of less than 150 kg) directed toward national aviation authorities.

In addition to ICAO and the three European organizations just discussed, any nation's civil aviation authority (CAA) is free to promulgate its own aviation rules and regulations for operations within their sovereign airspace, and until ICAO has created an overarching set of rules for UAS operations among its member states, operators of UASs must be sensitive to the rules and regulations of the contracting state that is providing air traffic services in international airspace. Another cross-national organization, the Joint Authorities for Rulemaking on Unmanned Systems (JARUS), is made up of a group of experts from the National Aviation Authorities (NAA) and regional safety organizations. This group's task is to recommend a single set of technical, safety, and operational requirements for the certification and safe integration of UASs into airspace and aerodromes. Their objective is to provide guidance material to facilitate each NAA's efforts in writing their own requirements and to avoid duplicate efforts.*

5.3 Standards versus Regulations

The FAA exercises its statutory mandate by making rules and regulations. Those efforts are often supplemented or enhanced by published standards that are created by industry organizations and approved by the FAA. Standards developers work with engineers, scientists, and other industry personnel to develop nonbiased standards or specification documents that serve industry and protect the public. These developers can be private concerns, trade organizations, or professional societies. Standards providers are distributors of codes, standards, and regulations. They may also provide access to a database of standards. The supplier may or may not be the developer of the standards distributed.

These organizations are essentially professional societies made up of industry representatives, engineers, and subject matter experts who provide advisory support to federal agencies such as the FAA. They make recommendations that may become a formal rule by adoption or reference. Engineering codes, standards, and regulations all serve to ensure the quality and safety of equipment, processes, and materials. The three most prominent of those advisory organizations playing a role in the evolution of unmanned aviation are the Society of Automotive Engineers (SAE), the Radio Technical Commission for Aeronautics (RTCA), and ASTM International (originally the American Society of Testing and Materials).

Aeronautical engineering codes are enforced by the FAA and are critical to developing industry practices. Whereas engineering regulations such as those found in the FARs are government-defined practices designed to ensure the protection of the public as well as uphold certain ethical standards for professional engineers, engineering standards ensure that organizations and companies adhere to accepted professional practices, including construction techniques, maintenance of equipment, personnel safety, and documentation. These codes, standards, and regulations also address issues regarding certification, personnel qualifications, and enforcement.

Manufacturing codes, standards, and regulations are generally designed to ensure the quality and safety of manufacturing processes and equipment, and aviation regulations are no exception. Manufacturing standards ensure that the equipment and processes used by manufacturers and factories are safe, reliable, and efficient. These standards are often voluntary guidelines, but can become mandatory by reference in the FARs. Manufacturing regulations are government-defined and usually involve legislation for controlling the practices of manufacturers that affect the environment, public health, or safety of workers. Aircraft manufacturers in the United States and European Union are required by law to produce aircraft that meet certain airworthiness and environmental emissions standards.

* www.jarus-rpas.org.

The FAA has supported and sponsored four domestic committees dedicated to developing standards and regulations for the manufacture and operation of unmanned aircraft. RTCA's Special Committee 203 Unmanned Aircraft Systems (SC-203) began developing minimum operational performance standards (MOPS) and minimum aviation system performance standards (MASPS) for UASs in 2004: "SC-203 products will help assure the safe, efficient and compatible operation of UAS with other vehicles operating within the NAS. SC-203 recommendations will be based on the premise that UAS and their operations will not have a negative impact on existing NAS users." SC-203's efforts ended in 2013, and new Special Committee, SC-228, Minimum Operational Performance Standards for Unmanned Aircraft Systems, was created shortly thereafter, in 2013. Its task is to develop MOPS for detect-and-avoid technology as well as command-and-control (C2) data link MOPS seeking L-Band and C-Band solutions. This committee's work in part builds upon earlier standards published by the North Atlantic Treaty Organization, STANAG 4586, "Standard Interfaces of UAV Control System (UCS) for NATO Interoperability, 3rd Ed."

ASTM's F-38 Unmanned Air Vehicle Systems Committee addresses issues related to design, performance, quality acceptance tests, and safety monitoring for unmanned air vehicle systems. Stakeholders include manufacturers of unmanned aerial vehicles and their components, federal agencies, design professionals, professional societies, maintenance professionals, trade associations, financial organizations, and academia.

SAE's G-10U Unmanned Aircraft Aerospace Behavioral Engineering Technology Committee was established to generate pilot training recommendations for UASs civil operations and has released their recommendations.

By Order 1110.150, signed on April 10, 2008, the FAA created a Small Unmanned Aircraft System (sUAS) Aviation Rulemaking Committee (ARC) according to the FAA administrator's authority under Title 49 United States Code (49 U.S.C.) § 106(p)(5). The committee's term was 20 months and was made up of representatives of aviation associations, industry operators, manufacturers, employee groups or unions, the FAA and other government entities, and other aviation industry participants, including academia. The committee delivered its formal recommendations to the FAA associate administrator in March 2009. The FAA's Air Traffic Organization simultaneously convened a safety risk management committee that was charged with describing the UAS systems under review, identifying hazards, analyzing risk, assessing risk, and treating risk to arrive at a safety management system (SMS) for UASs that would be coordinated or integrated with the ARC's recommendations. This process follows a number of FAA policies that require oversight and regulation of aeronautical systems that may impact safety in the NAS (FAA Order 8000.369 Safety Management System Guidance; FAA Order 1100.161 Air Traffic Safety Oversight; FAA Order 8000.36 Air Traffic Safety Compliance Process; FAA Order 1000.37 Air Traffic Organization Safety Management System Order; ATO-SMS implementation Plan Version 1.0, 2007; FAA SMS Manual Version 2.1 of June 2008; Safety and Standards Guidance Letter 08-1; and AC 150/5200-37 Introduction to SMS for Airport Operations).

The ARC's recommendations for regulations pertaining to small UASs were adopted in part and led to a published notice of proposed rulemaking for a new Part 107, Operation and Certification of Small Unmanned Aircraft Systems, released on February 15, 2015 (discussed later). This is the first set of regulations to be proposed by the FAA dealing specifically with UASs.

One previous attempt to address a narrow category of remotely piloted aircraft was AC 91-57, published in 1981. This AC was in reality an effort to regulate by not regulating the recreational modeling community, outlining and encouraging voluntary compliance with safety standards for model aircraft operators. The document's content was taken off the FAA's website, but it has not been revoked, so it remains as the operative standard for model aircraft operations within certain designated areas and under the authority of a voluntary organization, the Academy of Model Aeronautics (AMA). The AMA created its own set of standards and restrictions for its members, compliance with which is a prerequisite for the group insurance coverage for which its members are eligible.

Although AC 91-57 was specifically directed toward recreational modelers, the circular has at times been relied upon by commercial UAS operators and developers to make a claim that they can fly their small UASs under 400 feet AGL without communicating with the FAA and running afoul of the FARs. Policy Statement 05-01 and Guidance Document 08-01 both refer to AC 91-57 as the official policy with respect to recreational and hobbyist aero modeling, which is that those activities do not fall under the intent of FARs and are thus excluded. However, by inference, the FAA believed that it had the statutory power to regulate recreational models because they fell under the definition of *airplane* found in 14 CFR 1.1, but chose not to do so as a matter of policy. Section 336 of the FMRA has superseded that policy.*

5.4 How the Process Works

The sUAS ARC, and the subsequent NPRM discussed earlier, is an example of one of the FAA's processes for creating the rules, regulations, circulars, directives, and orders that it employs to bring some order to the aviation industry, which is one of the most heavily regulated industries in the United States and elsewhere. The FAA's rulemaking authority is derived from either executive order (from the Office of the President) or the U.S. Congress, through specific mandate or by delegation of Congress' lawmaking powers as conferred by the U.S. Constitution (Article 1, Section 8). The FAA relies on those two sources as well as recommendations from the National Transportation Safety Board, the public, and the FAA itself to initiate rulemaking. Ultimately the FAA makes rules to serve the public interest and to fulfill its mission of enhancing safety in the aviation environment.

The Administrative Procedures Act of 1946 and the Federal Register Act of 1935 govern the process of rulemaking. These two statutes combined were intended to ensure that the process is open to public scrutiny (that federal agencies do not make rules or impose regulations in secret or without full transparency). This is accomplished by procedural due process and publication requirements. This "informal rulemaking" is a four-step process that follows what often involves months or even years of industry rulemaking committee effort, internal FAA review and analysis, and interagency negotiation. Once the proposed rule has achieved a sufficient level of maturity, it will be published in the *Federal Register* as a "Notice of Proposed Rulemaking." The notice provides an opportunity to the general public to comment on the proposed rule within a certain period of time. The comments from the public must be resolved in some fashion before the final rule document is published, which should respond to the comments and provide an explanation of purpose and basis for rule as well as the way in which the comments were resolved. The last step is

* Public Law 112-95 (see note 15, and discussed below).

implementation, and the effective date must be at least 30 days after publication of the final rule unless it is interpretive, a direct rule, a general policy statement, an emergency rule, or a substantive rule that grants an exemption to an existing rule or requirement. Some agency rules or policies may be exempted from this process, such as interpretive rules or general policy statements, or if the agency can demonstrate that the notice-and-comment process would be impractical, unnecessary, or contrary to public interest (showing "good cause").

Rules that have gone through this informal rulemaking process have the same force and effect as a rule or regulation imposed by an act of Congress. Thus, the FAA is empowered to enforce those rules as if they were laws enacted by Congress. The rules are typically referenced to or codified in the Code of Federal Regulations (CFRs). There are exceptions to this, however. Direct final rules are implemented after a final rule is issued while still providing for a period of notice and comment. The rule becomes effective after the specified period if there are no adverse comments. The difference in this process from the notice of proposed rulemaking procedure is that there is no proposed rule published before the final rule is released. This is used for routine rules or regulations that are not anticipated to generate comment or controversy. Interim rules are usually effective immediately and are issued without prior notice, often in response to an emergency. A final rule may be issued based upon the interim rule after a period of comment. The status of an interim rule as final or amended or withdrawn is always published in the *Federal Register*. Last, interpretive rules may be issued to explain current regulations or its interpretation of existing statues or rules. This tool is not commonly used by the FAA, but may be useful when a rule is repeatedly misinterpreted, resulting in chronic compliance issues.

The point of this complex, sometimes cumbersome, and time-consuming process of rulemaking is to advance the cause of safety and harmonization so that all users and others affected by the aviation environment are protected from undue risk of harm. Further it is to ensure that all entities operate under the same set of rules and regulations and have abundant opportunity to engage in the process so that the outcomes may be influenced by multiple points of view. Each step of the process requires a series of reviews by other agencies of the federal government, such as the Office of the Secretary of Transportation, the Department of Transportation, the Office of Management and Budget, the General Accountability Office and the Office of the Federal Register. A flow chart of the aviation regulatory process would demonstrate at least a 12-step process, with multiple interim steps embedded in most of the broader categories. If a proposed rule were to be subjected to each and every possible review step, the list would include no fewer than 35 stops along the way. For example, due to the many and equally influential stakeholders that could be involved in an effort by the Department of Defense to create a new restricted area for UAS operations, testing, and training, it is commonly estimated that it would take 5 years to accomplish the goal. As an example, the implementation of an aviation safety device for air transport aircraft known as TCAS (traffic alert and collision avoidance system) required over 15 years from inception to implementation, and it took an act of Congress to mandate the use of TCAS in commercial airliners.

In addition to formal rules and regulations, the FAA issues orders, policies, directives, and guidance documents. The FAA routinely issues policy statements and guidance documents to clarify or explain how the FAA interprets and enforces the regulations. A policy statement gives guidance or acceptable practices on how to find compliance with a specific CFR section or paragraph. These documents are explanatory and not mandated. They are also not project specific. Practically speaking, this means that they are not enforceable in formal compliance proceedings, but they do provide guidance to users and the public on how best to comply with the FARs. Guidance documents are similar in nature and are likewise explanatory rather than mandatory.

The FAA's website contains links to all historical and current policy statements, guidance documents, orders, directives, circulars, and regulations. Binding orders and regulations are published in the *Federal Register* and are accessible on the Electronic Code of Federal Regulations (e-CFR) government website.*

5.5 Current Regulation of Unmanned Aircraft

As discussed earlier, until recently there was no specific reference in any of the Federal Aviation Regulations to unmanned aircraft, pilots/operators of unmanned aircraft, or operations in the national airspace of unmanned aircraft. A literal reading of the definitions listed in 14 CFR 1.1 would include all unmanned aircraft in the description of *aircraft*. There was no case authority, nor is there a rule or regulation that says that unmanned aircraft of any size or capability are *not* regulated. This conceivably would include radio-controlled model aircraft. In recognition of the reality that radio-controlled aircraft are aircraft but not of the type that the FAA is inclined to regulate, AC 91-57 was published in 1981. This AC encourages voluntary compliance with safety standards for model aircraft operators. The circular also acknowledges that model aircraft may pose a safety hazard to full-scale aircraft in flight and to persons and property on the ground.† Modelers are encouraged to select sites that are sufficiently far away from populated areas so as to not endanger people or property, and to avoid noise-sensitive areas such as schools and hospitals. Aircraft should be tested and evaluated for airworthiness and should not be flown more than 400 feet above ground level. If the aircraft is to be flown within 3 miles of an airport, contact with local controlling authorities should be initiated. And, above all, model aircraft should always give way to, or avoid, full-scale aircraft, and observers should be used to assist in that responsibility.‡

FAA policy statement AFS-400 UAS Policy 05-01 was issued on September 16, 2005, in response to dramatic increases in UAS operations in both the public and private sectors.§ The policy was intended to provide guidance to be used by the FAA to determine if UASs may be allowed to conduct flight operations in the U.S. NAS. AFS-400 personnel are to use this policy guidance when evaluating each application for a certificate of waiver or authorization (COA). Due to the rapid evolution of UAS technology, this policy is to be subject to continuous review and updated when appropriate.¶ The policy was not meant to be a substitute for any regulatory process, and was jointly developed by, and reflected the consensus opinion of, AFS-400, the Flight Technologies and Procedures Division, FAA Flight Standards Service (AFS); AIR 130, the Avionics Systems Branch, FAA Aircraft Certification

* Electronic Code of Federal Regulations: http://ecfr.gpoaccess.gov/cgi/t/text/text-idx?&c = ecfr&tpl = /ecfrbrowse/Title14/14tab_02.tpl.

† AC 91-57.

‡ AC 91-57.

§ FAA AFS-400 UAS Policy 05-01, September 16, 2005.

¶ FAA AFS-400 UAS Policy 05-01, September 16, 2005.

Service (AIR); and ATO-R, the Office of System Operations and Safety, FAA Air Traffic Organization (ATO).[*]

The 05-01 policy recognized that if UAS operators were strictly held to the "see and avoid" requirements of 14 CFR Part 91.113, "Right-of-Way Rules," there would be no UA flights in civil airspace.[†]

> The right-of-way rule states that "...when weather conditions permit, regardless of whether an operation is conducted under instrument flight rules or visual flight rules, vigilance shall be maintained by each person operating an aircraft so as to see and avoid other aircraft. When a rule of this section gives another aircraft the right-of-way, the pilot shall give way to that aircraft and may not pass over, under, or ahead of it unless well clear."[‡] The FAA's policy supports UA flight activities that can demonstrate that the proposed operations can be conducted at an acceptable level of safety.[§]

Another collision avoidance rule states "no person may operate an aircraft so close to another aircraft as to create a collision hazard."[¶] The FAA also recognizes that a certifiable "detect, sense and avoid" system, an acceptable solution to the see-and-avoid problem for UA, is many years away.[**]

Through the implementation of this policy, the FAA gave civil UAS developers and operators two choices: (1) they could operate their systems as public aircraft and apply for a COA that will permit operation of a specific aircraft in a specific operating environment with specific operating parameters and for no more than one year at a time; or (2) they could follow the normal procedures set forth in the CFRs to obtain a special airworthiness certificate for their aircraft,[††] operate the aircraft in strict compliance with all airspace regulations set forth in 14 CFR Part 91, and have them flown by certificated pilots.[‡‡] The policy also references AC 91-57, Model Aircraft Operating Standards, published in 1981, as it applies to model aircraft, and states "UA that comply with the guidance in AC 91-57 are considered model aircraft and are not evaluated by the UA criteria in this policy."[§§]

The FAA furthermore declared in this policy that it would not accept applications for civil COA, which means that only military or public aircraft were eligible.[¶¶] A *public aircraft* is defined in 14 CFR Part 1.1 as follows:

Public aircraft means any of the following aircraft when not being used for a commercial purpose or to carry an individual other than a crewmember or qualified non-crew member:

1. An aircraft used only for the U.S. government; an aircraft owned by the government and operated by any person for purposes related to crew training, equipment development, or demonstration; an aircraft owned and operated by the government of a State, the District of Columbia, or a territory or possession of the United States or a political subdivision of one of these governments; or an aircraft

[*] FAA AFS-400 UAS Policy 05-01, September 16, 2005.
[†] 14 CFR 91.113.
[‡] 14 CFR 91.113(b).
[§] FAA AFS-400 UAS Policy 05-01, Supra Note 43.
[¶] 14 CFR 91.111(a).
[**] FAA AFS-400 UAS Policy 05-01, Supra Note 43.
[††] 14 CFR 21.191.
[‡‡] FAA AFS-400 UAS Policy 05-01, Supra Note 43.
[§§] FAA AFS-400 UAS Policy 05-01, Supra Note 43.
[¶¶] FAA AFS-400 UAS Policy 05-01, Supra Note 43, § 6.13.

exclusively leased for at least 90 continuous days by the government of a State, the District of Columbia, or a territory or possession of the United States or a political subdivision of one of these governments.

i For the sole purpose of determining public aircraft status, *commercial purposes* means the transportation of persons or property for compensation or hire, but does not include the operation of an aircraft by the armed forces for reimbursement when that reimbursement is required by any Federal statute, regulation, or directive, in effect on November 1, 1999, or by one government on behalf of another government under a cost reimbursement agreement if the government on whose behalf the operation is conducted certifies to the Administrator of the FAA that the operation is necessary to respond to a significant and imminent threat to life or property (including natural resources) and that no service by a private operator is reasonably available to meet the threat.

ii For the sole purpose of determining public aircraft status, *governmental function* means an activity undertaken by a government, such as national defense, intelligence missions, firefighting, search and rescue, law enforcement (including transport of prisoners, detainees, and illegal aliens), aeronautical research, or biological or geological resource management.

iii For the sole purpose of determining public aircraft status, *qualified non-crew member* means an individual, other than a member of the crew, aboard an aircraft operated by the armed forces or an intelligence agency of the U.S. government, or whose presence is required to perform, or is associated with the performance of, a governmental function.

2. An aircraft owned or operated by the armed forces or chartered to provide transportation to the armed forces if—

i The aircraft is operated in accordance with title 10 of the United States Code;

ii The aircraft is operated in the performance of a governmental function under title 14, 31, 32, or 50 of the United States Code and the aircraft is not used for commercial purposes; or

iii The aircraft is chartered to provide transportation to the armed forces and the Secretary of Defense (or the Secretary of the department in which the Coast Guard is operating) designates the operation of the aircraft as being required in the national interest.

3. An aircraft owned or operated by the National Guard of a State, the District of Columbia, or any territory or possession of the United States, and that meets the criteria of paragraph (2) of this definition, qualifies as a public aircraft only to the extent that it is operated under the direct control of the Department of Defense.*

In summary, the FAA mandated that one intending to operate an unmanned aircraft in national airspace must do so either under the permission granted by a COA (available only to public entities, which includes law enforcement agencies and other government entities), or with an experimental airworthiness certificate issued pursuant to relevant parts of Title 14 of the CFSs. Specifically proscribed were operations that are of a

* 14 CFR 1.1.

commercial nature, without the protection of a COA, but ostensibly under the guidelines set forth in AC 91-57.

In recognition that some commercial for-hire UAS operators were flying their systems in national airspace under AC 91-57 guidelines, the FAA published a second policy statement on February 13, 2007.[*] This notice was a direct response to increasing efforts by U.S. law enforcement agencies and some small UAV manufacturers to introduce systems into operational service on the back of model aircraft guidelines. The policy stated that the FAA would only permit UAV operations under existing certificate of authorization and experimental aircraft arrangements. The policy states:

> The current FAA policy for UAS operations is that no person may operate a UAS in the National Airspace System without specific authority. For UAS operating as public aircraft the authority is the COA, for UAS operating as civil aircraft the authority is special airworthiness certificates, and for model aircraft the authority is AC 91-57.
>
> The FAA recognizes that people and companies other than modelers might be flying UAS with the mistaken understanding that they are legally operating under the authority of AC 91-57. AC 91-57 only applies to modelers, and thus specifically excludes its use by persons or companies for business purposes.
>
> The FAA has undertaken a safety review that will examine the feasibility of creating a different category of unmanned "vehicles" that may be defined by the operator's visual line of sight and are also small and slow enough to adequately mitigate hazards to other aircraft and persons on the ground. The end product of this analysis may be a new flight authorization instrument similar to AC 91-57, but focused on operations which do not qualify as sport and recreation, but also may not require a certificate of airworthiness. They will, however, require compliance with applicable FAA regulations and guidance developed for this category.

The gap that is created by these policies was a consistent definition of a "model aircraft," and, as discussed in previous sections of this chapter, some individuals and agencies took advantage of this gap to operate small (and not-so-small) UAVs with cameras and other sensing equipment on board, clearly for either a commercial or law enforcement purpose, without having applied for a COA or a special airworthiness certificate.[†]

This all changed as a result of two significant events. The first was the passage of the FAA Modernization and Reform Act of 2012 (FMRA), signed into law by President Obama on February 14, 2014.[‡] The second was the release of the aforementioned Notice of Proposed Rule Making for Operation and Certification of Small Unmanned Aircraft Systems" on February 15, 2015.[§]

The FMRA firmly stated Congress' intention to create (or recognize) a new class of aircraft that was to be regulated by the FAA and offered three definitions of unmanned aircraft, small unmanned aircraft, and public UASs.[¶] The Act mandated a timetable for integration of civil UASs into the national airspace.[**] The Act specifically exempts aircraft flown strictly for hobby or recreational use from regulation.[††] The Act further mandated establishment of six test ranges, approval of commercial operations in the Arctic,

[*] 72 FR 6689, Supra Note 40.

[†] For example, see website for Remote Controlled Aerial Photography Association: http://www.rcapa.net/.

[‡] Public Law 112-95 126 Stat. 11.

[§] Docket No.: FAA-2015-0150; Notice No. 15-01.

[¶] FMRA Sec. 331 Definitions.

[**] FMRA Sec. 332 Integration of Civil Unmanned Aircraft Systems Into National Airspace System.

[††] FMRA Sec. 336.

coordination of operational approvals with public agencies, the creation of regulations, standards and requirements for civil unmanned system operations, and a number of other methods to achieve full integration of UASs into the national airspace.[*] The statute requires the Secretary of Transportation (parent agency of the FAA) to determine if certain UASs may operate safely in the NAS before the completion of the plan and rulemaking required in Section 332.[†]

Section 333 of the Act literally opened the floodgates for civil commercial UAS operators and entrepreneurs to petition the FAA for authority to conduct a wide variety of commercial operations, including aerial photography, precision agriculture, power line and pipeline infrastructure inspection, news gathering, building inspections, insurance adjusting, and many more, in low risk, controlled environments. "333 Petitions," as they have become known, have spawned a cottage industry in the legal profession whereby law firms and practitioners have begun holding themselves out as experts in the field and have inundated the FAA with petitions. As of this writing, nearly 500 § 333 petitions have been granted, and hundreds more are in line for review and approval. This process is intended, by the words of the Act, to be an interim path to operational approval for civil UAS operators pending the finalization of the small UAS rule. The criteria for approval are stringent, and the proponents are expected to demonstrate that they can safely fly their UASs in a manner that follows all applicable operational rules in the FARs and does not endanger the safety of persons or property in the air or on the ground.

Civil operators also have the option of obtaining a Special Airworthiness Certificate in the experimental category for civil aircraft performing research and development, crew training, and market surveys, or may go through the UAS type and airworthiness certificate in the restricted category process under 14 CFR § 21.25(a)(2) and § 21.185 for a special purpose, or a type certificate for production under 14 CFR § 21.25(a)(1) or § 21.17.

The other significant event was the long-awaited release of the Small UAS NPRM (Notice of Proposed Rule Making) for the Proposed Part 107, released February 15, 2015. The comment period closed on April 24, 2015, and the FAA received approximately 4,700 comments that must be resolved and published before the rule becomes final.[‡] At the same event, the FAA announced three "Pathfinder" initiatives wherein the FAA will work with industry partners on three focus areas: visual line-of-sight operations in urban areas, with CNN; extended visual line-of-sight operations in rural areas, with PrecisionHawk; and beyond visual line-of-sight in rural/isolated areas, with BNSF rail system.

The key elements of the NPRM are that the aircraft must weigh less than 55 lbs., visual line-of-sight operations only, no operations directly over people, daytime operations only, yield right-of-way to other aircraft, maximum airspeed of 100 mph, maximum altitude of 500 ft. AGL, minimum weather visibility of 3 miles, no operations in Class A, other classes except E with ATC permission, no private pilot certificate but some testing and qualifications required, and many more too numerous to list here. The important thing to remember is that none of the provisions of this proposed rule has become enforceable until the rule is finalized, a process that could take a year or more after the comment period has closed, but in any case long after this book is published. And the final rule may read significantly different than the proposed rule, so the only alternatives for commercial UAS operators now are the § 333 petition process or special "Pathfinder" type arrangements, short of going through the full type and airworthiness certificate mechanism that has

[*] FMRA Sec. 332 and 334.

[†] FMRA Sec. 333 Special Rules for Certain Unmanned Aircraft Systems.

[‡] Comments by FAA Administrator Michael Huerta at AUVSI Press Conference, Atlanta, Georgia, May 6, 2015.

been available all along. The prudent entrepreneur or UAS business person will be best served by working with experts well versed in the intricate maze of FAA rules, regulations and policies before launching any size or configuration UAS for commercial purposes. The FAA's website offers a current look at all relevant regulations and guidelines, and the § 333 petitions are all on line, with some proprietary elements redacted or excluded from public view.

5.6 The FAA's Enforcement Authority over Unmanned Aircraft Systems

The FAA has two issues to face with respect to its enforcement authority over UAS operations. First, it must determine what it *can* regulate, and, second, it must decide what it *will* regulate. The answer to the second challenge largely depends upon the resolution of the first.

The FAA issues six types of regulations: mandatory, prohibitive, conditionally mandatory, conditionally prohibitive, authority or responsibility, and definition/explanation.[*] Mandatory and prohibitive regulations are enforceable. The other four types represent exceptions or conditions. A thorough analysis of the applicability of a regulation to a particular situation will include answering the following questions: (1) to whom does the regulation apply? (2) what does it say in its entirety? (3) where must the regulation must be complied with? (4) when must it be accomplished? (5) how does it apply to the situation in question? and (6) are there are any special conditions, exceptions, or exclusions?[†]

Since unmanned aircraft are now defined by statute as "aircraft," and there is no exception found elsewhere in the regulations that excludes UAVs from the definition (except for recreational or hobby aircraft), the most current interpretation would be that the FAA has full regulatory authority over all aircraft that are capable of and do fly in the national, navigable airspace. "*Navigable airspace* means airspace at and above the minimum flight altitudes prescribed by or under this chapter, including airspace needed for safe takeoff and landing."[‡] Minimum safe altitudes are prescribed at 1000 feet above the ground in a congested area, with a lateral separation from objects of 2000 feet, and an altitude of 500 feet above the surface, except over open water or sparsely populated areas. In those cases, the aircraft may not be operated closer than 500 feet to any person, vessel, vehicle, or structure.[§] The exception is when it is necessary for takeoff or landing, in which case the navigable airspace goes to the surface (and along a designated approach path or airport landing pattern).[¶] The 400-foot AGL altitude limit for model aircraft contained in AC 91-57 was probably an observance of the 500-foot minimum safe altitude for manned aircraft operating anywhere except in Class G (uncontrolled) airspace,[**] providing a 100-foot "buffer," in addition to the recommendation to not operate within close proximity to an airport. The actual FAA policy history of AC 91-57 is not available for confirmation, but the

[*] Anthony J. Adamski and Timothy J. Doyle, *Introduction to the Aviation Regulatory Process*, 5th ed. (Plymouth, MI: Hayden-McNeil, 2005), 62.

[†] Adamski and Doyle, Introduction.

[‡] 14 CFR 1.1.

[§] 14 CFR 91.119.

[¶] 14 CFR 91.119.

[**] 14 CFR Part 71.

foregoing is the commonly held belief of FAA officials and individuals familiar with the history of model aviation.*

The vast majority of the FARs are intended to provide for safe operations of aircraft that carry people, both for the protection of the crew and passengers, and for people and property on the ground. Although unmanned aircraft have been on the aviation scene for over 90 years, there is no evidence in any of the preambles to regulations or other historical documents currently available for review that the authors of any regulation before about 2005 contemplated application of a specific regulation to unmanned, remotely piloted aircraft. Moored balloons and kites,† unmanned rockets,‡ and unmanned free balloons,§ categories of objects or vehicles that are intended to occupy a place in the airspace and are unmanned, are specifically covered by existing regulations, but there was nothing similar for other types of unmanned aircraft.

The FAA has long maintained that it has enforcement authority under existing airspace regulations 14 CFR §§ 91.111 and 91.113, which require that an operator of an aircraft be able to safely operate near other aircraft and observe the right-of-way rules, but the more difficult issue is whether such aircraft must meet certification requirements for the systems and the qualification standards, with appropriate certificates, for pilots, sensor operators, mechanics, maintenance personnel, designers, and manufacturers.

As of the publication of the First Edition of this book, there had been no formal legal challenge to the FAA's enforcement authority over unmanned aircraft and their operations. Government contractors, Customs and Border Protection, the U.S. military establishment, and other public aircraft operators had, for the most part, followed the guidelines of AFS-400 UAS Policy 05-01, Interim Operational Approval Guidance 08-01 and AC 91-57. Likewise, there was no anecdotal evidence that the FAA had initiated any enforcement activity against anyone who was, or was perceived to be, operating a UAS outside of these guidelines. Until a robust set of regulations that specifically addressed the unique characteristics of unmanned aircraft was implemented, there was always the chance that someone would fly a commercial UAS in such an open and notorious manner that the FAA would be compelled to respond with more than a "friendly" warning letter or telephone call.

The FAA's public position on this issue, as evidenced by the February 13, 2007, policy statement published in the *Federal Register*, was that any unmanned aircraft to be operated in the national airspace, with the exception of radio-controlled models, must comply with the requirements for a COA if it is a public aircraft, or for a special airworthiness certificate if it is a civil aircraft. Thus, the agency had consistently answered the second question (what it will regulate) with a broad statement of policy that it had the responsible authority over airspace and aviation.

The next question, until recently, was even if the FAA exercises its declared authority over airspace and aviation and attempts enforcement against an operator of a "small" (model size) UAS who is using the system for some arguably commercial purpose, without an airworthiness certificate or a licensed pilot in control, just what regulation would be enforced, and what sanction would be appropriate to deter further violations?

There have been many entrepreneurs and developers around the world whose presence and activities in the civil small UAS market (the UASs are small, the market is not) were

* Benjamin Trapnell, Assistant Professor, University of North Dakota, Lifetime Member of the Academy of Model Aeronautics.

† 14 CFR 101.11 et seq.

‡ 14 CFR 101.21 et seq.

§ 14 CFR 101.31 et seq.

putting pressure on the FAA to take the lead in UAS rulemaking. If a farmer or other commercial agriculture concern were to acquire a small system and fly it over fields in what could be characterized as "sparsely populated" areas, at an altitude where possible conflict with manned aircraft could occur, is there in place a regulatory mechanism to stop this activity? Or, if a commercial photographer were to operate a small UAS equipped with a camera over a similar area for the purpose of photographing the land for advertising or some similar purpose, could the FAA prevent the operation? Congress answered those questions with the enactment of Public Law 112-95, FMRA of 2012, as discussed above.

The issue for the FAA in the foregoing scenarios is what tools are in the FAA toolbox to enforce whatever regulations it may deem enforceable. The vast majority of these systems do not have an airworthiness certificate. The FAA's central mission is to promote compliance with safety standards.[*] FAA Order 2150.3A acknowledges that civil aviation depends primarily upon voluntary compliance with regulatory requirements, and only when those efforts have failed should the agency take formal enforcement action.

A certificate holder cannot be deprived of "property" (the certificate) without due process.[†] Congress has given the FAA authority not only to make the rules[‡] but also to enforce them through a number of methods, including issuance of "an order amending, modifying, suspending, or revoking" a pilot's certificate if the public interest so requires.[§] Any other certificate issued by the FAA can be "amended, modified, suspended or revoked" in the same manner. The problem with the aforementioned scenarios is that the "pilot" in all likelihood will not be an FAA-certificated pilot, because it is not required for such operations, and the aircraft and its systems will not be certified as airworthy, again because it is not required. As long as the operator/pilot does not interfere with the safe operation of a manned aircraft or otherwise enter a controlled airspace (such as in an airport environment) without permission, there may be no violation of any existing regulation.

Taking the scenario a step further, if the pilot/operator inadvertently allows the UAS to come close enough to a manned aircraft to force the latter into an evasive maneuver (not an unlikely event even in a sparsely populated agricultural region), a possible violation of 14 CFR §91.111 (Operating Near Other Aircraft) could ensue. In this situation, the FAA has no certificate to revoke, and thus no statutory or regulatory authority to proceed with a formal enforcement proceeding pursuant to 49 U.S.C. §44709(b).

This leaves one other mechanism: the civil penalty the administrator may impose against an individual "acting as a pilot, flight engineer, mechanic, or repairman."[¶] The FAA is authorized to assess a civil penalty for violations of certain regulations, up to $400,000 against large entities or companies and up to $50,000 against individuals and small businesses.[**] The relevant section of the U.S. Code defines *pilot* as "an individual who holds a pilot certificate issued under Part 61 of title 14, Code of Federal Regulations."[††] Again, an argument could be made that a non-certificate holder would not be subject to even the civil penalty provisions of the U.S. Code, thus leaving the FAA with no effective or realistic enforcement power over "unauthorized" civil unmanned aircraft operations.

These questions all came to convergence with an enforcement action initiated by the FAA in 2012 against an aerial photographer/entrepreneur named Raphael Pirker. The

[*] FAA Order 2150.3A.

[†] *Coppenbarger v. FAA,* 558 F. 2d 836, 839 (7th Cir. 1977).

[‡] 49 U.S.C. § 44701(a).

[§] 49 U.S.C. § 44709 (b); *Garvey v. NTSB and Merrell,* 190 F. 3d 571 (1999).

[¶] 49 U.S.C. §46301 (d)(5)(A).

[**] 49 U.S.C. §46301 et seq.

[††] 49 U.S.C. §46301 (d)(1)(C).

FAA's "Order of Assessment" dated June 27, 2013, stated that on April 13, 2012, Pirker was assessed a civil penalty of an amount of $10,000 for his unauthorized operation of a Ritewing Zephyr powered glider aircraft in the vicinity of the University of Virginia campus. The allegations stated that Pirker did not have a pilot certificate, that he operated the flight for compensation, that he operated the aircraft in a manner at excessively low altitudes over vehicles, buildings, people, streets, and structures between 10 and 400 feet AGL, all in a careless and reckless manner so as to endanger the life or property of another, and many more details set forth in eleven numbered paragraphs. The specific FAR that was allegedly violated was 14 CFR § 91.13(a), which states that no person may operate an aircraft in a careless or reckless manner so as to endanger the life or property of another.*

Predictably, Pirker retained counsel and opposed the Order of Assessment. In a hard-fought legal standoff, Pirker filed a Motion to Dismiss the Order, essentially on the grounds that there was an absence of a valid rule for application of FAR regulatory authority over model aircraft flight operations. The FAA contested the motion, and the Administrative Law Judge who was assigned to the case ruled in Pirker's favor on the grounds that the Zephyr was a "model aircraft" to which § 91.13(a) did not apply. There were other arguments made with regard to the FAA's policy-making process and whether published policies, in this case one in particular (the 2007 Policy Memorandum referenced in footnote 22), were enforceable as rules or regulations. The FAA appealed the ALJ's order to the full NTSB (the first step in the appeal process from an FAA enforcement order), and the NTSB reversed the decision, finding in favor of the FAA on the narrow issue of whether the plain language of 49 U.S.C. § 40102(a)(6) and 14 C.F.R § 1.1 definitions of "aircraft" include any aircraft, manned or unmanned, large or small. They found that the prohibition on careless and reckless operation in § 91.13(a) applies with respect to the operation of any "aircraft"other than those set forth in parts 101 and 103. The case was remanded to the law judge for further a factual hearing on whether Pirker actually violated § 91.13(a), and Pirker and the FAA thereafter settled their differences for a penalty reportedly significantly less than the original $10,000 assessment.

This case, although interesting for the factual and legal issues it presents, provides no binding legal authority to anyone other than the parties to that particular case, and since the FAA essentially prevailed on the legal argument regarding its enforcement authority, it may be presumed that they will continue to invoke § 91.13(a) and other relevant FARs to deal with alleged violators of the rules subject to any restrictions contained in the FMRA.

5.7 The Way Forward: The Future of Unmanned Aircraft System Regulations

The foregoing discussion suggests that the FAA's enforcement toolbox may have been lacking in substance when dealing with ignorant (of existing FAA policy), uncooperative, or openly defiant UAS operators. The day finally came when the FAA's hand was forced to deal with a UAS operator, pilot, manufacturer, or business entity that was willing to take the FAA to task on its enforcement powers and push the envelope to see how far it can go before a judicial showdown takes place (the aforementioned Pirker case). As market forces create greater opportunities for developers and entrepreneurs to invest capital into

* Docket No. 2012EA210009 "Order of Assessment."

more sophisticated systems and bring the industry closer to solving the sense-and-avoid problem, there has been ever-increasing pressure on the FAA to put into place a regulatory structure that will allow the agency to reclaim its "ownership" of the airspace. This necessarily includes implementing reasonable operational and engineering standards through the rulemaking process that will allow the industry to grow while not negatively affecting the overall safety of the aviation environment.

The first task is to define the scope of what the FAA can and should regulate. Now that there is a definition of *model aircraft* that is precise enough to give notice to the public of the exact nature of the aircraft that will remain unregulated, no longer can the clever entrepreneur claim to legally fly a commercial UAS under the authority of AC 91.57. The definition of "model aircraft" should include such factors as size, weight, speed, performance capability, and kinetic energy, but it appears now to be dictated by operational intent. A commercially available and popular quadcopters drone that anyone can purchase from the Internet and learn to fly in a few hours can be "recreational" one day (filming the family dog romping through the woods) and a commercial flight the next day while taking photos of homes for compensation from a real estate business. A "size, weight, etc." approach would describe the physical attributes of the aircraft and its systems. In addition, there should be a precise description of the locations and altitudes where model aircraft can be flown. The FMRA does not specifically set forth design criteria other than a weight limit of 55 pounds and otherwise invokes "community-based safety guidelines" for operations, merely requiring coordination with ATC if flown within five miles of an airport. If modeling enthusiasts want to create increasingly larger and faster models that could easily overtake and possibly bring down a small general aviation aircraft, they must know where those aircraft can be legally operated and under what conditions.

The civilian UAS community needs to have standards by which admission to the airspace can be assessed and authorized. There must be a workable definition of a "commercial" UAS operation so that there is no confusion about flying a commercial UAS mission as a model aircraft. A non-enforceable AC such as 91-57 is of little assistance to the FAA as it attempts to deal with commercial, for-hire UAS operators who believe that they are exempt from any certification requirement and understand that ACs are not regulatory and are not rules, nor are FAA policy statements binding on anyone other than the FAA. It remains to be seen whether the FMRA and the inevitable judicial interpretations of the statute and the final Part 107 rule will deal with these questions.

The only real alternative left for the FAA is to engage in the rulemaking process, subject to the inevitable lengthy comment and revision schedule, and that is what is in play as this book goes to press. The outcome of this "sausage making" process is not clear or readily predictable. There are many potential gaps in the proposed rule, as evidenced by the sheer number of comments to the NPRM with which the FAA is presently dealing. One approach that was considered was simply to amend the current regulations to state that UASs are "aircraft" and that their operators are pilots for all purposes. An exception could have been delineated that would exclude the modelers, subjecting everyone else to the full spectrum of Title 14. This approach would require that all UASs be fully certified as airworthy, that their pilots and operators be properly certificated and rated, and that all airspace regulations be fully complied with. The FAA's system of certification is already in place, and all that is lacking are the standards and guidelines that must be met in each applicable category of regulation. This approach is now moot in light of the FMRA and the pending NPRM.

A second approach would be to systematically dissect each and every part and subpart of Title 14 of the CFRs and amend them as necessary, again through the rulemaking process, as required, to incorporate all known characteristics of unmanned aircraft. This was also considered and apparently abandoned in favor of chartering the sUAS ARC. Many regulations clearly would have no application to UASs (such as those under Part 121 pertaining to passenger seat restraints or flight attendant requirements), while a large portion of the remainder could have application by interpretation, and thus would be candidates for amendment. This process could conceivably take years, but if undertaken, the most logical place to start would be 14 CFR Part 91, Air Traffic and General Operating Rules; Part 71, Airspace; on to Part 61, Pilot and Crewman Certificates; and then to the aircraft design standards found in Parts 21 through 49.

The third alternative is to create an entirely new part to 14 CFR devoted exclusively to UASs, which would incorporate all the issues of "see-and-avoid" technology, airspace access, pilot qualifications, manufacturing standards, and airworthiness certification. The Part 107 NPRM is the first step toward that solution, although as currently written it does not deal with design and manufacturing standards or airworthiness certification. Those standards should evolve out of the efforts of ASTM's F-38 committee and that was the recommendation of the sUAS ARC. Unfortunately, although the FAA delegated the standards development process for small UASs to ASTM and the F-38 Committee and its subcommittees have been diligently working on those standards for over three years, as of this writing, the Part 107 NPRM makes no reference to ASTM or community-based standards, so the status of the standards that have been developed remains uncertain pending further clarification from the FAA.

In the meantime, pending the full integration of UASs into the aviation world, the FAA requires a tool to enforce its authority over the airspace and to carry out its mandate to promote public safety and to do no harm to the current system through lack of oversight or misguided oversight. This can best be accomplished by a rule that reinforces the FAA's authority over the airspace and provides for sufficient sanctions against violators who do not possess certificates to be revoked or suspended, or who are otherwise immune from civil penalty. It is the hope of this author that the final outcome of the Part 107 NPRM, combined with future rules that may evolve out of the statutory mandate of the FMRA, will be a comprehensive, yet user-friendly regulatory scheme for all UASs in all categories and sizes that can serve as a model to the rest of the world. This vision is necessary to promote and facilitate the extraordinary economic engine that has evolved in the few short years since the introduction of small multi-rotor and fixed-wing UASs into the marketplace, while preserving the outstanding record of aviation safety that the United States has enjoyed for many decades.

5.8 Conclusion

The aviation environment is complex, dynamic, and littered with pitfalls, landmines, and blind alleys, to mix several metaphors, and the designer, developer, operator, or user of a UAS seeking access to the NAS or in international airspace must proceed with caution to ensure that the rules of engagement are fully understood. The rulemaking and standards development processes for UASs are underway and are sure to be so for the foreseeable future. Active involvement by the industry and the user community in the process is not only encouraged but also absolutely mandatory for the industry to grow and evolve in an orderly fashion. The opportunities for technological advancement for unmanned systems, many of which will have a positive impact upon the rest of the world of aviation from a safety and efficiency perspective, are virtually unlimited. The greatest challenge for the FAA and other CAAs around the world is to arrive at coherent, rational, and enforceable policies, procedures, rules, and regulations governing the operation of remotely piloted aircraft, regardless of where they are deployed or for what purpose.

DISCUSSION QUESTIONS

5.1 Discuss the Federal Aviation Act of 1958.

5.2 What is the FAA's "toolbox?"

5.3 List and discuss the three tools the FAA uses to administer the FARs.

5.4 The FAA has supported and sponsored four domestic committees dedicated to developing standards and regulations for the manufacture and operation of unmanned aircraft. List and discuss each committee.

5.5 Discuss the initial intent of AC 91-57. Is this AC still applicable in light of recent developments in statutory, regulatory, and judicial law?

6

Human Factors in Unmanned Aerial Systems

Warren Jensen

CONTENTS

6.1 Introduction

While remotely piloted aircraft (RPA) are often referred to as "unmanned," it is clear that the humans involved and their actions are critical elements to safe operations. The goal of human factors is to provide operators and support personnel with the necessary knowledge, skills, and abilities and to achieve the overall goal of safe, effective, and efficient operations. The human has a great capacity to learn, perceive, integrate information, and execute complex decision making at a high skill level. Conversely, humans can be vulnerable to saturation, fatigue, spatial disorientation, and communication failures, among other issues. The challenge in the field of human factors is to address these issues through awareness of human traits and limitations, training programs, proficiency and currency, teamwork skill development, and effective workstation design. On the surface, it appears that the goal of human factors training is simply to minimize or eliminate errors. The overall goal is to optimize efficiency and effectiveness of humans managing RPA operations.

The purpose of this chapter is to provide a working discussion of some common issues seen when transitioning to and working with unmanned systems. This chapter must approach a wide range of operators and their experiences, operating equipment that can vary in its capabilities, and flying very different missions and conditions. In light of this concept, this chapter will focus on human performance challenges and strategies to approach those challenges.

6.2 The Enormity of the Scope

The large variety of remotely operated vehicles create physical and cognitive demands on human operators, as well as their capabilities. The aim of this chapter is to provide a broad discussion of human factors concepts and their application to the field of RPA operations. The topics are not specific to a single aircraft system or mission, as changes in systems and procedures are frequent. Many of these concepts are common to manned systems, or other professions operating in complex, dynamic settings (e.g., medicine and law enforcement).

6.3 A Caution Regarding Hindsight Bias

During the training phase of any career such as aviation, medicine, and stock trading, case studies provide excellent discussions for practitioners. Accident reviews and discussions are often a central part of aviation safety meetings. In these discussions, human factors-based errors are valuable learning opportunities, but they can be subject to hindsight bias on the part of the presenter and student. Hindsight bias can interfere with the ability to identify vulnerabilities in ourselves and our organizations, resulting in a failure to appreciate the risks and take appropriate corrective measures.

Hindsight bias is defined as the "belief that an event is more predictable after it becomes known than before it became known" (Roese and Vohs, 2012). In other words, once an error occurs, the reviewer can incorrectly surmise that event should have been more easily identified and prevented, but in reality, it is only due to the fact that the outcomes of the actions are known. For example, a pilot who lands an aircraft with the landing gear retracted will often face the question "What were you thinking?" or the statement "I would have never done that!" In a field that carefully examines mishaps for the purpose of shaping new training, designs or strategies, hindsight bias of personnel can impede progress toward more safe and efficient operations.

Several traits seen in hindsight bias are important to understand as an individual embarks on a safety-related career. First of all, individuals with hindsight bias feel the events that occurred were more easily predictable than could be reasonably explained. It can often be associated with an oversimplification of the events, with an emphasis on the error that occurred. Other events that influenced the error may be omitted from the discussion, due to memory distortion or desire to only select what we see as the salient events that contributed to the mishap.

The concern about hindsight bias stems from the fact that individuals, when transitioning to new aircraft, systems, or procedures, may look at the case files in a way that does not consider all the factors that can influence operations. As humans we prefer to link singular actions to outcomes, hoping to solve problems with a linear relationship. Even this warning for hindsight bias is likely to be ineffective. In an article by Paul Goodwin, the author states the danger of hindsight is that we "can hinder our learning from … past forecasting errors, limiting the extent to which we can improve our forecasting skills through experience" (Goodwin, 2010). Successful strategies to counteract hindsight include working to gain knowledge about the event and understand how an individual in the situation may think their choice was reasonable. It is important to realize that accident reports and operator-based insights from such events provide excellent learning experiences. The goal is to use them effectively and keep hindsight bias in check.

6.4 Human Perception and RPA Operations

Aircraft pilots transitioning to become RPA operators quickly note how the two systems differ in the perceptual inputs they receive during operations. The imaging technologies used in some RPA operations can provide valuable images of the nature of their operations (crop sensors, wildlife tracking, traffic management, etc.), but these images may have limited value for aircraft control and/or navigation. In addition, on-board RPA imaging systems (optical, infrared, etc.) may not provide protection in sensing and avoiding hazards, whether fixed or moving. The challenge is to understand how humans use visual systems to assess their surroundings and risks and how we attempt to compensate for the change in our perceptions and their assessments.

Currently, small RPAs must be flown within line of sight of the operator (which may also include ground observers). Normal visual acuity (20/20 for distant vision) of the operators will provide them with the ability to detect an object 2.3 meters in size one mile away. This ability to detect an object is dependent upon several factors, including contrast, color, shape, movement across the visual field of the observer, and the area (measured in angle) of sky the observer must search for the object (Williams and Gildea, 2014, pp. 6–7).

Visual displays from the RPA to guide operators for in-flight maneuvers can also be difficult to assess due to distortion (as compared to direct visualization) and loss of peripheral vision cues. Some systems using bore-sight visual displays for the assessment of landing performance require pilots to monitor visual image changes that differ from their prior experience. In addition, the changes they must monitor appear to be much more subtle in the critical phase of landing. These changes in perception can result in operators providing faulty control inputs which may result in mishaps.

Relative motion issues occur when an object does not move across an individual's peripheral vision. Motion in our peripheral vision is an attention cue, much like a car passing in a driver's peripheral vision on the highway. Objects that do not move across the peripheral vision are not easily detected. An example would be a small RPA flying toward a ground observer who is looking 30° to the right of the incoming RPA. Lack of motion across the peripheral vision may not draw the observer's attention to the incoming RPA.

Individuals scanning an "empty sky" in search of an RPA or airborne threats may focus their vision to a point closer than the distance to the vehicle or threat. This altered focus can impair the detection of those objects. In order to counter those effects, ground observers are advised to use short, fixed scans to areas of the sky, consciously pushing their focal length to the threat area. This strategy is used to counter the effect of relative motion and near-focus anomalies.

6.5 Attention

Charles Wickens' multiple resource theory was developed to explain human cognitive challenges of performing simultaneous and/or difficult tasks (Wickens, 2008). The theory describes brain resources as the brain's ability to perform tasks (e.g., visual perception, spatial processing, and motor skills) and also identifies the challenge of sharing a single resource for two separate tasks. For example, reading a book while driving a car would share visual perception resources and impair performance of one or both tasks. Attention is defined as the management of these resources. The resources can be managed several ways referred to as attention types. An understanding of the attention types can be valuable to understand each type's vulnerabilities, as well as strategies to optimize attention performance.

6.6 Selective Attention

Pilots, air traffic controllers, and RPA operators are required to monitor a number of information sources, such as aircraft attitude, speed, and position. When an operator gathers information through a sequential sampling of information sources (instruments, controls, or actual physical environment such as aircraft on runways), these operators are using selective attention. This process utilizes a systematic visual scan of information sources and involves the operator's skill to select and process these information sources. Pilots involved in instrument flying demonstrate a selective attention process, using a sequential scan to determine their speed, altitude, attitude, and other information.

The number of information sources the operator samples is referred to as load stress. With practice and proficiency, an individual's ability to manage load stress will improve. Currency, experience, and effective scan patterns improve performance in selective attention settings. Design of the information systems, to place instruments closer together, will also improve performance.

A key strategy in the management of selective attention is crew coordination of monitoring duties or for performing actions needed for operations. RPA operators may be able to coordinate team members to monitor aircraft position and/or alert others in a high load stress condition. For example, using an RPA in a lost child scenario would allow team members to monitor video images, RPA position, obstructions, and potential airborne conflicts simultaneously, while the demand of monitoring all these information sources could overwhelm a single operator.

6.7 Focused Attention

Directing brain resources to a single task is considered focused attention. A common example of focused attention would be an individual intently reading a book while unaware of his surroundings. Directing attention to an emergency checklist to the exclusion of other information is a similar situation known as channeled attention. Common problems experienced by individuals using focused, or channeled, attention are distractions and loss of situation awareness.

Watching a movie is more difficult with cell phones ringing, people talking, or other distractions. While some distractions, such as emergency alarms, are necessary for safe operations, they do interrupt the operators' ability to perform that task. The process of operating RPAs involves several settings in which focused attention is likely. The actions of an operator to look at the sensor operator's display, talk with onsite individuals, or refer to a manual may be necessary duties, but the operator must be aware of the negative effect a distraction can have on the performance of the primary task, such as airborne threat monitoring or manual RPA control.

Another issue of focused attention is the lack of situation awareness due to attention directed to only one aspect of the environment. Famous accidents have resulted when the crew turned to focused attention and lost overall situation awareness of the flight (NTSB). When one individual needs to use focused attention on a task, crew coordination is needed to divide tasks, such as monitoring aircraft systems and threats, for the crew to remain situationally aware. The primary concerns during focused attention are distraction management and loss of situation awareness prevention strategies.

Distraction management strategies include limiting the number of sources who communicate with vehicle operators during critical phases of flight and developing strategies to filter or prioritize information to curtail unnecessary distraction. Crew coordination would involve assigning tasks to team members during times when an individual must be in focused attention during an operation. This could include monitoring for airborne threats or systems malfunctions, during operations that would require operators to be using focused attention.

6.8 Divided Attention

Divided attention is the concept of performing two tasks simultaneously, such as talking to another individual while manually controlling an aircraft. Many tasks are done in this way in a very safe manner, but interference between these tasks can increase safety risks. Divided attention tasks are more difficult to perform when the tasks share resources. As mentioned earlier, resources are the cognitive skills we have to perform tasks. An example of sharing a single resource for two separate tasks would be seen when an operator is manually maneuvering a small RPA near obstructions while they are also observing visual displays for the optical images. If the operator cannot see the obstructions through the optical system, the process of alternating their visual focus between the RPA obstruction clearance task and the screen monitoring task, performance is likely to be negatively affected. In this case the demand for visual monitoring of both tasks can be overwhelming.

Even tasks that do not share resources, but are demanding of the operator, can have negative consequences. An example of this concept would be asking an individual to perform an unpracticed flight maneuver while engaging in conversation. While this combination of activities does not appear to share resources, it can affect performance if the cognitive demands of these tasks are high. Effective crew management involves assigning tasks that do not share resources.

6.9 Sustained Attention

Sustained attention is an attention process that monitors the environment for changes, which may or may not be foreseen. Examples of sustained attention would include an individual monitoring the progress of a flight or operation, looking to identify abnormalities or findings in which the operator must further investigate. This could be thought of as "guard duty" in which operators must maintain a high level of awareness. An example would be an RPA operator, monitoring an automated system for failures, flight performance, and navigation, all for the purpose of identifying performance that is out of pre-established parameters.

A common problem experienced during sustained attention is degraded monitoring skills after 20–30 minutes. Strategies described to improve sustained attention performance consider the design of monitoring tasks and alarm systems (proximity warnings for air traffic control), work–rest cycles, and motivation and comfort of the operators. Increasing the engagement of the operator in the monitoring task is also beneficial.

The above description is simplistic to the complex nature of cognitive function in complex, dynamic settings, where operators change attention types rapidly and without awareness. The point is that awareness of strategies can be beneficial to users to improve performance, as well as to understand vulnerabilities for the purpose of avoiding common errors.

6.10 Human Error

What is an error? Most individuals will agree that many events happen when driving a car, but what events are considered an error? All drivers tend to agree crossing the centerline to oncoming traffic is an error, but crossing over the centerline of a deserted highway is not held in the same level of agreement, as the likelihood of a negative outcome is decreased. Most people identify an error is an event where a negative consequence is likely to occur. Can the same action be an error in one setting and not another? If errors are only considered due to their consequence, this may be true.

Senders and Moray defined an error as an act that was "not intended by the actor, not desired by a set of rules or an external observer; or that led the task or system outside its acceptable limits" (Senders and Moray, 1991). Errors occur when an action results in an outcome that was not intended, even if the actions performed were intended. This author believes that identifying error for the purpose of improving performance should not be dependent upon the consequence related to the single occurrence. The saying "No harm, no foul" may not apply to RPA operations, in that the same error in the future may have significant consequences. To only identify and correct errors when negative consequences occur is inefficient at best and would more likely delay the identification and management of unacceptable risks.

It is evident that well trained and current operators commit errors. The nature of human error involves a wide range of activities, from overall strategy and planning errors, to inadequate monitoring of information systems, to poor technical execution of a skill. Some errors are inconsequential, which means that there is no negative outcome if they are not identified and corrected. Many errors are identified and corrected through monitoring actions of operators or other individuals. Anticipating, monitoring, identifying, and correcting errors are common tasks in the aviation industry.

Why can errors be difficult to manage? What reasons could explain our resistance to identify and effectively manage errors in ourselves and others? Part of this issue may be in the emotional responses we have to errors, whether they are due to the actions of others or ourselves.

When trained and qualified individuals commit errors, it is seen as an unexpected event. The operators may be unaware of their error, especially during routine actions. The unexpected nature of error can impair in an individual's recognition, acknowledgment, and correction of the error. In reality, errors are common and should be expected. Defensive driving strategies are based on the expectation of errors being committed by other drivers, and through anticipation of the error and immediate application of correction strategies, further errors or consequences of those errors can be avoided. Air traffic controllers are trained to expect errors in communication or understanding, for the purpose of immediate recognition and corrective action. Training professionals in RPA operations should include active monitoring and expectation of errors. Anticipation of errors can be crucial for timely correction as illustrated in the following interview.

> Following takeoff and climb-out from an uncontrolled airport, I radioed to the area that I was departing the airport traffic pattern to the east of the town, climbing through 3,000 feet to a cruise altitude of 5,500 feet. I received a call from another aircraft approaching that airport, who identified their position (location) to be "west of the town at 3,500 feet, over the high school, inbound for an GPS (global positioning system) approach." I recalled there was only one high school in the town and it was on the east side of town, directly under me. I immediately leveled the aircraft at 3,200 feet and saw the other aircraft pass overhead. I was fortunate to identify the error—they were east of town, not west—and avoid a collision (Interview with a pilot, 2011).

We can expect to experience distrust of other individuals who commit errors, which is likely to impair coordination and teamwork. The key is to manage errors as an expected event, with monitoring and correction strategies in place. Operators should be trained in error correction strategies.

Negative transfer is a common error when transitioning to new systems with similar layouts. Negative transfer is using a previous, well-established skill in a new setting in which the action is incorrect. Transferring from an aircraft to an RPA that uses the same joystick control, but has different functions for the buttons, would likely lead to an individual to perform incorrect actions. A driver who uses manual transmissions, now controlling an automatic transmission, may reach for controls in the wrong location in their new setting.

6.11 Threat and Error Management

Safety-based organizations have developed strategies to address human errors since the beginnings of aviation. Initially, rules and regulations were used to provide guidance to aircrews and air traffic controllers. Checklists aided operators to prevent memory errors. The 1970s saw the beginning of "Cockpit Resource Management" (CRM) to assist crew interactions with regard to communication, crew coordination, team building, and decision making (Helmreich et al., 1993). CRM training has been carefully evaluated and continues to evolve in its design and extent. The concept has extended beyond its original intent, with training involving cabin crews, maintenance teams, air traffic controllers, and individuals in law enforcement and medical fields.

The work of James Reason, Douglas Wiegmann, Thomas Shappell, and others further identified ways to classify errors and contributing causes, with the intent to identify trends and develop effective countermeasures. James Reason identified a model to identify defense strategies to manage errors (Reason, 2008, pp. 97–100).

In *Human Factors Analysis and Classification System*, HFACS, Shappell and Wiegmann (2000) provided a framework to identify and classify errors and contributing factors. Their overall goal was to develop a tool to create error management strategies based on the data obtained from the incident. The four layers of defenses were described as administrative procedures, supervisory controls, preconditions, and operator actions (Figure 6.1). Each layer uses designs or strategies to stop errors and, conversely, ineffectiveness of a defense layer may allow an error to continue. Several publications noted in the reference section of their work provide a thorough discussion of their assessment system, as well as applications to several aviation fields.

Organizational influences can range from formal policies, regulations, laws, and operating manuals. These documents provide guidance to the organization to properly conduct operations, similar to traffic laws influencing the safety and flow of traffic. Guidance from the administrative level can also lead to problems, such as an organization curtailing safety training for cost-cutting purposes.

Supervisory influence is the actions of other individuals to monitoring operators for the purpose of providing guidance and corrections. This aspect usually refers to a chain of command and appropriate monitoring and controls applied to an organization. First officers asserting themselves to correct errors would be considered an effective use of supervisory influence. On the other hand, some supervisory influence can be provided by other individuals, outside the operational group. Bystanders pointing out an obstruction or airborne threat can provide input to the operators as well.

Preconditions are considered any influence that would increase the risk of error. Gusty winds, poor visibility, approaching darkness are some of the common environmental preconditions in aviation operations. Other aspects of preconditions would be distraction, fatigue, or poor information and control display designs.

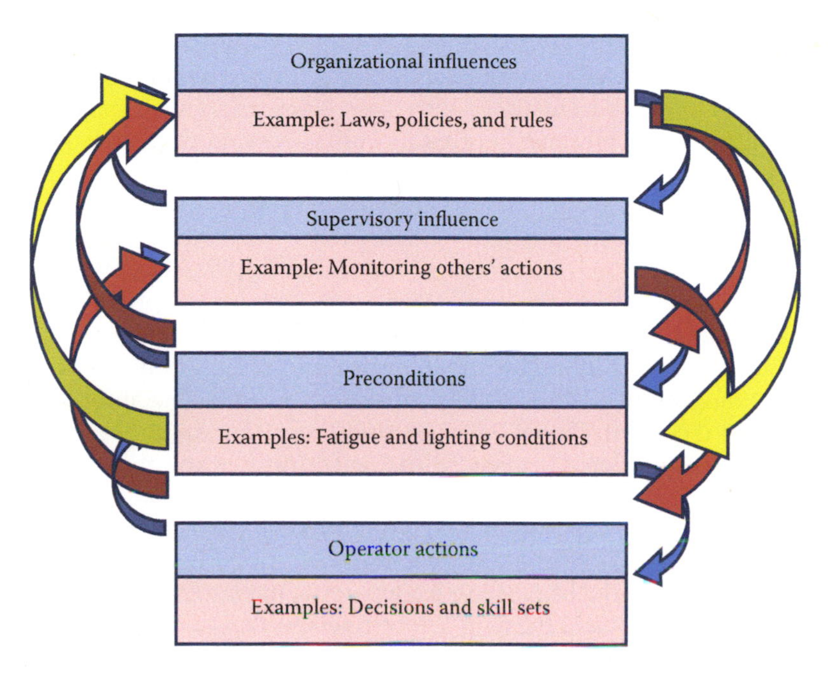

FIGURE 6.1
Layered error defenses. (Modified from Shappell, S. and Wiegmann, D. 2000. *The Human Factors Analysis and Classification System—HFACS*, DOT/FAA/AM-00/7.)

Operator actions include errors, such as decisional, skill-based and perceptual errors. Positive aspects of operator actions can also be seen, such as persistence, bravery, and highly developed skills to manage problems.

The arrows on the side of the diagram indicate the model is dynamic and actions can modify other defense layers. Operators can be influenced by policies of the administrative level and inputs and/or corrections from the supervisory level. Policies may be altered as a result of input from operators or supervisors. For example, would increasing frequency of airspace violations by RPAs influence the changing of administrative policies regarding operations? Would new policies on crew duty days decrease fatigue? Would decreasing medical certification requirements for operators result in a change of operator reliability? The defense layers can be changed (for better or worse) by influences at other levels.

The HFACS model presented in the literature identifies further factors to more clearly classify the issues related to each defense layer. The purpose for these evaluations is to identify factors and trends that could be corrected and/or studied further.

6.12 Crew Resource Management

Crew Resource Management was defined by John K. Lauber of the National Transportation Safety Board as "using all available resources—information, equipment and people—to achieve safe and effective flight operations" (Lauber, 1984). The concept of developing strategies to improve group performance through better coordination, communication, and decision effectiveness was fostered following a series of crew coordination incidents in the 1970s (NTSB). Since that time, coursework for the purpose of improving team management continued to expand, including cabin crews, airport ground personnel, air traffic controllers, as well as other industries such as medicine and law enforcement. A discussion of the history and development of this process is beyond the scope of this text, but a discussion of best practices, as they relate to unmanned systems, is appropriate.

Team concept principles stem from manned aviation operations, where the goal is to train pilots and other crewmembers to work together in a coordinated fashion, sharing information, decision-making, and actions (Weiner et al., 1993). In RPA operations, there are likely to be others outside the operations team that would provide input or direct operations. The fire chief at the site of major fire may suddenly need specific information that prompts a change in flight. Law enforcement and search and rescue personnel are likely to have changing needs for the RPA operators as well. The ability to clearly communicate needs and capabilities among these users will have an impact on the safety and effectiveness of the operation. Crew coordination training is an important component of effective RPA operations.

The initial goals of CRM training involved increasing participative management and assertiveness in the crews. While a clear chain of command is important in RPA operations, participative management refers to leaders who involve other crew members in the decision-making and actions of the team. Assertiveness was advocated to team members who were not in charge, but needed to clearly advocate their concerns to the group leader. Both skills are important in the development of teams and should be an integral part of training. This is especially important for teams that have never trained together previously.

Effective team management involves other attributes that contribute to the success of teams. A discussion of these cultural attributes can provide insight to groups regarding their interactions. This discussion involves the use of power distance, effective communication, ambiguity resolution, distraction management, and negative transfer. These are a few of the topics in advanced crew coordination coursework but are considered central issues in safe operations of unmanned systems.

Power distance is defined as the leader's consideration of opinions or concerns of individuals who are not in a position of authority (Hofstede et al., 2010). Leaders who do not ask for, or consider, others' opinions would be considered to have a high power distance. Conversely, leaders with low power distance allow others to provide their inputs to conduct of the operation. Low power distance is considered optimal for information exchange and effective decision making.

Effective communication within a group supports situation awareness, decision-making effectiveness, and error management. The use of standardized terminology is an important component of effectively transferring ideas. Directional guidance, position reporting, and requests for maneuvers all depend upon terminology which correctly and efficiently conveys concerns of the operator and support teams. The process of clearly encoding messages and providing unambiguous inquiries should go beyond classroom discussions to practice in simulations and operations.

Ambiguity is defined as a finding that can have two interpretations or two indications whose interpretations conflict (O'Brien and O'Hare, 2007). For example, is the image on the screen a fugitive with a rifle or a gardener with a rake? The other form of ambiguity would be an oil pressure alarm sounding, but the oil pressure gauge conflicts this finding, showing normal operational pressures. The state of uncertainty created by both forms of ambiguity is a challenge to situation awareness and safe operations.

Sometimes ambiguity is created by limitations of the mechanical systems (lighting, resolution of the optical system, etc.) or by crew interpretations or expectations of the information (engine noise, reflection of sound). Ambiguity may be apparent or unknown to the operators.

Ambiguity resolution involves strategies to improve situation awareness by gathering more information from sources, focusing to identify the issues or causes of the ambiguity, and evaluating the situation to determine the future implications of chosen actions. For instance, gathering more information regarding the position of a small RPA may involve asking other ground observers for information, utilizing global positioning information or referring to onboard camera output. Often ambiguity can be due to faulty assumptions made by team members. Everyone has a responsibility to identify and resolve ambiguity.

Distraction management involves identifying information or actions not pertinent to the operation, so as to prevent their influence on the operations. In commercial aviation, some non-critical alarms are silenced during critical phases of flight, such as take-off and landing. Increasing workload with non-relevant issues is considered a threat. The challenge is to understand what information is non-critical and, when in high workload operations, when distraction management strategies should be used.

For example, during launch procedures and obstacle avoidance procedures, it is wise to limit conversation unrelated to the mission or information that is not needed for that phase of the operation. Certain information, such as an engine performance abnormality, needs to be communicated even in a critical phase of flight. The difference can also be skills and experience of the operator to manage independent channels of information. Be mindful that high workload periods are not always predictable, nor is it simple to know whether incoming information is needed or if it creates a distraction. Careful observation and discussions with experienced operators are important training tools for distraction management.

6.13 Situation Awareness

Many professions (aviation, law enforcement, medicine) share the common challenge of maintaining awareness of the state of the operations, trends of changing conditions, and the identification and realistic appraisal of risks to the operators and the mission. The number of issues an operator can manage, while guarding against unnecessary distraction, can be a challenge and a skill worth developing. Situation awareness is critical in unmanned aircraft operations. In spite of the diversity of professions involved in situation awareness training, common cognitive processes are used.

Mica Endsley defined situation awareness as "an internalized mental model of the current state of the flight environment" (Garland, D.J. et al., 1999). Delta Air Lines training programs state it is "the ability to recognize events occurring to you and around you then reacting correctly to those events" (Captain Art Samson, personal communication, 2002). There are four aspects of situation awareness, seen as common steps in decision-making processes.

6.14 Vigilance

The task of vigilance requires an individual to direct their attention to information (e.g., aircraft position, instrument readings, airspace conflicts, structures) which has the most immediate impact on safety. Vigilance not only is managing the appropriate type of attention but also requires knowledge of the operations to identify and correctly assess the risks involved. For example, obstructions or low-contrast backgrounds could result in the loss of RPA visual contact during ground observation operations. Knowledge of this risk assists the ground observer's vigilance, allowing them to identify that potential problem prior to movement in that area.

Vigilance is commonly degraded by distractions. When using focused attention to provide inputs to the control systems, even momentary distractions can lead to significant problems. Distraction management strategies are designed to improve vigilance of tasks. These strategies include limiting conversation to operational issues during critical phases of flight. What about a police RPA operator who has risk to their own safety? Working in focused attention compromises an individual's ability to maintain situation awareness for the reason they can have difficulty in monitoring their surroundings while engaged in the focused attention task.

6.15 Diagnosis

When working in complex systems, correctly identifying and understanding a developing situation can be challenging. While we may be carefully observing the information systems we have, coming to a correct diagnosis of the situation can be dependent upon our training and familiarity with the issues, the nature of the human–machine interface, clarity of communication, ambiguous findings, our problem solving skills, and our bias in interpreting information. Spatial disorientation is a well-known accident cause in aviation and cases exist where pilots communicate their inability to interpret the instrument indications, even when the instruments are providing correct indications. This refers to an aspect of achieving a correct diagnosis of a problem.

Achieving the correct diagnosis involves many skills. As described by Gary Klein, the recognition-primed decision model depends upon the user's prior experiences and their ability to correlate their current situation (Klein, G.A., 1993). This process also requires the operator to ask questions or make observations to clarify the diagnosis. Klein describes the process as creating a story to explain the findings, using experience to shape the explanation. An individual's story will also lead them to consider successful alternatives to the problem and likelihood of success of each alternative.

For example, operators, flying piston-powered RPAs, are trained to recognize loss of power and quickly identify conditions or malfunctions that are responsible. Information is gathered and compared to patterns developed during training and experience. RPA trainees can spend a significant amount of time learning systems (fuel, electrical, etc.) and training with failure simulations to prepare for these contingencies. The goal is to develop mental models that can be recalled during the event, such as the loss of power. The operator, using this prior experience, can quickly identify what information is needed and create a story to explain the findings.

One challenge to make a correct diagnosis can lie in the clarity of the information presented to the operator. Incomplete, inaccurate, misleading, or conflicting information can impair this process. Klein describes an anomaly as a finding or information that does not fit our explanation (Klein). For example, when an engine tachometer shows decreasing performance, but there is no other confirming evidence and the vehicle is operating normally, the abnormal reading is considered an anomaly. The natural human reaction is to disregard the anomaly because it cannot easily be explained. As professionals, training emphasizes the need to consider abnormal readings that cannot be explained.

6.16 Risk Analysis

While this term has many meanings and applications, with regard to situation awareness, risk analysis asks "What is the likelihood and consequence of failure of each option I am considering?" Individuals who choose options that have a high likelihood of failure and/or significant consequences of failure may be demonstrating a loss of situation awareness. (However, it can be argued that heroic actions often have one or both of these aspects.) Granted, risk analysis of options we have not performed in the past can be difficult; our past experience of success in high risk actions may also mislead us when analyzing risk.

Vicarious learning is defined as an individual whose prior poor choices and safety risks have not resulted in negative outcomes and, as a result, the individual does not have a realistic assessment of the risk they are taking. For example, drivers who use excessive speed, but have no consequences such as accidents, speeding tickets, or negative comments from their passengers, may underestimate risks of driving at those speeds. Through their flawed risk analysis, they may develop a sense of invulnerability in those situations.

6.17 Action

Individuals who appropriately monitor their surroundings and correctly diagnose their situation may still fail to correct the problem appropriately. Students with writing assignments may delay beginning their work due to complacency. Individuals who commit errors may not be willing to identify and correct the error, as this will bring attention and possibly blame and liability. The hesitancy in both these examples would be considered a loss of situation awareness due to a lack of action.

6.18 Human–Machine Interfacing

The goal when designing workstations (control modules, information displays, etc.) is to create a system the operators can use efficiently and effectively. Often, the control systems are analogs of previously designed control systems, such as video gaming, aircraft controls, or computer keyboards.

6.19 Compatibility

In man–machine interfacing, compatibility refers to the consistency of the information and control systems to the operator's expectations. There are clear advantages to working with systems consistent with our prior experience. Laptop computers with similar layouts, commands, and software programs will result in faster learning, fewer errors, faster reaction times, less mental workload, and greater user satisfaction. Poor compatibility with inconsistent layouts can result in negative transfer errors. When transitioning to a new workstation, attention to compatibility can identify aspects that need modification or additional training to achieve compatibility.

6.20 Compatibility Types

Conceptual compatibility is the effective use of symbols, colors, sounds, or other indicators to convey information. Some symbols we see are cultural or related to specific profession, and their meaning may not be clear without specific training. Numbers on runways indicate magnetic heading rather than some other numbering strategy. Do the operators understand the meaning of the colors, symbols, or alarm sounds associated with their workstations? Are the keystrokes used for commands consistent with the operator's expectations?

Spatial compatibility refers to the organization of the information and control systems. Are the information systems located where the operator would expect them? Are the layers of menus consistent with expectations? New computer operating systems often challenge users when the organization is changed. Newer flat panel displays are easier for users to transition to when the locations are consistent with their prior experiences and therefore expectations.

Movement compatibility refers to the direction and sensitivity of the movement. Information systems can have movement compatibility with the deflection of instrument needles. A fuel gauge needle, for example, that moves in a direction opposite of the user's expectation can lead to an error in interpretation. The direction and sensitivity of controls can also lead to operator error.

Controlling a small, radio-controlled RPA through direct visual observation is a good example of movement compatibility challenges. While the aircraft flies directly away from your vantage point, the roll command of the control system consists of roll of the aircraft in the same direction. Control of the aircraft coming toward you may be more difficult to translate until you have a greater compatibility with the system.

Operators can learn systems (and become highly skilled) using systems that are initially poorly compatible. Be aware that systems that are poorly compatible will increase demands on the users.

Summary of points to be considered as follows:

1. Human error is common. Your job as a safety officer is to identify and manage error. Be alert and inquisitive to the activities occurring and anticipate errors. Learn and apply error management strategies in activities for the purpose of incorporating those actions as a natural process.

2. Consider the complete situation of another individual's error (as well as your own). Be insightful of factors that can lead to poor performance, such as lack of currency, poor communication, and distraction. Do not let hindsight bias interfere with the effectiveness of incident and accident reviews.
3. Situation awareness is difficult and fragile, at times. We need to work to achieve and maintain it. All four aspects of situation awareness discussed—vigilance, diagnosis, risk analysis, action—need to be actively processed to attain it. Distraction management strategies should be discussed and implemented with your team.
4. RPA operations can be perceptually challenging. Recognizing perceptual limitations to identify threats or the correct orientation of your RPA are necessary to correctly manage an operation. Spatial disorientation is a common concern in RPA operations due to the altered perceptions operators receive.
5. Attention is the ability to manage the inputs, processing, and outputs of your brain. The attention type will change rapidly and, at times, to the detriment of safety and effectiveness. Key strategies for improving task performance are to understand the challenges and employ corrective strategies of each attention type. For example, when operators are in focused attention while the RPA is negotiating obstacles, distraction management actions should be used.
6. Crew resource management development has included many useful team actions to identify and resolve problems in real-time operations. CRM training is highly recommended for RPA teams and groups in which they interact.
7. Human–machine interactions can be challenging. Proficiency, currency, and developing correct mental models of the systems you are using cannot be overemphasized.

Recommended Readings

1. A Summary of Unmanned Aircraft Accident/Incident Data: Human Factors Implications, Kevin Williams, USDOT, 2004.
2. Documentation of Sensory Information in the Operations of Unmanned Aircraft Systems, Kevin Williams, FAA, USDOT, 2008.
3. Defensive Flying for Pilots: An Introduction to Threat and Error Management, Merritt and Klinect, FAA, Human Factors Division, 2006.
4. *The Human Contribution*, James Reason, Ashgate Publishing, 2008.

References

Eastern Airlines, Inc, L-1011, N310EA, Miami, Florida, December 29, 1972, NTSB report AAR-73/14.

Garland, D.J., Wise, J.A., and Hopkin, V.D. (Eds). 2010. *Handbook of Aviation Human Factors*, Second edition, Taylor & Francis, Boca Raton, FL.

Goodwin, P. 2010. Why hindsight can damage foresight. *International Journal of Applied Forecasting*, www.forecasters.org/foresight, retrieved 5/29/2015.

Hofstede, G., Hofstede, G.J., and Minkov, M. 2010. *Cultures and Organizations: Software of the Mind*, third edition. McGraw-Hill, USA.

Klein, G.A. 1993. *A Recognition-Primed Decision (RPD) Model of Rapid Decision Making* (pp. 138–147). Ablex Publishing Corporation.

Lauber, J. 1984. Resource management in the cockpit. *Air Line Pilot*, 53, 20.

O'Brien, K.S. and O'Hare, D. 2007. Situational awareness ability and cognitive skills training in complex real-world task. *Ergonomics*, 50(7), 1064–1091.

Reason, J.T. 2008. *The Human Contribution*, Ashgate Publishing Limited, Burlington, VT.

Roese, N. and Vohs, K. 2012. Hindsight bias. *Perspectives on Psychological Science*, 7(5), 411–426.

Senders, J.W. and Moray, N.P. 1991. *Human Error: Cause, Prediction and Reduction*, Lawrence Erlbaum Associates, p. 25.

Shappell, S. and Wiegmann, D. 2000. *The Human Factors Analysis and Classification System—HFACS*, DOT/FAA/AM -00/7.

Shappell, S., Detwiler, C., Holcomb, K., Hackworth, C., Boquet, A., and Wiegmann, D. 2006. *Human Error and Commercial Aviation Accidents: A Comprehensive Fine-Grained Analysis Using HFACS.*

Weiner, E., Kanki, B., and Helmreich, R. 1993. *Cockpit Resource Management.* Academic Press, London, UK.

Wickens, C. 2008. Multiple resources and mental workload. *Human Factors*, 50(3), 449–455.

Williams, K.W. and Gildea, K.M. 2014. *A Review of Research Related to Unmanned Aircraft System Visual Observers*, DOT/FAA/AM-14/9, Office of Aerospace Medicine.

7

Safety Assessments

Eric J. Shappee

CONTENTS

7.1 Introduction

For years the aviation field has been rapidly advancing in technology. With all the change, aviation organizations and manufacturers have found themselves faced with new safety issues and ever-changing safety requirements. The unmanned aerial system/remotely piloted aircraft (UAS/RPA) field is no different. In fact, safety in this arena is more of a concern. With having no onboard pilot, complex operating systems, and ever-changing avionics as well as continuous software updates, safety appears to be one of the major hurdles for integrating UAS/PRA into the National Airspace System.

In this chapter, we will examine several safety tools and techniques such as the hazard analysis and its various forms. We will also look at the risk assessment process and provide some guidance on developing a risk assessment tool. And finally, we will discuss safety evaluations and provide some thoughts on UAS/RPA accident investigation considerations.

7.2 Hazard Analysis

Hazard analysis can take several forms. In this section, we will look at several common types of hazard analysis. The purposes/functions of the hazard analysis are all predicated on what stage of the operation for which you are applying it.

7.2.1 Purpose

Hazard analyses are common tools found in the system safety arena. Generally these tools are used throughout various stages of a product life cycle. In his book *System Safety for the 21st Century*, Richard Stephens identifies the various stages of a product life cycle. These stages or phases are concept, design, production, operations, and disposal. Although in UAS/RPA operations we do not look specifically at the development of a product throughout its life cycle, we do, however, look at its operational phase. We can subdivide the UAS/RPA operational phase into several general stages: planning, staging, launch, flight, and recovery. Applying the appropriate hazard analysis tool within each stage will allow for early identification and ultimately early resolution of safety issues.

7.2.2 Preliminary Hazard List

The preliminary hazard list (PHL) is just what it sounds like, a list. Simply put, it is a brain storming tool used to identify initial safety issues early on in the UAS/RPA operation. To get the most out of the PHL, you need to have a variety of input from the people familiar with each stage of the UAS/RPA operation and its phases. Figure 7.1 is an example of a PHL that can be used to aid in the process.

To use the PHL tool, we first need to have an in-depth understanding of the stage we are going to evaluate. At the top of the form, select the stage (planning, staging, launching, flight, and recovery) to be evaluated. Doing this helps to keep all the various sheets from team members organized in proper categories for ease of review. The next step is to list a tracking number (1, 2, 3, etc.) and potential hazards we see in the selected stage. For example, in the staging phase you may want to list items such as nearby terrain features (trees, power lines/poles, and antennas). After listing the hazards, we need to determine the probability and severity of the hazard. In the probability column we can enter frequent, probable, occasional, remote, or improbable. These probability levels are listed and defined in MIL-STD-882D/E and in the Appendix of this text.

The next column is severity. In this column you can use the categories of catastrophic, critical, marginal, or negligible. As the probability levels, the severity categories and definitions are also listed in MIL-STD-882D/E and in the Appendix. The last column of the hazard list is the RL or risk level. This is the point where we establish an initial RL value based on the probability and severity that we have identified. For instance, if we determined that launching the UAS/RPA at a field that has trees nearby would have a probability of impacting a tree to be Remote and the severity to be Critical, then using the risk matrix in MIL-STD-882D/E, we can determine that the risk RL for that hazard is a 10. Note

PRELIMINARY HAZARD LIST / ANALYSIS (PHL/A)

DATE: ____________ PREPARED BY: ____________ Page ____ OF ____

Operational Stage: ☐ **Planning** ☐ **Staging** ☐ **Launch** ☐ **Flight** ☐ **Recovery**

TRACK #	HAZARD	PROBABILITY	SEVERITY	RL	MITIGATING ACTION	RRL	NOTES

RL = Risk Level, RRL = Residual Risk Level Probability, Severity, and Risk Levels defined in MIL-STD- 882D/E

FIGURE 7.1
Preliminary hazard list/analysis (PHL/A).

that the higher the number, the lower the risk. If you decide to develop your own PHL/A, be careful since not all risk matrices are alike, some will be organized to have the lower number signify a lower risk.

7.2.3 Preliminary Hazard Analysis

Once the initial RLs have been identified, we now need to move into the analysis phase by looking at ways to mitigate listed hazards. This is fairly simple; here is where we ask what can be done to reduce or eliminate the hazard. When looking at mitigation we need to look at it in the terms of probability and severity. Concerning probability, we want to determine ways to eliminate or reduce the possibility of occurrence or better known as exposure. Let us say that we have determined that the field from which we want to operate has trees at the approach and departure end of the runway. In the mitigating action column we can list several solutions. The first can be relocating to another field with no trees, the second possible action could be to remove the trees, and the third, and probably the most reasonable, would be to establish or modify the launch and recovery procedures.

The next column is the RRL or residual risk level column. This time the question is whether we will lower the risk by implementing these mitigating actions. Just as when we determine the RL, we have to consider the probability and severity. You may find that one or both (Probability and Severity) may have changed. Changing any one of these factors can lower or increase the RL. Obviously if we increase the RL, then we do not want to implement that particular mitigating action.

The last column is labeled notes. This is fairly self explanatory. If we have any special concerns or instructions needed for the implementation of the mitigating actions, we will want to list them in this column in some detail. As we complete the PHL/A worksheet we must keep in mind that it is used as an initial hazard identification tool. Once the UAS activity is underway, an operational hazard analysis should be performed in order to evaluate the hazards after the mitigating action has been applied. Hazard analysis tools like the PHL/A are extremely useful when assessing the hazards of the UAS/RPA operational cycle. The main purpose for using the hazard analyses tool is to provide the user with a systematic approach to identifying, analyzing, and mitigating hazards early in the operation.

7.2.4 Operational Hazard Review and Analysis

Just like the PHL/A tool is used to identify initial safety issues early on in the UAS/RPA operation, the operational hazard review and analysis is used to identify and evaluate hazards throughout the entire operation and its stages (planning, staging, launching, flight, and recovery). This is a crucial part of the ongoing and continuous evaluation of hazards and provides the feedback necessary to determine that the mitigating actions employed have worked as expected.

Obviously we would want to continue monitoring the hazards we listed on the PHL/A, but there may be other hazards that appear during the UAS/RPA operation/activity that were not foreseen. Items that you should always consider with the OHR&A are in the area of human factors. These items are human interface with the equipment and operating systems as well as crew resource management (CRM). This can get complicated quickly, depending on the number of crew members and their specific tasks. Both human factors and CRM will be covered in more depth in later chapters but human factors issues and CRM must be continuously monitored.

The use of the Operational Hazard Review & Analysis (Figure 7.2) tool is very similar to the PHL/A. The main difference is the action review column. In this column we want to list whether the identified mitigating actions implemented from the PHL/A were adequate. If the actions were not adequate and the hazard has not changed, then list the hazard again. If the actions have modified the hazard, then list the modified one. At this point the rest of the OHR&A tool works like the PHL/A. To aid in keeping things organized I suggest the use of separate worksheets when it comes to hazard review and the evaluation of new operational hazards. I also suggest that the tracking numbers on the OHR&A sheet correspond with the ones listed on the PHL/A. Doing this will aid in keeping all the safety analysis and review information organized. Just as before, the probability levels, severity categories, and risk matrix are listed in MIL-STD-882D/E and in the Appendix.

7.2.5 Change Analysis

The change analysis serves a crucial role in the ongoing review and analysis of safety. What the change analysis allows you to do is review and examine any changes that have been made to the operation. For example, if we have a UAS/RPA system software change such as an upgrade for the UAS computers or operating systems, we will want to make an assessment of the changes and evaluate how these changes effect the overall operation. Another example would be a procedural change; you may have modified the launch procedure to get the vehicle in the air faster. This modification would also warrant an assessment of the changes made. To assess the change, use the OHR&A worksheet. List all hazards associated with the changes in the action review column and run the worksheet just as you would an OHR&A.

OPERATIONAL HAZARD REVIEW & ANALYSIS (OHR&A)							
DATE: ______		PREPARED BY: ______				Page ___ OF ___	
Operational Stage:	☐ Planning	☐ Staging		☐ Launch	☐ Flight	☐ Recovery	
TRACK #	ACTION REVIEW	PROBABILITY	SEVERITY	RL	MITIGATING ACTION	RRL	NOTES
RL = Risk Level, RRL = Residual Risk Level			Probability, Severity, and Risk Levels defined in MIL-STD- 882D/E				

FIGURE 7.2
Operational hazard review & analysis (OHR&A).

7.3 Risk Assessment

According to Maguire (2006), "Public perception of risk is the key to safety" (p. 47). I would dare to take it a step further and state that in the Unmanned Aircraft world, the public perception of risk is the first key to airspace integration and acceptance. The second being privacy concerns. How we approach and manage that risk is critical. One type of tool that has been used by military, airlines, and some flight training schools is a basic risk assessment matrix. The Risk Assessment tool in Figure 7.3 is a derivative of one that was developed for a flight training program. Risk assessment can best be defined as the evaluation of common operational hazards in term of severity and probability.

7.3.1 Purpose

The risk assessment tool serves two purposes. The first is it provides the UAS/RPA operator with a quick look at the operation before committing to the flight activity (a go/no-go decision). The second is that it allows safety and management real-time information needed to continually monitor the overall safety of the operation. This tool should be completed by the UAS/RPA operator before each flight and briefed to the crew. The briefing should consist of at least a review of the risks, hazards, and any concerns associated with the activity. This tool is meant to be an aid in the decision-making process and should not be the only means used to making the go/no-go decision.

sUAS RISK ASSESSMENT

Crew / Station: ______________ / ______ ______________ / ______

______________ / ______ ______________ / ______

Mission Type	SUPPORT 1	TRAINING 2	PAYLOAD CHECK 3	EXPERIMENTAL 4	
Hardware Changes	NO 1			YES 4	
Software Changes	NO 1			YES 4	
Airspace of Operation	Special Use 1	Class C 2	Class D 3	Class E, G 4	
Has PIC Flown This Type Aircraft	YES 1			NO 4	
Flight Condition	DAY 1			NIGHT 4	
Visibility	≥ 10 MILES 1	6 - 9 MILES 2	3 - 5 MILES 3	< 3 MILES 4	
Ceiling in feet AGL	≥ 10,000 1	3,000 - 4,900 2	1,000 - 2,900 3	< 1,000 4	
Surface Winds		0 - 10 KTS 2	11 - 15 KTS 3	> 15 KTS 4	
Forecast Winds		0 - 10 KTS 2	11 - 15 KTS 3	> 15 KTS 4	
Weather Deteriorating	NO 1			YES 4	
Mission Altitude in feet AGL		< 1,000 2	1,000 - 2,900 3	≥ 3,000 4	
Are All Crew Members Current	YES 1		NO 3 →	CURRENCY FLIGHT REQUIRED	
Other Range/ Airspace Activity	NO 1			YES 4	
Established Lost Link Procedures	YES 1			NO NO FLIGHT	
Observation Type	Line of Sight & Chase 1		Chase Only 3	Line of Sight Only 4	
UAS Grouping	GROUP I 1	GROUP II 2	GROUP III 3	GROUP IV 4	
Total					

RISK LEVEL			
20 - 30 **LOW**	31 - 40 **MEDIUM**	41 - 50 **SERIOUS**	51 - 64 **HIGH**

Aircraft Number: ______________ Aircraft Type: ______________

Flight Released By: ______________ Date ________ Time________

FIGURE 7.3
Small UAS risk assessment.

7.3.2 Development

The risk assessment tool shown in Figure 7.3 was designed for small UAS/RPA operations. As stated above, the risk assessment tool is meant to be an aid in the decision-making process. When considering developing a risk assessment tool you will want to tailor it for your specific operation. To get started, assemble those involved directly with the operation and discuss the operational factors such as weather, crew rest, and airspace. Also you should consider items listed on the PHL/A that would change per flight cycle.

Once you have developed the list, the next step is to identify how each factor can change in terms of probability and severity. At this point you will need to make a decision on whether or not to use a numeric ranking scale. If you choose not to, that is fine; the only caution I give, however, is that you may not have an easy way to identify the overall total RL (low, moderate, serious, and high). My recommendation would be to add some type of ranking system. The one in Figure 7.3 is a numeric system with a total value scale listed at the bottom of the sheet. Attached to each total value scale is a total RL category (low, moderate, serious, and high). These categories along with an example risk index are listed in MIL-STD-882D/E and in the Appendix. The numeric scale makes computer tracking and monitoring of overall operational risk easier. The overall risk category aids in briefing team members and gives them a meaningful RL for the operation.

The last item that needs to be mentioned and considered under development is the periodic reviewing and updating of the risk assessment tool to determine the effectiveness of the tool and make changes as necessary. You may find that some of the factors identified have changed. This could be due to a platform change or the operational factor or hazard was eliminated. Also review the OHR&A and change analysis to determine if any of the new hazards identified need to be considered.

7.3.3 Use

To use the risk assessment form in Figure 7.3, you will want to start by listing the crew and their position or station. Next, move to the matrix and start with the left-hand column where you see the first operational factor, mission type. From this point move right until you reach the type of mission. The choices listed are support which covers a broad range of activities such as disaster response, training, an example would be a new UAS/RPA operator, payload check which covers upgrades to payload or new payloads, and experimental which would be classified as a new vehicle or type or UAS/RPA operation.

Looking at the first row, if your mission type is training, the associated risk number would be 2 and you place a 2 in the far-right column. If your mission type is experimental, place a 4 in the far-right column. Continue down the operational factors list in the left column and move right to the associated RL that fits your flight, and place that number in the far-right column. As you can see, the farther right you go, the greater the associated RL. Once you have determined the RLs for each operational factor, add up the numbers in the far-right column to determine your total risk value.

Once the total risk value has been calculated, find which range your value falls within. For instance, if your total value is 26 the risk is low. The RLs of low, medium, serious, and high are derived from MIL-STD-882D/E. Below the RLs you will find spaces for aircraft number, aircraft type, flight released by, date, and time. All are self explanatory except the flight released by. This space should be reserved for someone with management authority such as the chief pilot and mission director. The idea behind this is to have management review each evaluated operational factor as well as the total overall risk value and sign for risk acceptance. Remember this is just a tool to help assess the risk and safety of the operation. This tool should not be the only means of determining a mission go or no-go.

Looking back at the matrix section, a few of the operational factors listed in the left column warrant further explanation. Hardware changes are items such as wing sets and engines. Items like operating system updates or new versions of software would fall under software changes. Under the operational factor of airspace of operation, you will find special use. A good example of special use would be restricted areas or an area with temporary flight restrictions (TFR). Also in the same row you will find Class C, D, E, and G airspaces. As the involvement of air traffic control is decreased you will notice that the RLs increase. Currently the two predominant airspaces for civil UAS/RPA operations are restricted areas and E and G—providing you are operating under some type of authorization or waiver such as a Certificate of Authorization (COA). When it comes to other range/airspace activities you have a choice of yes or no. The idea behind this operational factor is that if you have other aircraft in the vicinity/airspace or restricted area they could constitute a hazard and should be considered. The last item is UAS grouping. Detailed information on these groups can be found on the FAA website. In general, these groupings address a variety of items. Some of the items addressed are weight limits, speed limitations, and altitude restrictions.

7.4 Safety Evaluation

A major key to integrating UAS/RPV into the national airspace is its safety evaluation. The FAA recognizes that UAS/RPV will need to meet an acceptable level of risk. Doing this can be very challenging. This section will examine several ways to aid in the evaluation process of operational safety. Items that will be discussed will be risk assessments, flight test cards, and airworthiness.

7.4.1 Risk Assessment

As stated earlier, the purpose of risk assessment is twofold. First it provides the UAS/RPA operator with a quick look at the operation before committing to the flight activity (a go/no-go decision). Note that a risk assessment should be completed before every flight activity. Second it allows safety and management the means to review operational risks and continually monitor the overall safety. It is this review and continuous monitoring along with the completed risk assessment tools that provide the needed data to show an acceptable level of risk with the flight operation.

7.4.2 Flight Test Cards

Another key element of the safety evaluation is the flight test card. A flight test card is a set of tasks or functions that the UAS/RPA vehicle and/or ground station must be able to perform. These test cards are usually performed in some type of special use airspace such as a restricted area where an FAA authorization or waiver such as a COA is not required. After all, the whole purpose of the safety evaluation is to develop good safe practices and gain the needed safety data for FAA authorization to fly in the national airspace.

The test card shown in Figure 7.4 is the final flight test card that should be completed before the responsible organization will endorse the airworthiness certification. When developing test cards such as the airworthiness, auto land, or payload specific tests, you need to have an understanding of the equipment being used and its limitations. You also need to be familiar with the FAA requirements for UAS/RPA operations. If these test cards are developed properly they will be a great asset along with risk assessment. These two tools can go a long way in providing operational safety data for airworthiness certification.

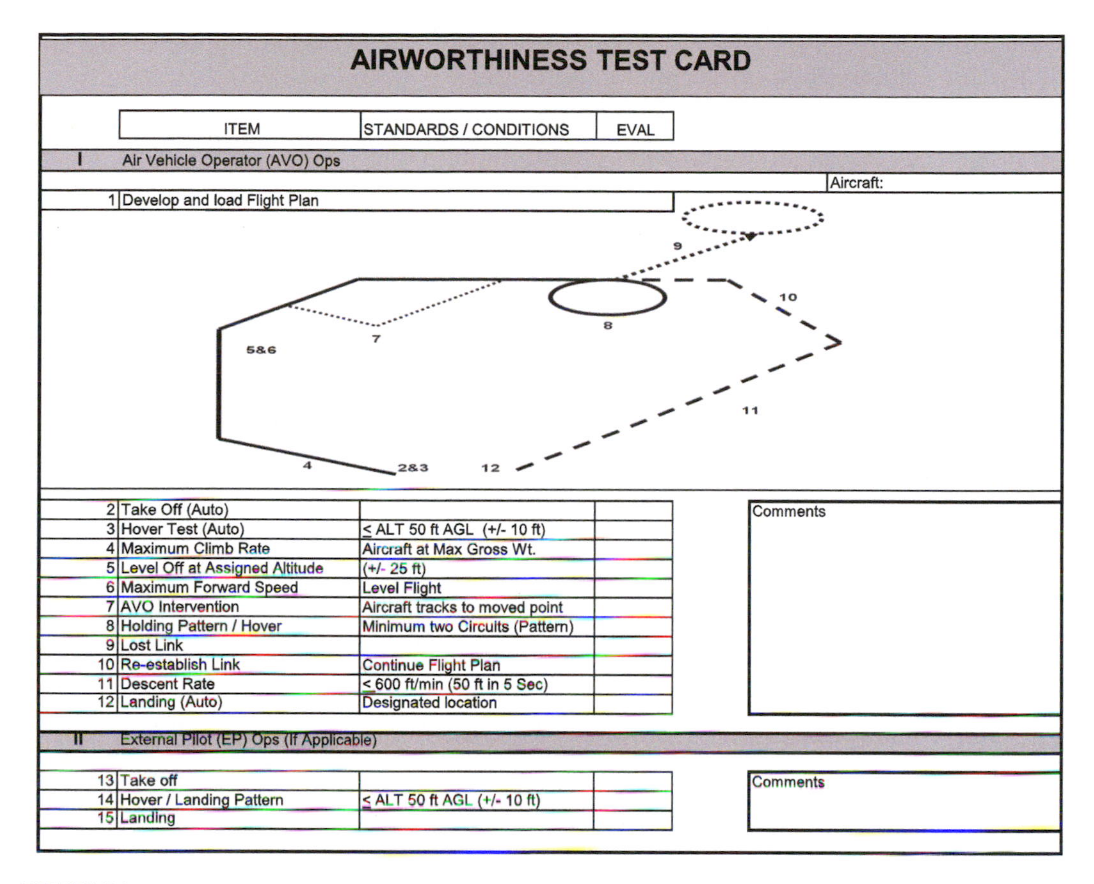

AIRWORTHINESS TEST CARD

	ITEM	STANDARDS / CONDITIONS	EVAL
I	Air Vehicle Operator (AVO) Ops		
1	Develop and load Flight Plan		
2	Take Off (Auto)		
3	Hover Test (Auto)	≤ ALT 50 ft AGL (+/- 10 ft)	
4	Maximum Climb Rate	Aircraft at Max Gross Wt.	
5	Level Off at Assigned Altitude	(+/- 25 ft)	
6	Maximum Forward Speed	Level Flight	
7	AVO Intervention	Aircraft tracks to moved point	
8	Holding Pattern / Hover	Minimum two Circuits (Pattern)	
9	Lost Link		
10	Re-establish Link	Continue Flight Plan	
11	Descent Rate	≤ 600 ft/min (50 ft in 5 Sec)	
12	Landing (Auto)	Designated location	
II	External Pilot (EP) Ops (If Applicable)		
13	Take off		
14	Hover / Landing Pattern	≤ ALT 50 ft AGL (+/- 10 ft)	
15	Landing		

Aircraft:

Comments

Comments

FIGURE 7.4
Airworthiness test card.

7.4.3 Airworthiness Certification

According to the FAA, public institutions have the option to self-certify airworthiness. To do this should involve a few more steps than just saying "everything looks good; we are airworthy; let's go fly." Using the tools, such as the risk assessments and flight test cards discussed in this chapter, will be very valuable in this process. However, we still need to consider operator and crew qualifications, air vehicle reliability, ground station reliability, and program/software capabilities before we even fly. When it comes to the operator and crew, we need to review their qualifications such as pilot certifications, experience, and competency using the system. As for air vehicle reliability, we need to consider structural integrity, power plant (engine) reliability, and aerodynamics and performance. When considering ground station reliability, we are asking, how reliable is the equipment? Is there a back up or contingency plan for equipment failures? The last consideration listed is program/software capabilities. Are the programs user friendly to minimize human factors issues? How reliable is the software and programs that are being used? Are there any backup systems? Are there any frequency issues/conflicts? As you can see the questions can be virtually endless. Just as critical as the items that are being evaluated is having some way of documenting this information, be it an application, checklist, or a combination thereof. These are but just a few basic areas that need to be examined along with some questions that would need answers in order to fly the air vehicle for further safety evaluation.

7.5 Accident Investigation Considerations

One subject about which there is very little information available is UAS/RPA accident investigation. Although many of the tools and techniques traditionally used in the manned aircraft type investigations will work in unmanned, there are, however, some unique differences.

7.5.1 Software and Hardware

Most of us are fairly familiar with how to operate the programs that are installed on our home computers or laptops. But, are we really familiar with the software? Do we really know all the features of the programs that we use? Unless you are a computer guru the answer is probably no. When it comes to software, many safety professionals, especially the system safety folks, know that software, if not compatible with other software or operating systems, can cause serious problems. Stephans (2004, p. 53) states, "A software specification error, design flaw, or the lack of generic safety critical requirement can contribute to or cause a system failure or erroneous human decision." When it comes to investigating UAS/RPA accidents you will want to take a close look at the software. To do this you will probably need someone who is very familiar with the specific operating system such as the software engineer or a programmer.

Like software, hardware is a critical area in accident investigation. The hardware components can be divided into two different categories. First is the hardware configuration. Here we would want to ask the question: "have all components been connected and physically checked for proper configuration?" Examples would be transmitters, backup power supplies, and antenna. The second category, and the one that could be the most problematic, is the interface between the hardware and software. Here is where you need to ask yourself, is the hardware and software compatible? Again you will need someone who is very familiar with the specific operating software and components.

As the person responsible for conducting UAS/RPA accident investigations it would be to your advantage to have someone with these special skills in this area on your team or party. Another benefit to having members who are very familiar with the software and hardware of the system is that most of the UAS/RPA systems record flight and operational data. With the expertise of these individuals, they should be able to extract and interpolate the data from the flight. They can also be beneficial in aiding you in the simulation to reconstruction of the entire flight. This type of information that is extracted is similar to the information that is retrieved from a flight data recorder.

7.5.2 Human Factors

Although the effect of human factors is covered in another chapter it warrants a special note in this section. As time goes on, you will see more and more studies concerning human factors and UAS/RPA operations. In this section, I offer some areas that you may want to consider when investigating. The first is crew coordination. Unlike the airlines with two or three cockpit crew, with UAS/RPA operations you can find significantly more crew than just the pilot and copilot, who in many cases is the payload operator. If your system is not equipped with an auto land or takeoff system, you will have added an external pilot with a control box (RC remote). In most operations you will need an observer or chase plane with pilot. This adds another level of complexity for crew coordination.

System complexity or user friendliness is another issue that you will want to consider. Many of the operating systems and associated software were designed by computer engineers with no aviation experience. What is simple for them may not be simple or even flow on a checklist for the UAS/RPA operator. A system not designed with human factors in mind could see a high increase in operator error. This error could occur anywhere from mission planning and programming to a situation in which time is critical, for example making quick decisions to avoid a collision or other disaster.

7.5.3 Suggestions

If you get tasked with investigating a UAS/RPA accident, I have a couple of suggestions. The first is do not try to tackle the investigation all on your own. You are going to need experts familiar with the field to help you get and analyze the information. Second, have a plan to get organized. Know what the major areas are for the investigation. One tool that I use is an investigation roster shown in Figure 7.5. This roster lists the major areas and provides space for assigning team members to specific tasks.

sUAS

INVESTIGATION ROSTER

IIC__________________ Case # __________ Start Date__________

	Section Chair					
Operations						
Pilot / Maintenance						
Safety Assessments						
Human Factors						
Software / Hardware						
ATC (If Applicable)						
Weather / Airfield						
Structures/ Perf						
Witnesses						
Other						

FIGURE 7.5
Investigation roster.

7.6 Conclusion and Recommendations

The information provided in this chapter is a good starting point for UAS/RPA safety and safety evaluations. The tools discussed in this chapter are tools that I have developed over the past several years and have used during the evaluation of UAS/RPAs we have flown and for which we have obtained Certificates of Authorizations. As the world of Remotely Piloted Aircraft grows, so will the need for safety. For those of you who are interested in broadening your knowledge of safety and jumping into this field with both feet, I would like to offer a couple of suggestions. First, take some safety courses. Take courses in the area of safety management, system safety, and safety management systems. I have found these courses to be invaluable when developing safety tools and evaluating safety of operations. Second, spend some time online or at the library reviewing some of the references listed from this chapter.

DISCUSSION QUESTIONS

7.1 List and discuss each of the UAS/RPA operational phases.
7.2 Define probability and severity.
7.3 Discuss the differences between the PHL/A and the OHR&A.
7.4 What are the two purposes of the risk assessment?
7.5 What is the purpose of the safety evaluation?
7.6 Discuss some of the differences between manned type accident investigation and unmanned.

Appendix

The following charts are excerpts from MIL-STD-882D/E.

Example Mishap Probability Levels

Description[a]	Level	Specific Individual Item	Fleet or Inventory[b]
Frequent	A	Likely to occur often in the life of an item, with a probability of occurrence greater than 10^{-1} in that life	Continuously experienced
Probable	B	Likely to occur several times in the life of an item, with a probability of occurrence less than 10^{-1} but greater than 10^{-2} in that life	Will occur frequently
Occasional	C	Possible to occur some time in the life of an item, with a probability of occurrence less than 10^{-2} but greater than 10^{-3} in that life	Will occur several times
Remote	D	Unlikely but possible to occur in the life of an item, with a probability of occurrence less than 10^{-3} but greater than 10^{-6} in that life	Unlikely, but can reasonably be expected to occur
Improbable	E	So unlikely, it can be assumed occurrence may not be experienced, with a probability of occurrence less than 10^{-6} in that life	Unlikely to occur, but possible

[a] Definitions of descriptive words may have to be modified based on the quantity of items involved.
[b] The expected size of the fleet or inventory should be defined prior to accomplishing an assessment of the system.

Example Mishap Severity Categories

Description	Category	Environmental, Safety, and Health Result Criteria
Catastrophic	I	Could result in death, permanent total disability, loss exceeding $1M, or irreversible severe environmental damage that violates law or regulation
Critical	II	Could result in permanent partial disability, injuries, or occupational illness that may result in hospitalization of at least three personnel, loss exceeding $200 K but less than $1 M, or reversible environmental damage causing a violation of law or regulation
Marginal	III	Could result in injury or occupational illness resulting in one or more lost work days(s), loss exceeding $20 K but less than $200 K, or mitigable environmental damage without violation of law or regulation where restoration activities can be accomplished
Negligible	IV	Could result in injury or illness not resulting in a lost work day, loss exceeding $2 K but less than $10 K, or minimal environmental damage not violating law or regulation

Example Mishap Risk Assessment Matrix (MRAM)

Severity Probability	Catastrophic	Critical	Marginal	Negligible
Frequent	1	3	7	13
Probable	2	5	9	16
Occasional	4	6	11	18
Remote	8	10	14	19
Improbable	12	15	17	20
Designed out	21	22	23	24

References

DOD. 2000. MIL-STD-882D Standard practice for system safety.

FAA. 2009. Small unmanned aircraft systems aviation rulemaking committee. Comprehensive set of recommendations for sUAS regulatory development.

Maguire, R. 2006. *Safety Cases and Safety Reports*. Burlington, VT: Ashgate Pub Co.

Shappee, E. 2006, March. Grading the go. *Mentor, 8*(3), 12.

Stephans, R. 2004. *System Safety for the 21st Century*. Hoboken, NJ: Wiley-IEEE.

8

Export Control and ITAR

Eric McClafferty and Rose Mooney

CONTENTS

8.1 Introduction

This chapter focuses on U.S. export control regulations related to unmanned aircraft systems (UAS), components of UASs, launch systems, payloads, and related equipment. Export controls not only regulate the movement of physical goods outside of the United States but they also cover releases of software and "know how" outside the United States that occur through a variety of mechanisms, including emails, phone calls, and the use of shared databases. Surprisingly to some, the rules cover releases of certain information to non-U.S. persons completely inside the United States. The regulations apply to U.S. person activities worldwide, such as when a U.S. citizen travels to another country. Violations of these rules undermine U.S. national security. Violations can also lead to criminal penalties, including jail time in some circumstances, to high dollar civil penalties, denial of export privileges, and a prohibition on selling to the government, not to mention reputational damage. Moreover, although U.S. export controls are enforced much more energetically than other countries' programs, because most of the controls discussed in this chapter originate in international agreements, many other countries have the same basic rules in place.

U.S. export controls require specific U.S. government export licenses for most exports of commercial and military UASs and certain vehicle control software. Specific, case-by-case export approvals are also required for releases in the United States of a good deal of "know how" to non-U.S. persons, especially of information related to product development.

This chapter will provide you with an understanding of the basic rules regarding export controls on military and commercial UASs and will familiarize you with the two principal U.S. government agencies that administer those rules. We will review the real-world consequences of not complying with the rules and describe how to develop a system to help avoid violations.

Export controls and economic sanctions have been around for thousands of years, such as when the city of Athens refused to trade with Megara in 432 BC,* and American history is full of examples of export controls and embargoes. For example, on October 20, 1774, during the first meeting of Congress in Philadelphia, the adopted Articles of Association stated that if Britain's Intolerable Acts were not repealed, the United States would boycott British products, and plans were also outlined for an embargo on exports to England.† In 1917, Congress passed the Trading with the Enemy Act, which was designed to restrict certain types of exports and activities during World War I. Certain parts of this law are still in effect.‡ The Export Control Act of 1940 was passed to address concerns that led to World War II and it was amended over time.§

More recently, the Arms Export Control Act of 1976 (AECA) created additional authority to control the import and export of defense articles and defense services, which includes certain types of unmanned vehicles.¶ Executive Order 11958 delegated the administration of the AECA to the U.S. Department of State, and its Directorate of Defense Trade Controls (DDTC) works closely with the U.S. Department of Defense (DoD) and other government agencies to establish policies, issue licenses and agreements, and enforce the regulations, as described in more detail below.** The AECA, which is implemented in the International Traffic in Arms Regulations (ITAR) at 22 C.F.R. § 120-130, requires overseas parties that receive defense articles from the United States to use them for legitimate self-defense and not transfer those articles without permission from the U.S. government. The U.S. State Department's export licensing decisions under the ITAR must consider whether the proposed shipments would help trigger an arms race, contribute to the proliferation of weapons of mass destruction, support terrorism, increase armed conflict, or undermine arms control and nonproliferation agreements, among other goals. These statues and regulations are the basis for controls on military UASs and associated items.

Nonmilitary export controls on "dual-use" items (commercial items that could also be used for a military, terrorist, or weapons proliferation purpose) are administered principally by the U.S. Department of Commerce's Bureau of Industry and Security (BIS), principally under the Export Administration Act (EEA) of 1979†† and the International Emergency Economic Powers Act, as amended.‡‡ BIS uses the Export Administration Regulations§§ (the EAR [properly pronounced "E" "A" "R", not the thing you use to hear]) to administer dual-use exports, including exports of certain UASs, software, know how, and numerous payloads, among other items. The dual-use controls emphasize protecting national security, while also trying to ensure that legitimate commercial activity is allowed to proceed without undue disruption. We will see that this is quite challenging because the level of export

* http://www.ancient.eu/Peloponnesian_War/.
† https://history.state.gov/milestones/1776-1783/continental-congress.
‡ http://www.gpo.gov/fdsys/granule/USCODE-2011-title50/USCODE-2011-title50-app-tradingwi.
§ http://www.gpo.gov/fdsys/granule/USCODE-2011-title50/USCODE-2011-title50-app-exportati-other.
¶ http://www.gpo.gov/fdsys/pkg/USCODE-2010-title22/html/USCODE-2010-title22-chap39.htm.
** http://www.archives.gov/federal-register/codification/executive-order/11958.html.
†† Export Administration Act of 1979, as amended (Pub. L. 96-72, 93 Stat. 503, 50 U.S.C. app. 2401–2420).
‡‡ International Emergency Economic Powers Act, as amended, (Pub. L. 95-223, 91 Stat. 1628, 50 U.S.C. 1701–1706).
§§ 15 C.F.R. § 730 et seq., https://www.bis.doc.gov/index.php/regulations/export-administration-regulations-ear.

controls on dual-use UASs are higher than you might expect, and so many exports of commercial systems require export licenses.

Remember that this chapter is an overview regarding this complex topic that should be treated as an introduction and not an endpoint. Export control rules are varied and detailed, so the specifics of any "real-world" activity that has any international twist to it need to be evaluated carefully under the applicable rules to be sure you and the people you are working, researching, and studying with do not violate the law. In addition, export control laws are dynamic and are revised regularly, so the regulations cited in this book need to be checked for continued applicability.

8.2 Glossary of Terms for Export Control Understanding

The technical definitions of the following terms, which can be lengthy and complex, differ between the ITAR and the EAR. To fully understand the meaning of the terms described *informally* below, you must consult the specific language of either the ITAR or the EAR regulatory definition (depending on whether you are shipping a commercial/dual-use item or a military item).* Note that the responsible agencies are working as this is written to harmonize key export control definitions as part of an overall U.S. export reform process, but that task is not complete. For the time being, we are providing the following informal glossary to help you get at the heart of how to think about key terms in the special world of export regulation.

Export—Think about a product, software, or know how (e.g., a drawing, an email with instructions on how to make something) crossing the U.S. border on its way to another country. That is a classic export. Another type of export occurs when a person or organization in the United States releases know how to a non-U.S. person inside the United States. That is "deemed" to be an export of that know how to that person's home country since they are assumed to actually or potentially take that know how back to their country or send it there.

Export license or agreement—Virtually all military items, controlled by the ITAR, require an export license from the Department of State for shipment outside the United States. Many exports of commercial/dual use UASs require export licenses from the Department of Commerce under the EAR. Exports of military know how, whether outside the United States or to a non-U.S. person in the United States, can require individual export licenses or Technical Assistance Agreements. Sharing commercial/dual-use technology subject to the EAR can also require export license approval. Also, in very limited circumstances, certain license exceptions are available from the Departments of Commerce or State.

Products—Think about a physical object, such as an entire UAS, or an imaging system that will become part of the payload for a UAS. Certain cameras and imaging systems are controlled under the ITAR. Others are principally designed for commercial end uses, but they may be sophisticated and potentially useful for military or terrorist end uses. If these commercial/dual-use items meet specific control

* The ITAR definitions are found in 22 C.F.R. § 120.1 and the EAR definitions are found in 15 C.F.R. Part 772.

criteria listed in the Commerce Department's EAR, they may require a license for shipment to certain countries.

Software—Software can be controlled for export on its own or while integrated with a physical product. Software that contains encryption functionality, as so much does these days, is subject to export control.

Technology (EAR) or Technical Data (ITAR)—Think about "know how" that is required to develop, produce or, sometimes, even use an export-controlled item. Sending one export-controlled item outside the United States without a license can create a serious violation of the law, but what about exporting the know how required to produce thousands of those items? Enforcement agencies are very worried about exports of controlled know how, which can occur in so many forms these days—in computer files, email attachments, phone conversations, and many other forms.

Import—Think about a military item that comes permanently from another country across the U.S. border into the United States We will not focus on these imported items in this chapter, but it is important to know that these items are also controlled. The Department of Justice's Bureau of Alcohol, Tobacco, Firearms and Explosives (sometimes known as ATF) regulates the import of certain military items under its own regulations.

Missile Technology Control Regime (MTCR)—An informal multilateral agreement among 34 countries that seek to limit the export and proliferation of missiles, software, and technology. Some UASs fit into the MTCR definition of missiles and so are covered by these rules. The MTCR Guidelines express an export policy that is applied to a list of items, the MTCR Equipment, Software, and Technology Annex. The Annex has two parts—Category I and Category II and it also addresses certain end-use controls.

"Category I items include complete rocket and unmanned aerial vehicle systems (including ballistic missiles, space launch vehicles, sounding rockets, cruise missiles, target drones, and reconnaissance drones) capable of delivering a payload of at least 500 kg to a range of at least 300 km, their major complete subsystems (such as rocket stages, engines, guidance sets, and re-entry vehicles), and related software and technology, as well as specially designed production facilities for these items. Pursuant to the MTCR Guidelines, exports of Category I items are subject to an unconditional strong presumption of denial regardless of the purpose of the export and are licensed for export only on rare occasions. Additionally, exports of production facilities for Category I items are prohibited absolutely."*

"Category II items include other less-sensitive and dual-use missile related components, as well as other complete missile systems capable of a range of at least 300 km, regardless of payload. Their export is subject to licensing requirements taking into consideration the nonproliferation factors specified in the MTCR Guidelines. Exports judged by the exporting country to be intended for use in WMD delivery are to be subjected to a strong presumption of denial."*

Re-export—Think about a U.S.-made product that is shipped from the United States under an export license naming ABCD Company in Germany as the end user. After the product has been used by ABCD Company for six months, the president of the company sells the product and ships it to EFGH Company in South Africa. This U.S. product has been re-exported from Germany to South Africa.

* http://www.mtcr.info/english/FAQ-E.html.

U.S. export control laws are different from most other country's laws because that second shipment is generally still controlled by U.S. licensing authorities and it may require a U.S. re-export license.

Transfer—Think about that product that went from the United States to ABCD Company in Germany. Instead of sending it to South Africa, what if the president sold it to GHIJ Company in another town in Germany? That would be a transfer. Such transfers can also be controlled under U.S. export control rules.

Unmanned Aerial Vehicle—The Commerce Department's regulation on commercial/dual-use items (15 C.F.R. Part 772) define a UAV as, "any 'aircraft' capable of initiating flight and sustaining controlled flight and navigation without any human presence on board."

> "In addition, according to section 744.3 of the EAR, unmanned air vehicles, which are the same as 'unmanned aerial vehicles,' include, but are not limited to, cruise missile systems, target drones, and reconnaissance drones."*

8.3 The Sources of Export Controls

While the United States has certain U.S.-only, unilateral export controls, many of the items controlled for export from the United States are also controlled for export by U.S. allies and by other countries. In resolution 1540 (2004), the United Nations Security Council decided that member States should not provide support to non-State actors attempting to use nuclear, chemical, or biological weapons and their means of delivery, in particular for terrorist purposes.† The resolution requires States to adopt and enforce appropriate export control laws.‡ Other multilateral agreements with U.S. allies, including the MTCR and the Wassenaar Arrangement,§ both of which address UASs, establish an international framework for a common set of export controls that are then expressed in each member country's legislation and regulations. Many people assume incorrectly that the U.S. stands alone in implementing export controls and placing the burden of complying with these complex regulations on its citizens and companies. But this is not accurate. As demonstrated by the 34 countries that implement MTCR controls and the 41 that implement the Wassenaar Arrangement controls on the UAS, many countries have rules very similar to those in the United States, including the specific rules that apply to UASs.

8.4 What Is Export Control?

Export control is a key method that the U.S. government uses to protect sensitive equipment, software, and technology. This is done to promote U.S. foreign policy and national

* http://www.bis.doc.gov/index.php/forms-documents/doc_view/838-772 page 42.
† United Nations, www.un.org/en/sc/1540/.
‡ United Nations, www.un.org/disarmament/WMD/1540/.
§ http://www.wassenaar.org/.

security interests in cooperation with certain other countries by regulating manufacturing, sales, and distribution of these items. Specifically, export controls regulate the transfer, shipment, or movement of defense articles, including data, outside of the United States. In 22 C.F.R. § 120.17, for example, ITAR defines export as follows:

- "Sending or taking a defense article out of the United States in any manner, except by mere travel outside of the United States by a person whose knowledge includes technical data; or
- Transferring registration, control or ownership to a foreign person of any aircraft, vessel, or satellite covered by the U.S. Munitions List (USML), whether in the United States or abroad; or
- Disclosing (including oral or visual disclosure) or transferring in the United States any defense article to an embassy, any agency or subdivision of a foreign government (e.g., diplomatic missions); or
- Disclosing (including oral or visual disclosure) or transferring technical data to a foreign person, whether in the United States or abroad; or
- Performing a defense service on behalf of, or for the benefit of, a foreign person, whether in the United States or abroad."* (As discussed, the ITAR definition is a bit different from the EAR definition.)

Traditionally there has been an unconditional strong presumption of denial (Category I) or a presumption of denial (Category II) for license applications for most military UASs, in line with multilateral obligations under the MTCR (see the section on MTCR in the glossary above). That policy was loosened somewhat in February 2015, when the U.S. State Department opened the door to allow additional exports of these items to certain allies under specified conditions. The new policy was implemented to provide enhanced "operational capabilities and capacity of trusted partner nations, increasing U.S. interoperability with these partners for coalition operations, ensuring responsible use of these systems, and easing the stress of U.S. force structure for these capabilities."† In addition, the new policy is designed to align with developments in the UAS industry, such as wider availability of systems and the emerging commercial UAS market. This will be discussed further in the following sections.

If you wish to export a UAS that qualifies as a defense article, or share know how relating to how to develop or produce that article, you must get an ITAR license or a special agreement from DDTC. The DDTC has been delegated ultimate authority to decide which items it controls under the ITAR and which are under BIS jurisdiction. Exporters are still permitted to self-classify their products as ITAR or EAR items so long as they do so strictly in line with the control language in the regulations. But they can also seek written guidance from DDTC in the form of a Commodity Jurisdiction (CJ) if they wish. So, if someone is not sure whether to classify a UAS as a military or dual-use item, he can ask DDTC for a written answer in a CJ application.

Because both the ITAR and the EAR are "strict liability" regimes, exporters are responsible for getting their classifications right. If they misclassify their products and ship without a required export license, they can be subject to substantial penalties, whether they

* 22 C.F.R. § 120.

† State Department, www.state.gov, Diplomacy in Action, U.S. Export Policy for Military Unmanned Aerial Systems.

knew exactly what the rules were or not. If an item is not under ITAR jurisdiction, then that almost always means that it is a dual-use item, and BIS, after consulting with other U.S. government agencies, is responsible for issuing export licenses and enforcing the rules related to those shipments.

As described above, in Figure 8.1, in addition to the State (DDTC) and Commerce (BIS) controls on exports, the United States implements certain very strict trade embargoes and economic sanctions, which are also part of the U.S. export control regime. For example, with the very narrow exception of certain informational materials (e.g., a newsletter) nothing can be sent to Cuba, Iran, North Korea, Sudan, or Syria without an export license from the Department of Treasury's Office of Foreign Assets Control (OFAC). Those licenses are very difficult, or impossible, to obtain, with the exception of those proposed for shipments of certain humanitarian items like agricultural products, food, and medicine.

Exports of military items are administered by the U.S. Department of State, which works closely with the DoD to administer export controls on military UASs through the ITAR (22 C.F.R. 120 et seq.), which implements U.S. controls and international controls like those in the MTCR. Exports of commercial/dual-use UASs and associated items are administered by the U.S. Department of Commerce through the EAR (15 C.F.R. 730 et seq.), which again implements U.S. and international controls like those in the MCTR and Wassenaar Arrangement. Licensing is determined at both agencies by the characteristics of the UAS and its function, who is receiving the item, and what they intend to do with it, among other factors. In summary, commercial/dual-use items are regulated through EAR and its

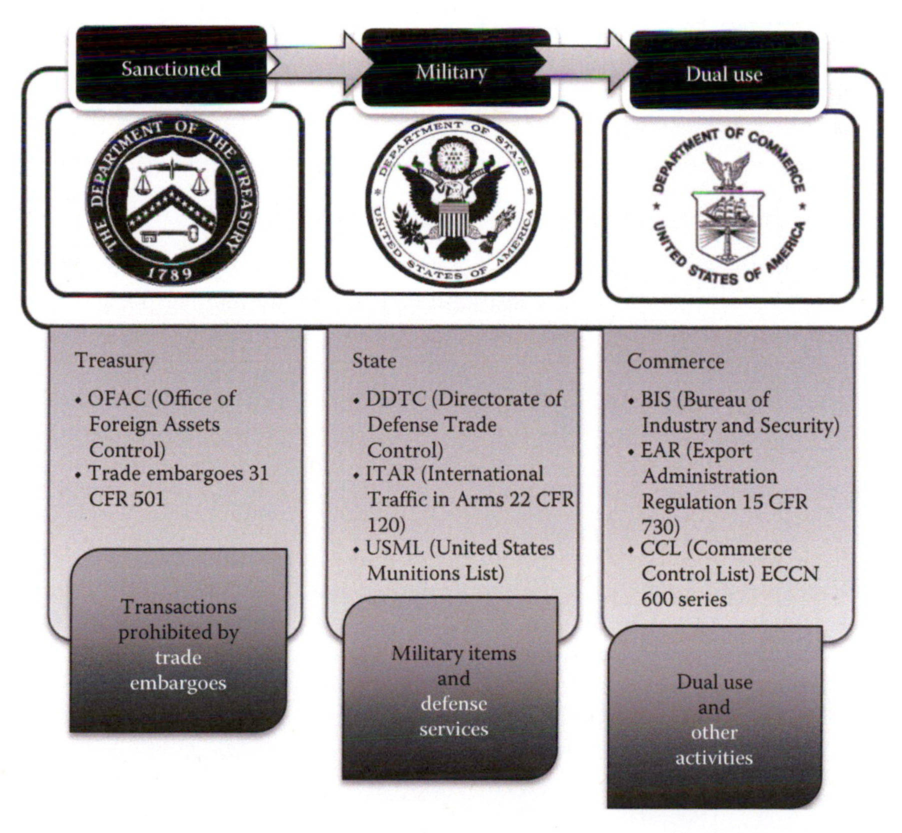

FIGURE 8.1
Export control agency responsibility.

classification system is called the Commerce Control List (CCL). Defense-related articles and services are regulated through ITAR and its classification system called the United States Munitions List (USML).

8.5 Where Do Export Controls Come From?

A multitude of statutes, executive orders, and regulations establish the U.S. export control system. The Arms Export Control Act mentioned above is important because it establishes the framework for exports of military items and it is the foundation for the ITAR. The foundation of the commercial/dual-use export control regime administered by the Department of Commerce is found in the EAA of 1979, as amended (50 U.S.C. app. 2401–2420), and the International Emergency Economic Powers Act (IEEPA), as amended (50 U.S.C. 1701–1706). IEEPA keeps the EAA in effect, even though the EAA has lapsed and is no longer formally law. The Missile Technology Controls Policy (50 U.S.C. app. 2402 (note)) is also relevant to UAS exports, as are a number of other statutes. Fortunately for those who have to deal with the alphabet soup of statutes passed by Congress, with provisions spread out in a variety of laws passed at different times, the key export control agencies, State and Commerce, each administer regulations that implement most of the statutes. Not surprisingly, given the involved legal background, the EAR and the ITAR are complicated. The next section breaks down some key components of the EAR and ITAR as they relate to UASs.

8.6 Export Administration Regulations

As mentioned earlier in the chapter, the EAR were developed to implement the EAA of 1979, which deals with controls on commercial/dual-use items. The U.S. Department of Commerce's BIS administers the EAR to control certain exports, re-exports, certain U.S. person activities, and end uses for items exported from the United States. If an item is not controlled as military under the ITAR, it likely is subject to the jurisdiction of the EAR. The EAR is hundreds of pages long, but for our limited purposes here we will focus just on a few sections that directly affect UASs. As described below, anyone who wants to export commercial/dual-use products, software or know how from the United States, needs to know what parts of the EAR affect their activities.

The EAR consists of a list of controlled items that are assembled on the CCL,[*] a country chart that shows where CCL listed controlled items can be shipped,[†] a set of general prohibitions that indicate what exporters cannot do,[‡] prohibitions on certain activities by U.S. persons,[§] end-use restrictions that apply to all items subject to the regulations, and a variety of other instructions. Not all parts of the EAR can be explained, or even referenced, in this short chapter. We will focus on some of the CCL categories that cover UAS, UAS

[*] See 15 C.F.R. Part 774, Supplement 1.

[†] Id. at Part 738, Supplement 1.

[‡] Id. at Part 736.

[§] Id. at part 744.

software, and UAS technology and some of the other EAR controls. Other issues are necessarily left for the reader to research.

8.6.1 Commercial Control List

The CCL is a classification system used to identify commercial items that are controlled for export as dual-use items under the EAR. It is divided into 10 sections, the most relevant of which for UASs are category 9 (Aerospace and Propulsion), 7 (Navigation and Avionics), 6 (Sensors and Lasers), 5 (Telecommunications and Information Security), 4 (Computers), 3 (Electronics), 2 (Processing Equipment), and 1 (Materials—e.g., composite materials). Other categories also come into play from time to time. Within the CCL are entries called Export Control Classification Numbers (ECCN), which provide technical control criteria. If a product, software, or technology meets the control criteria in an ECCN, it is controlled for export under that ECCN category. If an ECCN applies, the question then becomes what is the reason for export control assigned by the ECCN? Certain reasons for control create licensing requirements for certain countries. Knowing how a product is classified—Is it an ITAR item or a CCL item? What ITAR or CCL category does it fit into?—is absolutely critical for export control purposes. Without that information, you simply do not know how to behave. If you do not know the ECCN of a dual-use item, you do not know if you need an export license to send that product to Country A or not. And you do not know if you need a license to release information to an engineer from Country Y who is in the United States on a work visa. Even product discussions beyond what is available, public knowledge may require an export license.

CCL categories are broken up into five sections that contain ECCNs with different numbers. The "product" sections are A (systems, equipment, and components), B (test, inspection, and production equipment), and C (material). Section D covers software and E covers technology (know how). Within each of these sections are the ECCNs.

As an example, here is an edited version of the principal ECCN for unmanned aircraft on the CCL (there are many other relevant ECCNs, but this one is particularly important to understand):

ECCN 9A012 Non-military "unmanned aerial vehicles," ("UAVs"), unmanned "airships," associated systems, equipment and "components," as follows

License Requirements	
Reason for Control: NS, MT, AT	
Control(s) Country Chart	
(See Supp. No.	
1 to part 738).	
NS applies to entire entry	NS Column 1
MT applies to non-military unmanned air vehicle systems (UAVs) and remotely piloted vehicles (RPVs) that are capable of a maximum range of at least 300 kilometers (km), regardless of payload.	MT Column 1
AT applies to entire entry	AT Column 1

Items:

a. "UAVs" or unmanned "airships" having any of the following:

a.1. An autonomous flight control and navigation capability (e.g., an autopilot with an Inertial Navigation System); *or* a.2. Capability of controlled flight out of the direct visual range involving a human operator (e.g., televisual remote control);

b. Associated systems, equipment and "components," as follows:

b.1. Equipment "specially designed" for remotely controlling the "UAVs" or unmanned "airships," controlled by 9A012.a.

b.2. Systems for navigation, attitude, guidance or control, other than those controlled in Category 7, "specially designed" to provide autonomous flight control or navigation capability to "UAVs" or unmanned "airships," controlled by 9A012.a.

b.3. Equipment or "components" "specially designed" to convert a manned "aircraft" or a manned "airship" to a "UAV" or unmanned "airship," controlled by 9A012.a;

b.4. Air breathing reciprocating or rotary internal combustion type engines, "specially designed" or modified to propel "UAVs" or unmanned "airships," at altitudes above 50,000 feet (15,240 meters).

Note that 9A012 does not control model aircraft or model "airships." However, at the time of this writing, we are aware that pending revisions to this ECCN category are likely to focus more on whether a UAS is capable of certain flight times or flight in certain wind conditions. It would be prudent to check the current regulations before exporting!

In order to determine if a UAS you are considering exporting needs a license, you need to be sure it is not controlled by the ITAR. Next, you need to determine if it meets the criteria described in one of the 9A012 subcategories (or another ECCN category). If it does, then you need to determine the "reason for control," which is listed in the ECCN. In ECCN 9A012, two of the listed reasons for control are NS1 (for national security column 1) and MT1 (an MTCR control) that apply to certain items that fall into ECCN 9A012. After determining the reasons for control, consult the Commerce Country Chart to see if a license is required. A license is needed if the country chart shows an "X" in the country list/reason for control matrix. See Figures 8.1 and 8.2.

For example, say we wanted to ship an ECCN 9A012 UAS to Algeria. You see in Table 8.1, which is a part of the Commerce Country Chart that comes from Supplement 1 to Part 738 of the EAR, that there is an "X" in the NS1 box and the MT1 box after Algeria. That means the NS1- and MT1-controlled UAS needs a Commerce Department license before it can be shipped to Algeria.

Before you apply for a license, however, it is smart to search for a license exception in the EAR (none are listed as available in the ECCN category itself, but you can also look in part 740 of the EAR for others). Almost no license exceptions are available for UASs, however. If a license is required and no license exception is available, then you must apply for a Commerce Department export license. There is no government fee for the application process. Organizations may not, repeat *not*, ship the UAS without a required export license. Doing so could subject the organization to criminal charges and other severe penalties.

In addition to ECCN 9A012, certain UASs that come with or are designed for certain aerosol dispensing systems are controlled in ECCN 9A120. In addition, 9B010 controls "Equipment 'specially designed' for the production of 'UAVs' and associated systems, equipment and 'components,' controlled by 9A012."* UAS component and payload controls

* See 15 C.F.R. Part 774, Supplement 1.

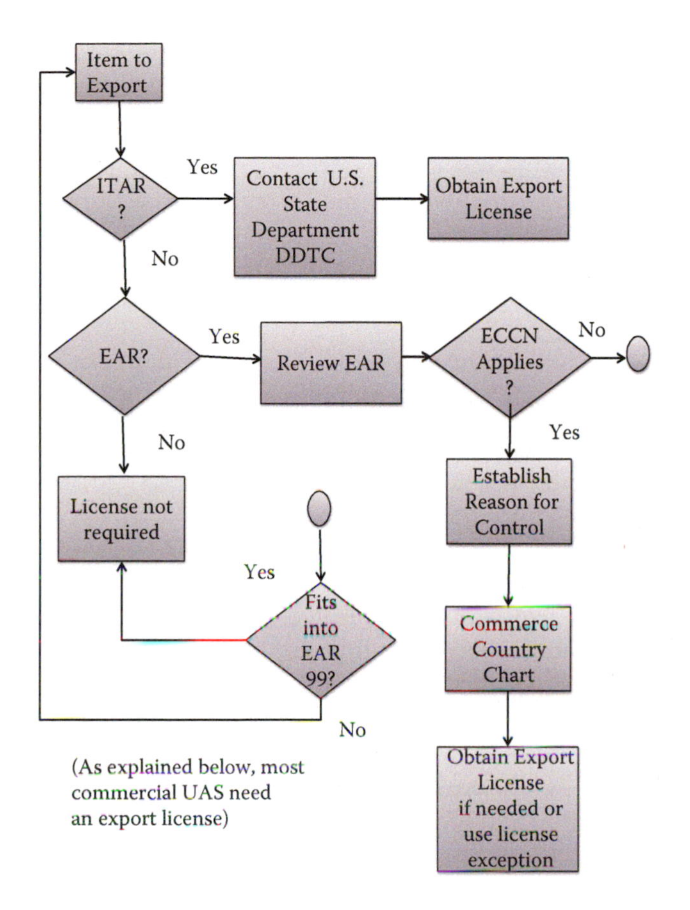

FIGURE 8.2
Export control flowchart.

TABLE 8.1
CCL Overview and Country Chart

Countries	Commerce Country Chart															
	Reason for Control															
	Chemical and biological weapons			Nuclear nonproliferation		National security		Missile tech	Regional stability		Firearms convention		Crime control		Anti-terrorism	
	CB 1	CB 2	CB 3	NP 1	NP 2	NS 1	NS 2	MT 1	RS 1	RS 2	FC 1	CC 1	CC 2	CC 3	AT 1	AT 2
Afghanistan	×	×	×	×		×	×	×	×	×		×		×		
Albania[2,3]	×	×		×		×	×	×	×							
Algeria	×	×		×		×	×	×	×	×		×		×		

are spread throughout the CCL, such as those that control guidance systems, composite materials, controllers for motors and other components, cameras, lasers, and sensors. One also finds cross references in the EAR to ITAR controlled items, such as those in ECCN 9A115, which controls "apparatus, devices and vehicles, designed or modified for the transport, handling, control, activation and launching of rockets, missiles, and unmanned aerial vehicles capable of achieving a 'range' equal to or greater than 300 km. (These items are 'subject to the ITAR.' See 22 CFR parts 120 through 130.)"

Also, as part of the export reform process introduced by the Departments of State and Commerce over the last few years, several parts and components of UASs have been moved from the ITAR to the EAR, including, but not limited to, the following items that fall under the newly created 9A610:

> t. Composite structures, laminates and manufactures thereof 'specially designed' for unmanned aerial vehicles controlled under USML Category VIII (a) with a range equal to or greater than 300 km.
> u. Apparatus and devices 'specially designed' for the handling, control, activation and non-ship-based launching of UAVs or drones controlled by either USML paragraph VIII (a) or ECCN 9A610.a, and capable of a range equal to or greater than 300 km.
> v. Radar altimeters designed or modified for use in UAVs or drones controlled by either USML paragraph VIII (a) or ECCN 9A610.a. and capable of delivering at least 500 kilograms payload to a range of at least 300 km.
> w. Hydraulic, mechanical, electro-optical, or electromechanical flight control systems (including fly-by-wire systems) and attitude control equipment designed or modified for UAVs or drones controlled by either USML paragraph VIII (a) or ECCN 9A610.a. and capable of delivering at least 500 kilograms payload to a range of at least 300 km.*

In addition, a variety of complex control provisions exist in ECCNs 9D and 9E for software associated with a MTCR controlled UAS and for a 9A012-controlled UAS. For example, 9D004.e controls "software" "specially designed" or modified for the operation of "UAVs" and associated systems, equipment, and "components," controlled by 9A012.† It is therefore vitally important that software and know how related to UASs should also be classified carefully under either the ITAR or the EAR and that a full licensing analysis be conducted before any exports occur across U.S. borders or before software or know how are released to non-U.S. persons in the United States.

Many UAS items that are subject to control under the EAR are subject to NS1 (National Security 1) or MT1 (Missile Technology 1) controls. Other than the controls on sanctioned countries, these are the most severe export controls that exist under the EAR. *It means there is a license requirement for virtually every country in the world.*

Perhaps counterintuitively, under export reform, certain items that transferred from the ITAR to CCL category 9A610 are eligible for a new, but complex license exception Strategic Trade Authorization (shortened in the regulations to STA) found in 15 C.F.R. § 740.20. Items in 9A610 may in some cases be exported without a license to countries that are eligible to receive products under this license exception, so long as all of the requirements of the license exception and the EAR are satisfied in full.

Note that the paragraphs above mention only selected ECCN categories that may apply to UASs. As indicated, under the EAR, there are a number of other categories that are not specifically mentioned, but should not be ignored.

If your item is a commercial/dual-use item (not ITAR), but it is not called out in a specific ECCN category on the CCL, it is still under the jurisdiction of EAR, but it may be classified as an EAR99 item. A validated EAR99 classification that is definitely not in a controlled ECCN category means that the commercial item generally does not require

* 15 C.F.R. Part 774, Supplement 1, ECCN 9A610.

† See 15 C.F.R. Part 774, Supplement 1 (note that terms in "parenthesis" in this part of the regulation are defined in Part 772 of the EAR).

a license to be exported or re-exported. *Except* that EAR99 items may not be exported from the United States to sanctioned countries (including, but not limited to Cuba, Iran, North Korea, Sudan, and Syria), *and* they may not be exported for prohibited end uses (e.g., certain missile, nuclear, chemical/biological weapons, or nuclear propulsion end uses),* and they may not be sent to "denied persons," such as U.S. government listed terrorists, narcotics traffickers, and others. Few commercial UAS items currently fall under EAR99, unless those items are for hobby purposes or they fall outside of the control category.†

Key export control policies on UAS are described in part 742 of the EAR, which addresses "Unmanned Aerial Vehicles (UAVs) of any type, including sensors for guidance and control of these systems, except model airplanes." The end-use controls referenced above are not based on the CCL and ECCN categories. Virtually any item proposed for export from the United States that fits into one of the prohibited end uses described in part 744 for UAS (there are several) requires an export license issued prior to shipment. For example, a pencil that is proposed for export from the United States for the listed prohibited end uses in certain countries would require a Department of Commerce export license based solely on the proposed end use. Individuals who are overly focused on ITAR and EAR CCL level item controls often overlook the end-use controls. If you are involved with exporting, researching, or studying UAS and/or working with non-U.S. persons on UAS issues, please read parts 742 and 744 of the EAR in addition to knowing the CCL controls described above.

8.6.2 Missile Technology Control Regime Annex

"The MTCR was originally established in 1987 by Canada, France, Germany, Italy, Japan, the United Kingdom and the United States. Since that time, the number of MTCR partners has increased to a total of thirty-four countries all of which have equal standing within the Regime"‡ The countries and the year they joined can be seen in Table 8.2. The proliferation

TABLE 8.2

MCTR Countries As of March 2016

Argentina (1991)	Greece (1992)	Republic of Korea (2001)
Australia (1990)	Hungary (1993)	Russian Federation (1995)
Austria (1991)	Iceland (1993)	South Africa (1995)
Belgium (1990)	Ireland (1992)	Spain (1990)
Bulgaria (2004)	Italy (1987)	Sweden (1991)
Brazil (1995)	Japan (1987)	Switzerland (1992)
Canada (1987)	Luxembourg (1990)	Turkey (1997)
Czech Republic (1998)	The Netherlands (1990)	Ukraine (1998)
Denmark (1990)	New Zealand (1991)	United Kingdom (1987)
Finland (1991)	Norway (1990)	United States (1987)
France (1987)	Poland (1998)	
Germany (1987)	Portugal (1992)	

* 15 C.F.R. § 744.

† The consolidated denied persons list can be found on the Department of Commerce, Bureau of Industry and Security website: https://www.bis.doc.gov/index.php/policy-guidance/lists-of-parties-of-concern.

‡ www.mctr.info/english.

of weapons of mass destruction initiated the MCTR to be formed to protect peace and provide security. The MCTR has an elected chair who is currently Ambassador Roald Naess of Norway.*

The entire MTCR Annex is listed in part 121.16 of the ITAR. That section states, "Some of the items on the Missile Technology Control Regime Annex are controlled by both the Department of Commerce on the Commodity Control List and by the Department of State on the United States Munitions List." 22 C.F.R. 121.16. Other MTCR items are on the CCL or the ITAR. To the extent an article in 121.16 is on the USML (121.1), a reference appears in parentheses in 121.16 listing the USML Category in which it appears. EAR items that are controlled under the MTCR are identified in the CCL with an MT reason for control listed in the ECCN category. The USML is spelled out in 22 C.F.R. § 121.1 and is discussed in the following section.

As it applies to UASs, Category I includes complete systems "capable of carrying a payload of 500 kg to a distance of 300 km and come with a strong presumption of denial for export" (Endnote 30). Category II is capable of a range of 300 km and may contain less-sensitive components and items regardless of their payload.

8.7 International Traffic in Arms Regulation

The Arms Export Control Act, which was passed on July 5, 1940, forbids the export of minerals, chemicals, and aircraft parts without a license. Three weeks later the ban added aviation fuel, iron, and scrap metal. "22 U.S.C. 2778 of the Arms Export Control Act (AECA) provides the authority to control the export of defense articles and services, and charges the President to exercise this authority. Executive Order (EO) 11958, as amended, delegated this statutory authority to the Secretary of State" (Endnote 32). EO 11958 was enacted on January 18, 1977. The EO states "Coordination. (a) In addition to the specific provisions of Section 1 of this Order, the Secretary of State and the Secretary of Defense, in carrying out the functions delegated to them under this Order, shall consult with each other and with the heads of other departments and agencies, including the Secretary of the Treasury, the Director of the United States International Development Cooperation Agency, and the Director of the Arms Control and Disarmament Agency, on matters pertaining to their responsibilities. (b) In accordance with Section 2(b) of the Act and under the directions of the President, the Secretary of State, taking into account other United States activities abroad, shall be responsible for the continuous supervision and general direction of sales and exports under the Act, including but not limited to, the negotiations, conclusion, and termination of international agreements, and determining whether there shall be a sale to a country and the amount thereof, and whether there shall be delivery or other performance under such sale or export, to the end that sales and exports are integrated with other United States activities and the foreign policy of the United States is best served thereby."†

* Department of Commerce, www.bis.doc.gov, Commerce Control List (CCL).

† Executive Order 11958.

The following ITAR categories deal with UASs:*

Category VIII—Aircraft, Space, and Associated Equipment

A few highlights of the very broad scope of Category VIII export controls on military UASs include, but are not limited to:

> "(a)(5) Unarmed military unmanned aerial vehicles (UAVs) (MT if the UAV has a range equal to or greater than 300 km);
>
> (a)(6) Armed unmanned aerial vehicles (UAVs) (MT if the UAV has a range equal to or greater than 300 km);
>
> (a)(13) Controls Optionally Piloted Vehicles (OPV) (i.e., aircraft specially designed to operate with and without a pilot physically located in the aircraft) (MT if the OPV has a range equal to or greater than 300 km);
>
> (d) Controls "Ship-based launching and recovery equipment specially designed for defense articles described in paragraph (a) of this category and land-based variants thereof (MT if the ship-based launching and recovery equipment is for an unmanned aerial vehicle, drone, or missile that has a range equal to or greater than 300 km)."
>
> (h)(12) Controls "Unmanned aerial vehicle (UAV) flight control systems and vehicle management systems with swarming capability (i.e., UAVs interact with each other to avoid collisions and stay together, or, if weaponized, coordinate targeting)."
>
> Associated technical data and software are included in this category as well, as indicated in the sections above.

Category XI—Military and Space Electronics

A wide variety of electronic items and associated technical data for UASs potentially fit into this category. Examples include Category XI (a)(3) covering certain radar equipment, (a)(4), which includes electronic combat parts and components (e.g., sensors that detect weapons launches); and (a)(5) command, control, and communications (C3); command, control, communications, and computers (C4); command, control, communications, computers, intelligence, surveillance, and reconnaissance (C4ISR) items. This complex section covers many other electronic items for UASs.

Category XV—Spacecraft Systems and Associated Equipment Aircraft

A number of potentially relevant UAS articles and associated items can also be found in Category XV.

Other USML Categories Also Having the Potential to Include Items Relevant to USML Controls

Here is a full listing of the USML categories with a few examples of where UAS-related items might appear:

Category I (Firearms, Close Assault Weapons, and Combat Shotguns), II (Guns and Armaments) (*armed vehicles would have Category II items*), III (Ammunition/Ordnance) (*again, armed vehicles would have these items*), IV (Launch Vehicles, Guided Missiles, Ballistic Missiles, Rockets, Torpedoes, Bombs, and Mines) (*certain armed vehicles would have these*),

---

* 22 CFR § 121.1 The United States Munitions List.

V (Explosives and Energetic Materials, Propellants, Incendiary Agents and Their Constituents (*some UAS or UAS armaments use USML-listed propellants*), VI (Surface Vessels of War and Special Naval Equipment, VII (Ground Vehicles), VIII (Aircraft and Related Articles), IX (Military Training Equipment and Training) (*military training related to military UASs is often controlled under the ITAR*), X (Personal Protective Equipment), XI (Military Electronics), XII (Fire Control, Range Finder, Optical and Guidance and Control Equipment), and XIII (Materials and Miscellaneous Articles), XIV (Toxicological Agents), XV (Spacecraft and Related Articles), XVI (Nuclear Weapon Related Articles), XVII (Classified Articles, Technical Data, and Defense Services Not Otherwise Enumerated), XVIII (Directed Energy Weapons), XIX (Gas Turbine Engines and Associated Equipment [certain UASs use gas turbine engines, *turbine components and equipment to make or repair components* that are controlled under the ITAR *(or in some cases the EAR)*]), XX (Submersible Vessels and Related Articles), and XXI (Articles, Technical Data, and Defense Services Not Otherwise Enumerated).*

A number of other USML categories are also relevant to UASs and payloads, but those categories must be determined on a case-by-case basis. ITAR is administered by the DDTC through licenses or special agreements.

8.8 How Do Export Control Issues Come Up in Real Life?

Here is a story.

The ink still wasn't dry on the undergrad degree, and it was an hour into Pat's first day on the job at "X" Inc., a small UAS manufacturer. Pat was hired to be the director of engineering and there was a lot to learn. In fact, Pat's head was already swimming with new information, while Kelley, the boss, gave a tour of the production facility and the offices. "About 50% of the company's products are exported, and that includes full systems, autopilots, cameras, and software," Kelley explained. "Aside from the manufacturing operation here in the U.S., there's a second production and engineering facility in India, and we're working with the team there to develop a radically improved design. One thing we don't have a great grasp on, however, are the export control reg's," Kelley declared. "I don't look good in stripes Pat, if you know what I mean, so make that priority one—and make sure you get it right, ok?" Kelley was smiling when saying this, but Pat could see the underlying anxiety in Kelley's eyes.

It was pretty clear that the products weren't military items, but they were designed to fly out of the operator's visual range using a camera system and a tablet controller and they had over 90 minutes flight capability in normal operation because of excellent batteries and a lightweight design. Even with just that basic information, Pat was already pretty sure that the products were controlled for export under the EAR and that turned out to be a good first impression. That meant that BIS export licenses were needed to ship the product to every country except Canada, but Kelley had said the company was not getting any export licenses. That was a problem because each shipment was a violation of federal law. Also, if the product under development with India was controlled for export under the EAR (which they turned out to be), a license was needed to share information with the India facility about how to develop it. Another problem.

* 22 CFR § 121.1 The United States Munitions List.

The first step would be to make absolutely sure how the products were classified under the BIS CCL. If licenses were indeed needed for exports, the company would have to apply for them starting now and there could be no support of any kind for prior exports that violated the law unless the government said that was OK. There were probably 30 steps after that, but those could be addressed in logical fashion without too much panic . . . hopefully. This job was going to be more interesting, dramatic, and busier than Pat had thought at 8 a.m.

8.9 How to Protect Export-Controlled Products and Information ("Know How")?

The most important part of protecting export-controlled products and information, or "know how," is to understand what is controlled by classifying it, train your personnel to recognize controlled know how, and set up a systematic compliance program. Whether an entity is dealing with ITAR or EAR items, software, or information, compliance is important. A compliance program should include the following: a written corporate commitment and policy statement from senior management, a clear organizational control structure that identifies responsible personnel and contact information in the event an issue is spotted, classification, identification (e.g., labeling), receipt and tracking of controlled items, a technology control plan to deal with know how, a licensing program to obtain approvals for exports, re-exports/transfers, solid recordkeeping for a specified period, internal monitoring and auditing, customer screening, regular training tailored to fit assigned responsibilities, a plan to address potential violations and penalties, and a plan to address contacts by regulating agencies and enforcement personnel, among other steps.*

A good place to start in developing an ITAR-specific export compliance program is with the DDTC's publication *Getting Started with Defense Trade*. This publication will help answer the question "Does defense export control apply to me?" It does this with the following steps:

1. Check to see if the item to export is on the USML that is found in 22 CFR 121 in ITAR.
2. Unsure if the item is covered then file a CJ request.
3. If it is on the USML then you must be registered with the DDTC.
4. After registration apply for an export license. This can be done through D-Trade. D-Trade is an electronic export licensing system in DDTC.
5. For basic questions call the DDTC Response Team.†

Once an item is determined to be export controlled there are different types of approvals that may be required before certain technical discussions can be had with a potential overseas customer or before product can be exported. Under the ITAR, for example, there are basic export licensing approvals, and there are also marketing, brokering, and product/software/data export licenses. There are also warehousing and technical data exchange agreements for certain foreign manufacturing and technology exchanges. There are also a complex set of export license exceptions, approvals for Foreign Military Sales that the U.S. DoD helps organize, and other types of licensing depending on what is being exported and

* Department of State, www.pmddtc.state.gov, Compliance Programs.

† Department of State, www.pmddtc.state.gov, Getting Started with DDTC.

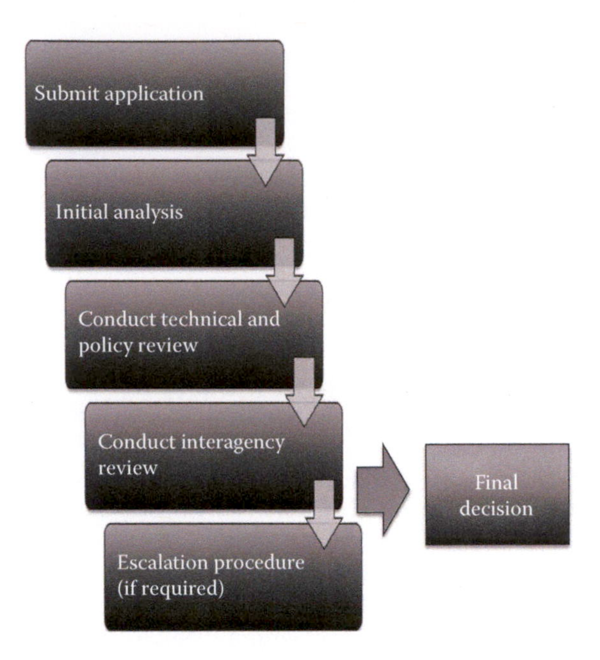

FIGURE 8.3
Department of Commerce License Review Process.

to whom. The U.S. Department of Commerce has a similar export license application and review process, but it has fewer license types (see Figure 8.3). Commerce does also have a variety of license exceptions, but it is critically important to be sure that all conditions of a license exception (ITAR or EAR) are met perfectly before using one at either agency.

Once it is determined that the item is export controlled, it is essential to protect the item and associated know how. It is often confusing how to do that most effectively and in compliance with the regulations. It is important to develop a comprehensive export compliance program, but here are some basic tips to keep protected materials safe. Do not permit foreign nationals any access to controlled information including through emails, presentations, conversations, viewing of computer screens, printed materials, or in any other way. The following are a few helpful example tips that can be incorporated into your compliance program:

- Do not discuss export-controlled projects in group meetings unless everyone is cleared to participate
- Do not leave workstations with export-controlled data visible; set screens to lock within a short time period if the computer is unattended
- Monitors should face away from doors and windows
- All data should be locked up when not in use
- Shred all export-controlled materials before disposing of them
- Only discuss export-controlled projects in protected areas
- Isolate data files on secure media and servers, avoiding non-U.S. personal information technology system access to the data (especially administrative access)
- Qualify other universities and subcontractors on export-controlled projects*

* National Contract Management Association.

8.10 What Are Export Control Violations?

Violations of the ITAR or the EAR can create serious criminal liability for organizations or for individuals. People have been imprisoned for violating the ITAR and EAR and there have been many criminal cases involving UASs and associated equipment. In testimony before Congress, Assistant Secretary for Export Administration at the Department of Commerce Kevin J. Wolf stated that

> BIS's Export Enforcement team, along with the Department of Homeland Security's Bureau of Immigration and Customs Enforcement, and the Federal Bureau of Investigation, enforce controls on dual-use exports. These agencies, through investigations of suspected violations of law and regulations, and the interdiction of suspected illicit shipments, have provided the necessary evidence to successfully prosecute both criminal and civil cases on export violations. Our multilateral controls also provide a strong framework for cooperative enforcement efforts overseas when such efforts call for an international approach.*

He highlighted several UAS enforcement cases, including

> Aviation Services International
>
> On September 24, 2009, Aviation Services International BV (ASI), an aircraft supply company in the Netherlands, Robert Kraaipoel, director of ASI, Neils Kraaipoel, sales manager of ASI, and Delta Logistics pled guilty in U.S. District Court in Washington, DC, to charges related to a conspiracy to illegally export aircraft components and other U.S.-origin commodities to entities in Iran, via the Netherlands, the United Arab Emirates, and Cyprus. Between October 2005 and October 2007, the defendants received orders from customers in Iran for U.S.-origin items, including video recorder units for end use in Unmanned Aerial Vehicles, then contacted companies in the United States and negotiated purchases on behalf of the Iranian customers. The defendants provided false end-user certificates to U.S. companies to conceal the true end-users in Iran. The defendants caused U.S. companies to ship items to ASI in the Netherlands or other locations in the United Arab Emirates and Cyprus, which were then repackaged and transshipped to Iran. In a related case, ASI, Robert Kraaipoel and Niels Kraaipoel settled administrative charges with BIS that included, in part, ASI and Robert Kraaipoel being placed on BIS's Denied Persons List for seven years. Niels Kraaipoel agreed to a three-year denial of his export privileges that would be suspended pending no future export violations.
>
> ARC International
>
> On February 3, 2010, Harold Hanson (Hanson) and Nina Yaming Qi Hanson (Qi) were sentenced in U.S. District Court in the District of Columbia. Qi was sentenced to 105 days in jail with credit for time served, placed on one year of supervised release, ordered to pay a fine of $250 and a $100 special assessment fee and ordered to attend a U.S. Department of Commerce-sponsored export education training program. Hanson was sentenced to 24 months' probation, required to pay a fine of $250 and a $100 special assessment fee, ordered to perform 120 hours of community service, and also ordered to attend a U.S. Department of Commerce-sponsored export training program. On November 13, 2009, Hanson and Qi pleaded guilty to making false statements. On March 12, 2009, a federal grand jury in the District of Columbia returned an indictment

* Department of Commerce, www.bis.doc.gov/index.php/forms-documents/doc_view/730-testimony-by-assistant-secretary-wolf-before-the-national-security-and-foreign-affairs-subcommittee.

charging Qi, her husband Hanson (an employee at Walter Reed Army Medical Center), and a Maryland company, ARC International, LLC, with illegally exporting miniature Unmanned Aerial Vehicle (UAV) Autopilots controlled for national security reasons to a company in the People's Republic of China.

Mayrow General Trading

In September 2008, a federal grand jury in Miami, FL, returned a Superseding Indictment charging eight individuals and eight corporations in connection with their participation in conspiracies to export U.S.-manufactured commodities to prohibited entities and to Iran. They were charged with conspiracy, violations of the International Emergency Economic Powers Act and the United States Iran Embargo, and making false statements to federal agencies in connection with the export of thousands of U.S. goods to Iran. Charges against defendant Majid Seif, also known as Mark Ong, and Vast Solutions alleged that Seif and Vast exported radio control devices and accessories used in Unmanned Aerial Vehicles from a Singapore firm to Malaysia. The radio control devices were then shipped to Iran.

Landstar/Yi-Lan Chen

On February 3, 2010, Yi-Lan Chen, also known as Kevin Chen, was arrested on charges of illegally exporting commodities for Iran's missile program. According to the affidavit filed in support of the criminal complaint, Chen caused dual-use goods to be exported from the United States, including P200 Turbine Engines, which the investigation revealed were for end users in Iran. "The P200 Turbine Engines are designed for use as model airplane engines but can also be used to operate Unmanned Aerial Vehicles and military target drones."*

Do not be lulled into a sense of complacency by these serious criminal cases. There are any number of other serious civil penalty cases involving releases of a wide variety of commercial/dual-use know how or product exports that caused companies, universities, and individuals many headaches and a lot of money. Even though the rules are very complex, the U.S. government puts the obligation on all U.S. persons to know and apply them properly. If you play any part in the UAS industry or in many academic settings, you need to be familiar with export control rules to protect yourself and your organization from a violation.

Moreover, as noted before, export controls are a strict liability regime. You can be punished severely even if you do not know the rules. So it is no defense to bury your head in the sand and say that the subject is too complex or hope that someone else will deal with it. This is critically important to understand, because when dealing with UASs, as you have seen from above, even the simplest commercial system, payload item, or data file can be controlled by the EAR, not to mention the ITAR. As an example, the civil penalty for a violation of the EAR can be $250,000 per shipment or per data release to the wrong non-U.S. person, or twice the value of the item subject to the violation, whichever is higher!

8.11 How Do We Perform Work Outside the United States?

With all the complexities surrounding export controls, some people in the industry or in academia who deal with UASs might be tempted to throw their hands in the air and just focus on domestic sales or activities. Even if they do, they still have to be worried about

* Department of Commerce, www.bis.doc.gov/index.php/forms-documents/doc_view/730-testimony-by-assistant-secretary-wolf-before-the-national-security-and-foreign-affairs-subcommittee.

domestic data releases to non-U.S. persons. Do not make the mistake, however, of restricting hiring in your program just to U.S. persons without taking further compliance steps. That could lead you into a claim of discrimination on the basis of national origin. You must carefully ensure that you are complying with labor and employment rules, and with U.S. export control rules.

All is not lost in pursuit of international business and international academic cooperation. In fact, many companies in the industry are looking outside the United States for sales because of the many restrictions on flying even small commercial UASs in the U.S. national airspace, notwithstanding the FAAs recently started rulemaking process in that area. Many companies, universities, and others effectively deal with export control regulations that affect their day-to-day activities. In addition to getting products, software, and technology classified properly, in almost any situation where you are dealing with UASs, you need to put a written export control policy and compliance program in place. DDTC has a short guide to what is needed in a compliance system that can be found at this link: https://www.pmddtc.state.gov/compliance/documents/compliance_programs.pdf.

While you need to adapt these basics to your organization's specific circumstances and products, this basic approach is a good starting spot for organizations dealing with both military and commercial items. As indicated in the guidance, the best approach is to create a written policy and guidelines, identify responsible individuals, conduct regular training, audits, proper recordkeeping, and a standard approach to dealing with potential violations. These are all effective steps described in the DDTC guidance. Another best practice is to create and require the use of systematic procedures (e.g., checklists) that help people identify and react properly to export control issues in relevant situations. That way people do not have to rely on their memory of these complex rules. This is critical as ignorance is not an accepted excuse when it comes to export control. Building export control steps into existing business or academic practices is a smart way to attack compliance. For example, including a line in the checklist used for all international shipments that determines whether a license has been obtained would be an effective step to help prevent violations.

It is also vitally important to have the right message from leadership—"these are important rules and we are going to comply." And it is critical to have a point person who is responsible for the program that people can contact with questions. If that person needs some help, which is common because of the complexity of this area, they should have access to the regulations and be able to seek expert guidance from time to time. Violations need to be handled intelligently. In our experience, the best approach is to be honest and bring violations you spot to your export control officer or superior. Of course, the best option is to avoid violations in the first place and seek help and guidance whenever you have a question, and certainly before the item in question is exported. When in doubt, ask!

DISCUSSION QUESTIONS

8.1 What is the purpose of export controls? What is the underlying purpose or justification for enacting export control regulations? What is the purpose of the export control license? Who must obtain an export control license? Which federal agency issues such licenses? Describe in a very general way, AECA and ITAR regulations. Do these apply to UASs and associated technologies? Generally, describe the differences between ITAR and EAR regulations? Which agency controls exports under the latter regulation?

8.2 Are EAR and ITAR terms currently harmonized? Explain your answer. Review and discuss the terms and definitions provided in Section 8.2 and tell how you believe these might apply to UASs and associated components, subsystems, and intellectual properties.

8.3 Is it true that the United States is one of only a handful of countries that restrict trade through export control regulations? Explain your answer.

8.4 Describe what is meant by the term, "export" under ITAR regulations. What was the justification for the U.S. State Department loosening export controls somewhat in February 2015? What two options exist if you wish to export a UAS that qualifies as a defense article or share information related to development or production of that article?

8.5 Where do export controls come from?

8.6 Describe in detail EAR regulations. List and describe those UAS technologies which have been removed by the Departments of State and Commerce from EAR and ITAR control. Why are export controls described as "strict liability" laws? Describe the MTCR and corresponding annex in detail.

8.7 Describe, in detail, ITAR, including those categories and sections that refer to a UAS.

8.8 Describe what is meant by a compliance program and tell why it is important. What should such a program include?

8.9 List and discuss examples of export control violations. What could have been done differently to avoid these problems?

8.10 Individuals working in industry or in academia who deal with UASs should focus exclusively on domestic sales or activities to avoid the complications associated with performing work outside the United States. Do you agree with this statement? Support your answer with a convincing explanation.

9

Unmanned Aircraft System Design

Brian Argrow

CONTENTS

9.1 Introduction: Mission-Derived Design

Since their conception in the early twentieth century through the first few years of the twenty-first century, the design and development of unmanned aircraft systems (UAS) have been driven primarily for military missions. In addition to risk reduction, in the latter years it was anticipated that an uninhabited aircraft would result in a significant reduction in system cost, compared to manned aircraft that performed the same or a similar mission. While UASs have sometimes provided opportunity for cost savings against competing manned aircraft, this is not universally true, and the view that UASs are competing against manned systems often ignores the broader complementary or supplementary role that the UAS might provide in mission design. As discussed in the 2003 UAS report [SAB03], the cost picture often remains ambiguous, "… because of limited UAV experience and because procurement numbers are so low that per-unit costs have remained high." Regardless, as of this writing, military customers continue to drive UAS development. The past decade, however, has witnessed the emergence of a civilian demand for small UAS (sUAS) creating a market pent-up by a worldwide regulatory environment unprepared to integrate any UAS, particularly sUAS, into the airspace system.

Because there are no UAS-specific airspace regulations, a UAS must be operated in compliance with manned aircraft regulations. Specifically, this means that a UAS must possess a see-and-avoid capability and demonstrate the level of safety equivalent to that exhibited by a manned aircraft. These requirements have severely limited the commercial UAS market with the result that in the United States most civilian UAS operations to date have been carried out by public agencies that can obtain a Federal Aviation Administration (FAA) certificate of authorization or waiver (COA). The 2012 FAA Modernization and Reform Act is currently enabling limited operations of an increasing number of small civil (commercial) UASs in the National Airspace System (NAS). As of this writing, the FAA has granted approximately 450 exemptions to allow the operation of sUAS in the NAS. Most larger UASs (>55 lb) used for civilian applications, such as for homeland security or for atmospheric and oceanic research, are converted military systems or were designed for civilian applications based on an original military design. For the brief survey that follows, several fixed-wing UASs are selected to represent performance characteristics across the spectrum of size, speed, endurance, and range of systems in operation today. Not included is the class of multi-rotor or multi-copter designs that have proliferated in recent years, and are now being deployed in increasing numbers for both hobby and professional purposes. These multi-copters have had a major impact on the public perception of sUAS such that today when someone speaks of a "drone," this now connotes a small multi-copter—usually carrying a camera, rather than the large military UAS that were originally referred to as drones.

UASs have been traditionally used, according to the frequently used phrase, for missions that are too "dirty, dangerous, or dull" for a human to be physically present in the aircraft. However, advances in communications and control systems have enabled increased autonomy of operations, thus enabling more specific mission-driven UAS designs where, from the start, the purpose is not to physically remove a human pilot from the aircraft, but to directly design the aircraft as a platform or tool for specific applications. This explains the growth in UAS designs not derived from manned aircraft, including small sUAS, where habitation of the aircraft is impossible. The absence of humans aboard the aircraft opens the design space compared to that for manned

FIGURE 9.1
UAS airframes. (a) NASA Global Hawk [http://www.nasa.gov/centers/armstrong/aircraft/GlobalHawk/], (b) NASA Ikhana [http://www.nasa.gov/centers/armstrong/multimedia/imagegallery/Ikhana/ED14-0341-09.html], (c) RECUV Tempest UAS (University of Colorado/NOAA), and (d) DJI Phantom 2 Vision Quadcopter (Amazon) [http://ecx.images-amazon.com/images/I/61x8yJcesfL_SX425_jpg]

aircraft. This affects the design choices from the airframe materials and construction techniques, to the quality of servos, to system redundancies, to recovery systems (e.g., no landing gear).

Figure 9.1 displays the range of size and variety of UASs in operation today. In the following sections, we will discuss the typical UAS subsystems. Considering the incredible range of size, performance, and applications of UASs, it might be surprising to consider that most UASs are made up of the same subsystems. However, these subsystems obviously vary tremendously in the number, size, weight, power requirements, and complexity of the components.

9.2 The Design Process

Raymer [RAY12] presents the design wheel, shown in Figure 9.2, and an overview of the design process. The wheel illustrates the iterative nature of the process where requirements are set by preceding trade studies, concepts are derived from requirements, the design analysis may generate new concepts, and the cycle may repeat.

Raymer also notes that those involved in the design process can never agree on where the process begins. Most conventional aircraft designs, however, generally have a starting point anchored to previous designs for similar purposes. Today, one might argue that the design process for a UAS differs and that it usually begins with the mission requirements, leading to the mission-derived design, which is discussed in the previous section. The absence of humans in the aircraft opens the design space, enabling designs to be more mission-driven than in the case of the design of conventional manned systems. As a result, current UAS designs range from those for which the airframe might appear externally similar to a comparably sized manned aircraft, to those such as the recent multi-copter designs that have no conventional counterpart. Even in the case of the UAS sized on the order of conventional manned aircraft, unless the UAS airframe is actually a re-purposed manned airframe, or an "optionally piloted" aircraft (where there is an option to have a human pilot on board), there are usually features that make them obviously uninhabitable, such as no obvious cockpit or windows.

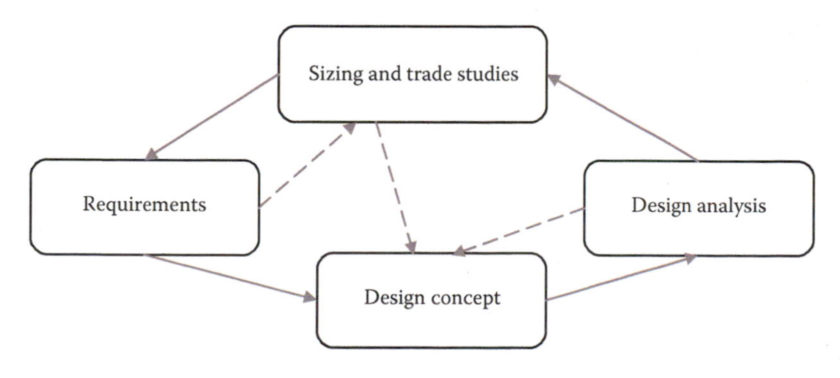

FIGURE 9.2
The design wheel. (From Raymer, D.P. (RAY12) 2012. *Aircraft Design: A Conceptual Approach*, 5th ed., AIAA, Chap 1 and Chap. 17.)

9.3 Unmanned Aircraft Subsystems

9.3.1 Design Tools

The methods for aircraft design layout and design analysis used in the industry tend to be proprietary and highly computerized [RAY12], and these methods are directly applicable to the design of airframes for large UASs. Many of the methods and tools developed for conventional, large aircraft design are not directly applicable to the design of many sUAS airframes. This is particularly true for tools based on the empirical relations or databases of manned aircraft, since sUAS generally operate at flight speeds and Reynolds numbers much lower than for manned aircraft. In addition, many sUAS use electric propulsion systems powered by batteries, fuel cells, or solar cells, which rely on energy storage or conversion systems not typically modeled in conventional aircraft design tools. Because nonproprietary conventional aircraft design tools are generally well established and readily available, the following discussion will not emphasize those, but will instead focus a bit more on the tools available for sUAS design, many of which have origins in model aircraft design.

9.3.2 Airframe

Aerodynamic databases such as the University of Illinois Urbana-Champaign (UIUC) Airfoil Data Site [UIUC15] have expanded wind-tunnel airfoil data to include Reynolds number values in the range of interest to model airplane hobbyist and sUAS airframe designers. The emergence of high-fidelity computational fluid dynamics (CFD) tools, such as those included in commercial multiphysics modeling packages which include ANSYS Fluent [ANS15], Star-CCM+ [CCM15], and COMSOL [COM15], enable "virtual wind-tunnel" testing of airframe components as well as entire airframes. Simpler panel method-based tools such as AVL [AVL15] compute stability derivatives as well as aerodynamic forces and moments on simplified geometries, and are also particularly useful for preliminary airframe design.

Materials for a manned aircraft airframe are generally selected to provide maximum strength and stiffness with minimum weight. Strength and stiffness, and knowledge of how those properties vary in the material undergoing structural loading and deformation, are required to maintain aerodynamically optimized shapes for maximum design efficiency, or simply to anticipate shape changes so that reliable control laws for aircraft stability can be designed. The materials for UAS airframe construction are typically those used for a range of aircraft from those in manned aviation to hobby aircraft, thus ranging from aerospace-grade metals and composites, supported by a vast database of material properties and construction standards for certified manned aircraft, to balsa and Monokote (a thin plastic film used for the outer skin of many hobby aircraft), to various forms used for hobby aircraft construction. The databases for hobby construction materials that might be used for sUAS airframes are sometimes only supported by the shared knowledge and goodwill of the user community, and the material properties that manufacturers choose to make public. This lack of materials data and construction standards for sUAS airframes constructed from hobbyist materials remains a significant obstacle to their integration into the NAS.

Figure 9.3a shows the payload weight capability for a range of large-to-small UASs compared to the overall per-airframe cost (airframe cost includes the communication and control systems) for a representative legacy U.S. military systems. The trend-line fit to these

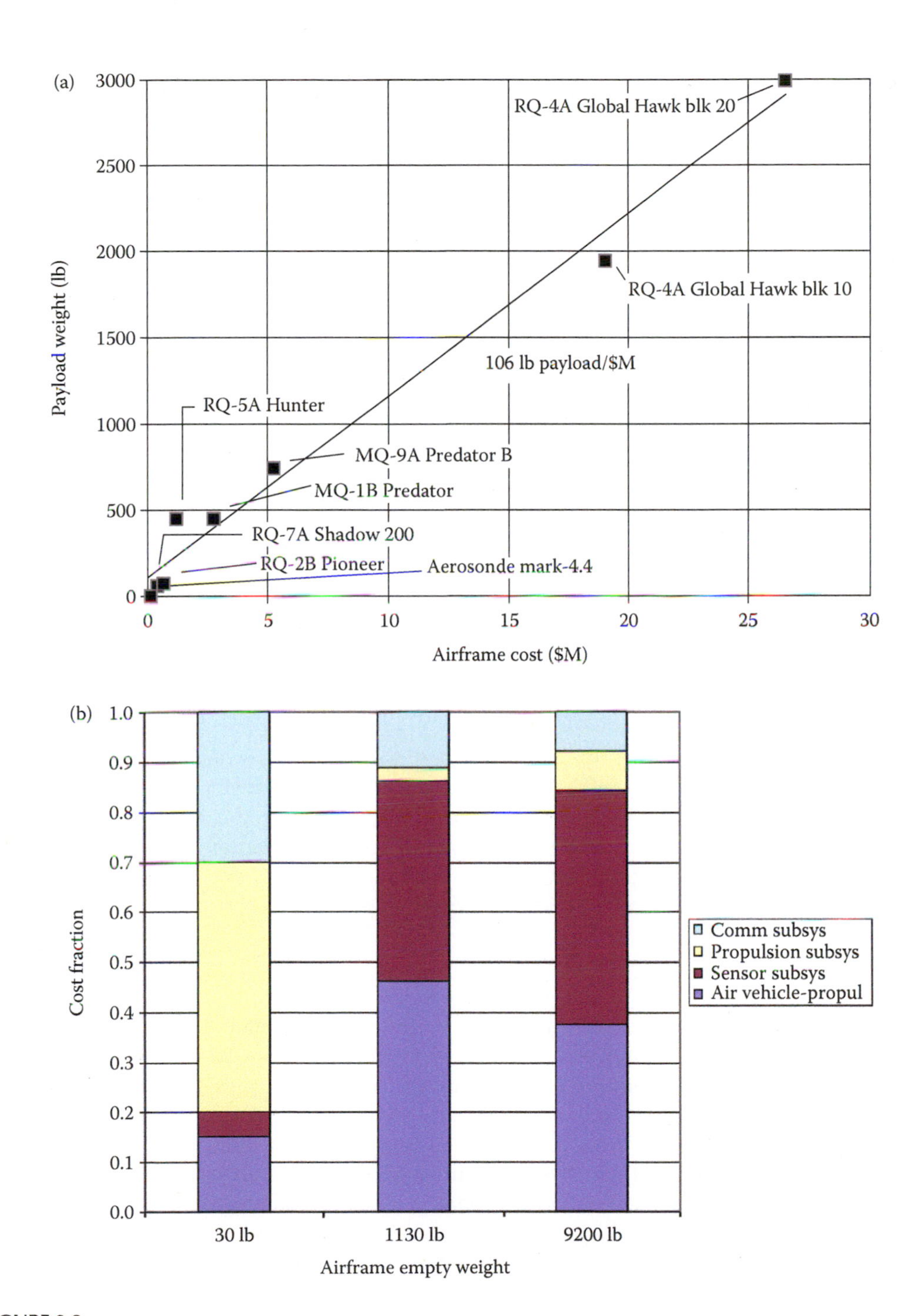

FIGURE 9.3
(a) Trend for airframe cost relative to payload capacity for representative legacy systems; (b) Fraction of system cost versus airframe empty weight for three representative legacy systems.

data shows that the airframe cost is a bit over U.S.$1M per hundred pounds of payload. While it is no surprise that larger payloads require larger, more expensive airframes, it shows that sUAS enable the airframe to continue to be sized according to the payload, and do not reach a lower threshold of airframe size necessary to meet the requirement of safely carrying a human pilot. Figure 9.3b shows the fractional cost of subsystems for three high-performance legacy UASs, categorized by airframe empty weight (the data are from the year 2003). Figure 9.3b shows that the cost of the high-performance sUAS is dominated by the specialized propulsion system, and for the large UAS the largest fraction of the cost is for the sensor package. The purpose of Figure 9.3 is to illustrate trends that are only notional compared to the civilian systems in development today.

9.3.3 Propulsion System

Figure 9.4 illustrates the range of performance of several contemporary UASs. These are primarily military systems [OSD05] with data supplemented by the 2009 UVS Yearbook [VBL09] and product brochures.

Unmanned aircraft performance is most clearly defined by the propulsion system, with the greatest capability provided by gas turbine engines, followed by internal-combustion engines and battery-powered electric motors. Similar to manned aircraft, UAS propulsion systems are chosen based on the mission requirements, where excess power required for subsystems and payloads might be an important driver for the selection of propulsion system.

Equations to estimate the range and endurance performance of aircraft propelled by piston-driven, that is, internal-combustion (IC), engines or gas turbines are well established (e.g., [AND99], [MCC95], [RAY12]). Recently, Traub [TRA09] developed relations for estimating the range and endurance performance of battery-powered aircraft, which includes the effects of discharge rate and voltage drop on the effective battery capacity. Tools such as MotoCalc [MOTO15], originally developed for hobbyists, are readily applicable for preliminary propulsion system design for battery-powered sUAS.

9.3.4 Flight Control System

The autopilot was a key to enabling aircraft to be reliably controlled without a human pilot on board. Elmer Sperry is generally credited with the development of the first true autopilot used in a UAS. Building on his experience in developing gyrostabilizer systems for submarines and then manned aircraft, he worked with Glen Curtiss, another aviation pioneer, to produce the first controlled unmanned aircraft. The flight of the Curtiss–Sperry Aerial Torpedo on March 6, 1918 marked the first successful flight of a powered unmanned aircraft, and unmanned aviation's counterpart to the Wright Brothers' flight 14 years earlier [NEW04].

Today, only sUAS operating within close visual range are typically flown through full wireless remote control (RC). Additionally, many small rotorcraft employ some type of stability augmentation, such as gyros, to assist the RC pilot in controlling the aircraft. Other UASs are generally flown with some level of automation or autonomy, particularly if they are flown beyond visual range. This includes large UASs such as the General Atomics Predator family of UASs and the Northrop Grumman Global Hawk, pictured in Figure 9.1. In the case of the Predator, the pilot controls the aircraft with a joystick, with a variable amount of direct joystick control that depends on the mission segment (e.g., takeoff, cruise, and landing). The Global Hawk is operated at an increased level of autonomy,

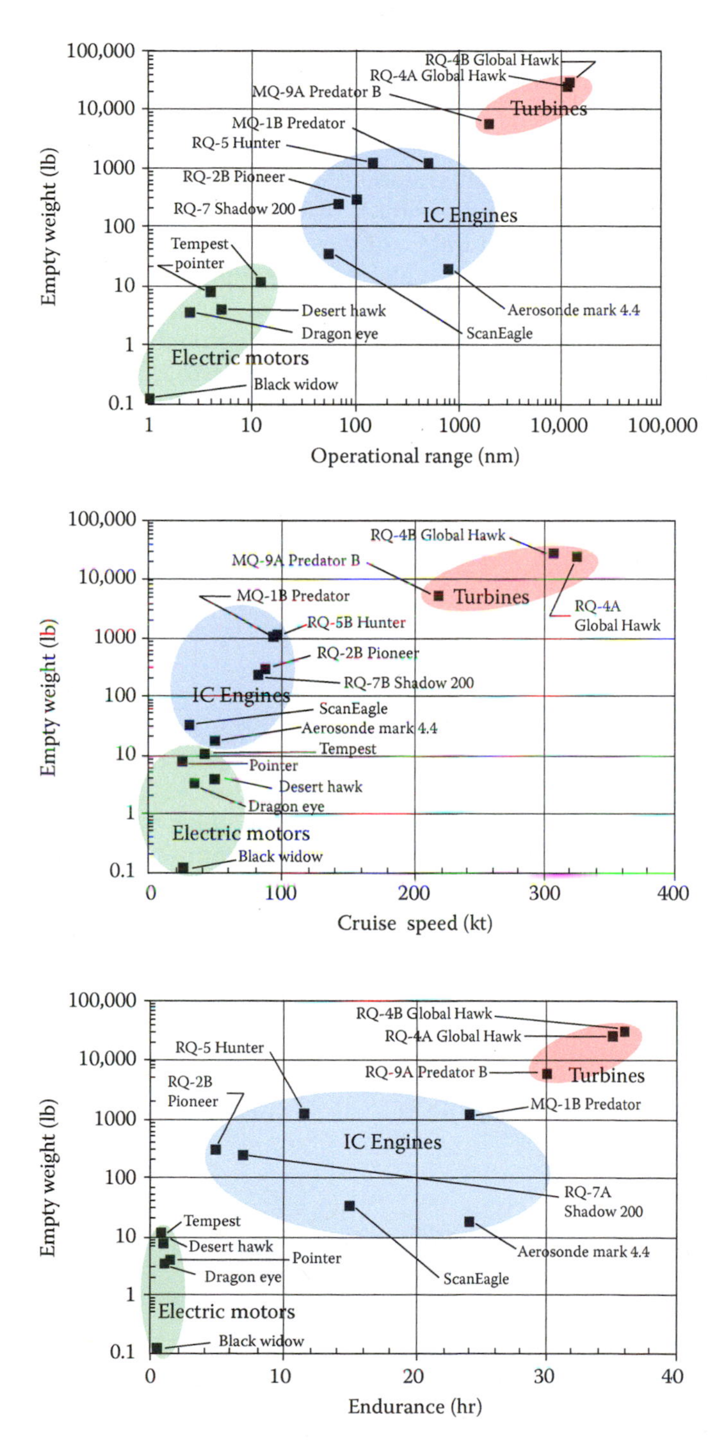

FIGURE 9.4
Illustration of the performance of representative UASs over a range of sizes.

with automated takeoff and landing, and flight plans where the autopilot system controls the trajectory of the aircraft between waypoints. In this case, the role of the remote pilot is primarily to monitor the aircraft and to potentially update the flight plan or to handle contingencies.

For the UAS not flown in RC mode or stability-augmented RC mode, a flight control system (often referred to as an autopilot, the term that we will use) is employed to control the aircraft. The autopilot system is typically composed of a microprocessor or computer that runs algorithms designed to control the aircraft over a preplanned flight path, or to augment control of the aircraft that is receiving steering commands (commands such as heading changes, not associated with maintaining stability) from a remotely located pilot. Typically, an inner control loop receives high-frequency sensor data to manage the aircraft attitude, while an outer-loop controller manages the aircraft position while following a flight plan.

In some UASs, a separate flight computer might run independently of the autopilot computer, with algorithms that issue high-level commands to the autopilot outer control loop. The degree to which such commands are based on "decisions" made by the flight computer algorithms without direct human interaction is often a measure of the system's level of autonomy.

9.3.5 Control Station

The ground control station, or more simply the control station (CS), is the part of the UAS that provides the control interface to the unmanned aircraft (UA) through which the pilot/operator steers the UA or otherwise manages the mission. Commands issued through the CS may range from a pilot sending real-time joystick altitude-control commands in RC mode to manually fly the UA, to an operator monitoring and interfacing with a highly autonomous system through high-level commands. Figure 9.5 contrasts the CS for the NASA Ikhana UAS (Figure 9.5a) compared to a recent version of the Black Swift Technologies SwiftPilot CS (Figure 9.5b) where the operator interface is through a tablet computer.

The most important feature of the CS is the interface that enables the human pilot/operator to interact with the UAS flight control system. The sensory inputs available to the pilot/operator through the CS interface generally cannot be reproduced to realistically represent the experience the pilot would have on board the UA. Therefore, it is no surprise that a CS "cockpit," based on the design of a manned aircraft cockpit, might

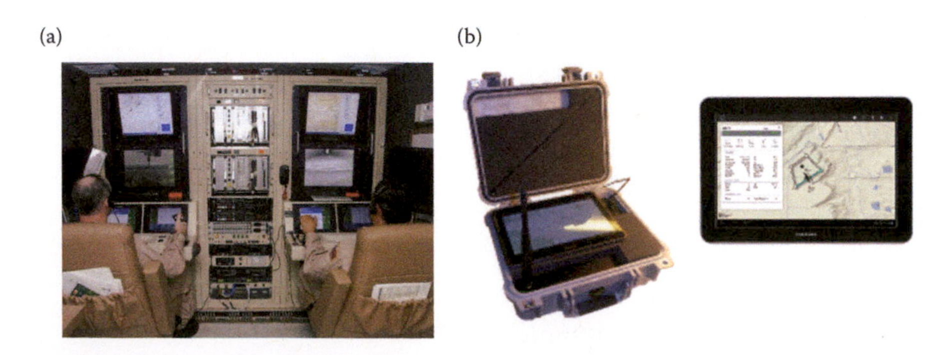

FIGURE 9.5
(a) Pilots at the controls in the Ikhana ground control station (NASA Photo/Tony Landis); (b) The Black Swift Technologies SwiftPilot ground control station and tablet interface for small UASs.

not provide the optimal interface for remotely operating a UAS. As an example, in an interview published by Wallace [WAL09], when discussing his experience with the NASA Ikhana UAS (a civilian version of the General Atomics Predator-B), an Ikhana pilot had quoted: "I don't think that any pilots were there on the days GA [General Atomics] designed this setup." Another is quoted: "The hardest part is learning the keyboard and menu navigation. When they say 'IFF ident,' good luck. It's like three or four keystrokes to do something like that," and another pilot stated: "You get saturated pretty quickly with this system…". These quotes are cited to illustrate that the UAS pilot/operator experience is different from that in a manned aircraft, and that it is important for CS interface design to consider the "human factors" that produce a different experience than that encountered in an onboard cockpit.

9.3.6 Payloads

Figure 9.3b shows the fractional cost breakdown of three representative high-performance UASs. In the case of the two larger UASs, the payload (sensor subsystem) contributes significantly more to the total cost than for the small UAS. This reflects the trend that sUAS are more likely to be used to carry lower cost, but not necessarily less capable, commercial-off-the-shelf (COTS) payloads, such as cameras, meteorological sensors, etc. The larger airframes are generally designed for higher reliability and to carry the more expensive, more power-demanding sensors for military and specialized civilian applications, such as EO/IR (electro-optical/infrared) sensors, lidar, synthetic aperture radar, etc.

Miniaturization of sensor packages and increasing reliability in small airframes and subsystems has enabled some mission capabilities, originally only achievable by relatively large and expensive UASs, to be scaled to smaller UASs. Again, the flexibility of not having to design the airframe to include human occupation enables more specific mission-derived design, so that the aircraft size and performance might be more optimized to meet the mission requirements. The DJI Phantom Quadrotor with an integrated video camera, shown in Figure 9.1d, is an example of how sensor miniaturization combined with small airframes and advanced battery technologies has led to the proliferation of sUAS designs available to the general public and professionals alike.

9.3.7 Communications, Command, and Control (C3)

When writing "C3", the order of the words communications, command, and control in the definition is important. For current UAS design, the requirements for safe operations in the NAS dictate that design starts with the communications system to enable the operator to remain on the control loop as a monitor or in the control loop to carry out manual tasks. Commands can then be sent to the aircraft through the uplink, while the downlink enables telemetry to report the UAS health and status or a payload link to deliver sensor data. Commands issued from the control station by the operator over the uplink might enable direct control of the aircraft. The frequency (how often updates are required) of these commands determines the requirements for availability and latency in the wireless C3 links.

Figure 9.6 shows a schematic of the RECUV Networked UAS Communications, Command, and Control (NetUASC3) architecture that was developed to support operations of the Tempest UAS for the second Verification of the Origins of Rotation in Tornadoes Experiment (VORTEX-2) [ELS11]. The various arrows show the lines of communication that link the aircraft and control station and also simultaneously enable communication with

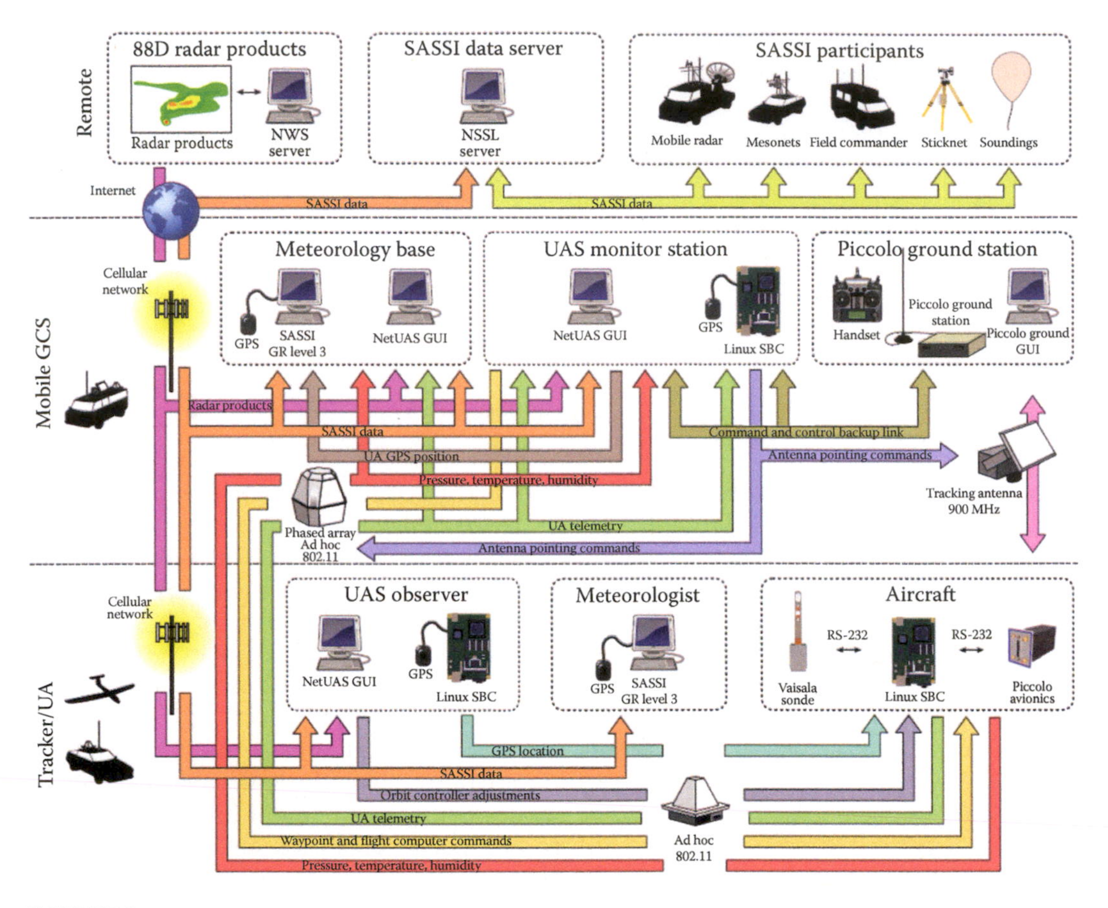

FIGURE 9.6
Schematic of the RECUV Networked UAS Communications, Command, and Control (NetUASC3). (From Elston, J. et al. 2011. *Journal of Field Robotics*, Vol. 28, No. 4, pp. 461–483.)

various data servers and other clients sending and receiving data. This diagram clearly illustrates how, in this case, the aircraft is but one of many components that can make up an unmanned aircraft system.

9.4 Standards for UAS Design, Construction, and Operations

While many large UASs might be designed to incorporate components used in type-certified manned aircraft, as previously discussed, even these aircraft do not meet the requirements for FAA type certification, and there is no expectation that any sUAS designed with components manufactured for the RC hobby community will ever meet FAA standards for type certification.

Several engineering standards organizations have taken up the task to develop standards for UAS design and operations. The ASTM International Committee F38 on Unmanned Aircraft Systems is developing one of the most comprehensive sets of UAS design and operations standards, and guidance materials. The scope, as stated by the Committee, is to develop technical publications that include [ASTM15]

1. Minimum safety, performance, and flight proficiency requirements.
2. Quality assurance—to install manufacturing controls that will assure vehicles conform to design criteria.
3. Productions acceptance test and procedures assuring that the completed vehicle systems meet reported performance as demonstrated in the prototype vehicle system. This includes limits such as empty weight and center of gravity, performance specifications, controllability and maneuverability trim, stability, stall speed and handling characteristics, engine cooling and operation characteristics, propeller limits, systems functions, and folding or removable lifting surfaces.
4. A baseline plan for continued airworthiness systems, including methods for monitoring and maintaining continued operational safety, and processes for identifying, reporting, and remedying safety-of-flight issues.

The Radio Technical Commission for Aeronautics (RTCA) is chartered by FAA as a Federal Advisory Committee. The RTCA Special Committee 203 (SC-203), Unmanned Aircraft Systems was established in 2004 (sunsetted in 2013) to develop standards and certification criteria, and procedures for sense-and-avoid systems as well as protocols to be used for the certification of command, control, and communications systems [RTCA203]. The RTCA SC-228, Minimum Operational Performance Standards for Unmanned Aircraft Systems was established in 2013 and is developing Minimum Operational Performance Standards (MOPS) for detect-and-avoid equipment and command and control (C2) data link MOPS establishing L-Band and C-Band solutions [RTCA228]. One of the earlier UAS standards that was published is "Standard Interfaces of UAV Control System (UCS) for NATO UAV Interoperability, 3rd Ed." [NSA07]. Other organizations that are developing UAS standards include the Society of Automotive Engineers (SAE) and the American Institute of Aeronautics and Astronautics (AIAA).

9.5 UAS Design Verification and Mission Validation

Decades of academic and industrial investment in Modeling and Simulation (M&S) tools has not only facilitated the design process, but has also led to M&S tools that can be used in the verification of a UAS design and the preparation for mission validation. Depending on the level of autonomy designed into the UAS, software testing and verification can be the major cost, measured in schedule and development costs.

Hardware-in-the-loop (HIL) testing is widely used for the verification of the aircraft control system, sensor/payload integration, and the CS interface. Typically, a simulation creates a synthetic flight environment where simulated sensor inputs are sent to the autopilot to create a telemetry stream to the CS that is identical to what is measured in flight, so the CS interface exactly replicates what the UAS operator would see or experience in an actual flight. From an observer's point of view, an HIL test can be interesting to observe, since one might see the control services moving on the stationary aircraft exactly as they would be moving to control the aircraft in a real flight. Therefore, this type of testing does not only put the autopilot and flight computer software through their paces but may also verify the function of the servos and control services prior to an actual flight. In some cases, the control software of the autopilot might be separately run on the simulation computer or another computer and not on the actual autopilot hardware. In this case, any latency that might be associated with running the algorithms on the autopilot hardware must also be simulated to realistically replicate the autopilot function. For a small UAS, a flight simulator for HIL testing is included with most COTS autopilot systems, designed in some cases to integrate with high-fidelity graphical interfaces originally developed for video games.

After systems verification, the simulated environment might be used to prepare for mission validation, or for pilot/operator training, where the tasks of an actual mission might be simulated. Depending on where the remote pilot/operator is located, it might be impossible for that person to distinguish between a simulated mission and an actual mission.

DISCUSSION QUESTIONS

9.1 In the introductory paragraphs, the author makes several statements. Answer the following questions based on that passage. What has "pent-up" the global demand for UASs? Why have UAV per-unit costs remained high? What entities have carried out most civilian sUAS operations and why? What design opportunities are attributable to the lack of a pilot on board the UAS?

9.2 Generalize the design process. Describe tools currently available to designers of sUAS airframes and powerplants and tell how these influence the design process.

9.3 What are the functions of the autopilot? What is the difference between the autopilot inner and outer control loops? Whom does the author identify as the developer of the first successful flight of a powered, unmanned aircraft?

9.4 Describe the purpose of the ground control station (GCS) or control station (CS). How does the CS of the General Atomics Predator differ from the pilot interface (i.e., cockpit controls) typically found on complex, high-performance manned aircraft?

9.5 What is meant by C3? Describe the difference between operator "in the loop" and operator "on the loop."

9.6 What is meant by aircraft certification standards? List organizations involved in developing UAS standards and the corresponding areas to which these would apply.

9.7 Describe hardware-in-the-loop operation. What is the difference between UAS design verification and mission validation? In which would HIL be most useful and why?

References

[AND99] Anderson, J.D. 1999. *Aircraft Performance*, McGraw-Hill, New York, Chap 5.

[ANS15] http://www.ansys.com/, accessed 4/1/2015.

[ASTM15] ASTM International Committee F38 on Unmanned Aircraft Systems, http://www.astm.org/COMMIT/SCOPES/F38.htm, accessed 4/1/2015.

[AVL15] http://web.mit.edu/drela/Public/web/avl/, accessed 4/1/15.

[CCM15] http://www.cd-adapco.com/, accessed 4/1/15.

[COM15] http://www.comsol.com/, accessed 4/1/15.

[ELS11] Elston, J., Roadman, J., Stachura, M., Argrow, B., Houston, A., and Frew, E. 2011. The tempest unmanned aircraft system for *in situ* observations of Tornadic Supercells: Design and VORTEX2 flight results, *Journal of Field Robotics*, Vol. 28, No. 4, pp. 461–483.

[MCC95] McCormick, B.W. 1995. *Aerodynamics, Aeronautics and Flight Mechanics*, Wiley, New York, pp. 378–385.

[MOTO15] MotoCalc, http://www.motocalc.com/, accessed 4/1/15.

[NEW04] Newcome, L. 2004. *Unmanned Aviation: A Brief History of Unmanned Aerial Vehicles*, AIAA, Chap. 3.

[NGC15] Northrup Grumman Corp., RQ-4 Global Hawk High Altitude, Long–Endurance Unmanned Aerial Reconnaissance System, http://www.northropgrumman.com/capabilities/rq4block10 globalhawk/documents/hale_factsheet.pdf, accessed 4/1/2015.

[NSA07] NATO Standardization Agency. 2007. *Standard Interfaces of UAV Control System (UCS) for NATO UAV Interoperability*, STANAG 4586, Ed. 2.5.

[OSD05] 2005. *Unmanned Aircraft Systems Roadmap: 2005–2030*, Office of the Secretary of Defense Memorandum for Secretaries of the Military Departments.

[RAY12] Raymer, D. P. 2012. *Aircraft Design: A Conceptual Approach*, 5th ed., AIAA, Chap 1 and Chap. 17.

[RTCA203] SC-203 Unmanned Aircraft Systems (UAS), http://www.rtca.org/, accessed 4/1/2015.

[RTCA228] SC-228 Minimum Operational Performance Standards for Unmanned Aircraft Systems, http://www.rtca.org/, accessed 4/1/2015.

[SAB03] Unmanned Aerial Vehicles in Perspective: Effects, Capabilities, and Technologies, 2003, Vol 1: Summary, SAB-TR-03-01, USAF Scientific Advisory Board, AF/SB, Washington, Sep.

[TRA09] Traub, L.W. 2009. Range and endurance estimates for battery-powered aircraft, *Journal of Aircraft*, Vol. 48, No. 2, pp. 703–707.

[UIUC15] UIUC Airfoil Data Site, http://m-selig.ae.illinois.edu/ads.html, accessed 4/1/2015.

[VBL09] van Blyenburgh, P. 2009. *Unmanned Aircraft Systems: The Global Perspective 2009/2010*, Blyenburgh & Co., Paris, France, May.

[WAL09] Wallace, L. 2009. Remote control: Flying a predator, *Flying*, http://www.flyingmag.com/pilot-reports/turboprops/remote-control-flying-predator, accessed 4/1/2015.

10

UAS Airframe and Powerplant Design

Michael T. Most

CONTENTS

10.1 Introduction

Of necessity, the design of any given unmanned aircraft system (UAS) will correspond directly to the intended mission of the platform—in the case of unmanned aircraft (UA), form truly follows function. Any UA that does not conform to this basic principle will fail to optimize mission goals. For example, infrastructure inspections (e.g., bridges, monuments, flue gas stacks, wind turbine blades and towers) may demand stable, low vibration platforms to obtain the high resolution images necessary to ensure the integrity of the structure under examination. In these applications, a high degree of endurance and range may not be a major requirement, and a multi-rotor (quad-, hexa-, or octocopter) UA, powered by an electric motor, may be best suited to the mission. On the other hand, while covering larger areas searching for pipeline failures, abandoned mine shafts, or a lost hiker may not require high definition imagery, the mission will dictate greater range and endurance. In this latter example, a conventionally configured fixed-wing aircraft, having a high fineness ratio fuselage and wings of high aspect ratio powered by a gasoline, reciprocating engine may be best suited to the mission. In the civilian world, for the foreseeable future, obtaining remotely sensed data as one step in delivering a finished data product is often the raison d'être of the UA platform and its associated systems. Thus, the mission is currently a significant factor in determining the selection or configuration and design of a UAS.

Due to the multifaceted functionality of UASs and the diversity of operations suited to the application of this technology (as documented throughout this book), unmanned aircraft run the design gamut between small UAs that fit in the palm of one's hand (e.g., IAI Malat Mosquito and the Prox Dynamic Black Hornet Nano, which weighs approximately 16 g, or just over one-half ounce) to large fixed-wing aircraft, such as the Boeing SolarEagle, designed by the Phantom Works to incorporate a wing spanning more than 120 m, and the Titan Solara, both of which will loiter above 18,500 m (~60,000 feet) for periods of five years and beyond. Between these design extremes are small UASs, or sUAS, such as the vectored thrust, ducted fan vertical takeoff and landing (VTOL) craft, the Honeywell RQ-16 T-Hawk (named for the Tarantula Hawk, a waspish insect found in the Southwest desert), helicopters, multi-rotors and fixed-wing aircraft comparable in size to corporate jets (at about 14.5 m, the wingspan of a General Atomics Predator approaches that of many Learjets, Citations, etc.), or even larger aircraft (the wingspan of a Northrop Grumman Global Hawk/Euro Hawk7 surpasses that of a Boeing 727 or Airbus A320). The choice of UAS powerplants is similarly diverse, ranging from turboprop and turbofan gas turbines to two-stroke heavy fuel engines, or HFEs, with electric, Wankel (aka, rotary), and Otto cycle reciprocating engines filling niches between the extremes of the design spectrum.

Given such diversity, providing an in-depth, detailed discussion of all design considerations is far beyond the scope of this chapter, or even of a book of this size. Rather, the author's intention is to adumbrate design considerations as influenced by mission goals and performance and payload constraints, while providing a general overview of common designs and the reasons behind the decisions resulting in the selection of these over others in a diverse field of alternate platforms.

10.2 A Few Observations Regarding UAS Design

What factors should most influence the process of UAS design or the selection, for a particular mission, of an existing platform? Useful load? Maximum sustainable speed? Airframe stability? Freedom from vibration? Safety? Yes, safety is a concern but likely not to the degree necessary for passenger-carrying manned aircraft—it is not possible to remove all risk from flight operations and attempts to completely eliminate it are often counterproductive, excessively restricting operations, research and development to produce a catatonic state wherein paralysis inhibits innovation and impedes progress. Reliability? Maintainability? Expected life cycle? Endurance and range? (It should be noted, parenthetically, that these last two characteristics are closely related, but the terms are not strictly interchangeable. For example, if an aircraft flies very, very slowly while consuming small amounts of power, it may exhibit limited range, but high endurance. Conversely, an aircraft that flies very fast while consuming large amounts of power, may have great range, but limited endurance. Although the two terms are often used interchangeably, care should be exercised to ensure that equivalence is warranted.) The answer to the question posed at the outset of this paragraph is that the consideration of each of these design attributes is certainly important, but none, by itself, is substantially overriding.

However, two preeminent factors do exist. As previously mentioned, the nature of the intended mission should take precedence above all other factors in a ground-up design or in the decision to select an existing platform for a particular application. The second factor of paramount importance is cognitive and it lies at the heart of the design process. It is the realization that any UAS platform should be chosen or designed according to the rule of interdependent synergism, a phrase alluding to the interrelationship of all subsystems and components wherein these function together in a way that maximizes system performance to optimize the mission goal. Only when the entire design is considered as a synergistically holistic construction of individual components with sometimes competing system demands and often diametric operational constraints, can it be realized that the design of any given UA system is a tradeoff—something of a linear programming problem wherein the optimization of the entire system (to achieve the mission goal) may only result from the sometimes suboptimal performance of the individual components and those subsystems that are integrated to form the UA system.

For example, perhaps the endurance of an electric aircraft is insufficient to complete its intended mission of surveying an area of specified dimensions. Increasing the size of the battery (in terms of ampere-hour capacity) will allow the aircraft to stay aloft longer, but doing so will also diminish useful load and, therefore, decrease the choice of payloads available to gather data for the mission. Or, alternately, necessitate the installation of a smaller, lighter, but disproportionately more expensive sensor package. Installing a larger, heavier battery may also alter the center of gravity and negatively affect other performance and aerodynamic characteristics. Or, perhaps a hand-launched electric airplane is difficult to get airborne on a windless day. Choosing a more powerful motor and installing a larger propeller having less blade angle may improve launches, but will also reduce endurance unless a higher capacity (larger and heavier) battery is installed, which will, again, reduce flexibility in payload selection, and alter other performance characteristics as the domino effect associated with a simple alteration manifests itself. The point is, and it is an important one, that the aircraft platform and all subsystems and components should function synergistically as a system to optimize overall operational capability of the UAS in achieving mission goals—however, maximizing mission effectiveness is often only achieved at the expense of suboptimal performance at the component level. Maximizing overall design capability can be an extremely complex process, often, at once, both art and science.

10.2.1 Form Follows Function: The Best Place to Begin the Design Process

Because "form follows function," the most likely "best place" to begin the design process is to consider the intended mission of the unmanned aircraft. Keep in mind that the UAS is exactly that—a systemic entity comprised of components and subsystems, and any design change to any element of the system has the potential to affect the entire system. At the most basic level, the UA exists only as a platform to carry the payload necessary to successfully accomplish mission goals. The mission, in turn, will determine factors such as the type of payload (number, weight, and configuration of sensors), endurance and, possibly, aircraft structure and configuration. Payload will, in turn, influence gross weight, powerplant selection, structural loading, CG, choice of nav/com and C2 design factors, amount of lift required, among other factors. Endurance requirements will affect choices regarding drag, airfoil and wing design, fuel load (and, consequently gross weight), powerplant, etc. The amount, type, and cost of UA ground support equipment (GSE) as well as the choice and design of launch and recovery subsystems will also be affected by the overall UAS design as determined by the mission. Consequently, because the mission affects every aspect of airframe, powerplant, and system architecture and integration, the most fundamental design consideration is how to best configure the UAS for its intended purpose.

10.2.2 Economic Influences on the Design Process

As a rule of thumb, the design process should begin with consideration of the type of mission that is to be flown, but economic factors cannot be ignored. In general, as the size of the platform increases, so do the costs associated with both manufacturing and operations (e.g., those expenses associated with maintenance, support, and operation). For example, the cost of operating a UAS not requiring substantial logistical support, investment in GSE, or a large, mobile ground control station (GCS) may be only 20%–40% of that necessary for a manned aircraft performing a similar mission, whereas, associated expenses will double for larger unmanned aircraft requiring greater end-to-end support (Austin 2010). In fact, although affording definite advantages in 3-D (i.e., dull, dirty, and dangerous) missions, the additional investment in technology and the extra operational and personnel costs associated with operating larger UASs (e.g., a Reaper or Global Hawk) may approach those associated with manned aircraft.

If the UA is to be more than a "one-off," mission-specific design (i.e., produced in quantity, particularly for commercial distribution), life-cycle costs become significant. Broadly, these fall into one of two categories: (1) nonrecurring costs (testing, research and development, or R&D, tooling, startup costs, etc.), which must be recouped over the anticipated life of the UAS, and (2) recurring costs over the life of the product. Life-cycle costs may be viewed from the perspective of either the manufacturer or the operator. The manufacturer's life-cycle costs are those encountered during the production life of the platform, including costs which are nonrecurring, whereas those incurred by the owner/operator are associated with the fulfillment of mission requirements during the life of the UAS. These are, nonetheless, related.

The design of both manned and unmanned aircraft must achieve some necessary threshold of performance and reliability while minimizing life-cycle costs. From the operator's perspective, life-cycle costs derive not only from operating costs, but also include the initial purchase price of the platform (acquisition cost). The two perspectives on life cycle are related through acquisition costs, which include a portion of the manufacturer's initial development and design costs. The two perspectives on life-cycle costs are further associated through operating costs. If reliability is low and maintenance, support, and fuel costs are high, then the manufacturer will likely sell fewer units over the production run which will, in turn, increase not only the operator's life-cycle costs but also those of the manufacturer. Thus, efficiency, which affects endurance, range, reliability, and maintainability are important design considerations for both the operator and the UAS manufacturer.

10.2.3 Exogenous Factors Affecting the Design of UASs

The builder or manufacturer cannot control all factors influencing design. Those design factors falling outside the designer's purview of control authority and yet which must be incorporated into the UA platform are referred to as exogenous design factors—that is, attributes dictated by other than economic, intended mission, aerodynamic, airframe or propulsions system considerations, but rather by entities or functionalities outside the immediate design process. Examples of exogenous design influences include industry standards which frequently inform, or even drive, FAA regulations. Sources of such standards, which provide guidance for all aspects of UAS operation, construction, maintenance and design, are manifold and include private entities (ARINC, ASTM, RTCA), trade organizations (AUVSI), professional organizations (SAE), and ad hoc committees of industry leaders. As part of the regulatory process, the FAA solicits input and recommendations from representatives of these organizations who constitute either full (ARAC) or partial (ARC) membership of the respective rulemaking committees; thus, the influence of these groups can be very significant. Just as occurred in manned aviation, increasing regulation of UAS flight is inevitable and without foreseeable end.

Although industry influence of the design (and operation) of UASs is closely bound to the regulatory process, the influence of industry standards may be strong without the undergirding force of the federal government. An example from manned aircraft is the adoption of the Air Transport Association (ATA) system of organizing technical and maintenance data for complex and turbine-powered aircraft. Although the ATA-100 coding system was never enacted as a regulation, the adoption by airlines, manufacturers, corporate flight departments, and maintenance operations has been complete. Because the major airframe and powerplant manufacturers providing parts, spares, components, and aircraft to the operators of corporate and transport category aircraft began using the ATA system, all entities in any way affiliated with the industry were forced to adopt and integrate these into all relevant aspects of their operations. In this way, the industry-developed standard of ATA coding came to carry the weight of regulation without regulatory action. It is not unlikely that a similar form of pervasive adoption of certain industry standards, as exogenous design influences, will also occur in the unmanned aircraft industry as it evolves and matures.

10.2.4 Selected Preliminary Comments Relevant to UAS Flight Dynamics and Physics

Regardless of size or configuration, that is, whether fixed wing or rotary wing, UASs share most design features associated with their manned counterparts. The reason is that unmanned aircraft are subject to the same physical laws and flight dynamics as manned aircraft. Both manned and unmanned aircraft operate in the viscous fluid of the atmosphere. The same four forces, the gravitational pull of the earth on the mass of the aircraft (weight), aerodynamic drag, thrust, and lift, all affect both manned and unmanned aircraft in flight. These forces are interrelated. Lift acts opposite (counteracts) aircraft weight. An increase in weight reduces range and endurance while increasing drag and the amount of thrust necessary for a desired level of performance (Hurt 1965). Thrust opposes drag, which, in part, develops from the creation of lift and from the production of thrust, and, yet, the generation of thrust contributes to total drag (e.g., nacelles, cowls, propeller components, and powerplant cooling systems are sources of parasite drag).

Drag is the sum of all forces opposing movement of an aircraft through the viscous medium of the atmosphere. Total aircraft drag develops from multiple sources. As stated, creating lift induces drag, the result of tip vortices. High pressure air spills from under lower wing surface, curling around the wingtips into the low-pressure area above to create a swirl of air, or vortex, trailing off the tip and behind and below the aircraft. The dissipation of energy in these vortices is the source of drag attributable to an airfoil generating lift. Not surprisingly, this is termed induced drag. The remainder of total drag is referred to as parasite drag. Components of parasite drag include skin friction, form drag, interference drag, cooling drag, leakage drag, and the strong influence of compressibility effects, sometimes referred to as compressibility drag or wave drag. Because airflows may pass through the speed of sound (Mach) as it accelerates over the cambered surfaces of airfoils and other curved surfaces of the airframe (e.g., the canopy) when the airspeed of the aircraft is around 75% of Mach (or, 0.75M), compressibility drag is a factor in the design of very few unmanned aircraft (e.g., the Lockheed Martin QF-16, the 3+ Mach D-21 developed at Lockheed's Skunk Works® and Boeing's scramjet-powered, hypersonic X-51 Waverider) (Figures 10.1 and 10.2).

The remaining sources of parasite drag, to some extent, all affect the performance (e.g., range, endurance, useful load, required thrust) of a UAS. Skin friction drag develops from the shearing stresses that dissipate energy in the thin boundary layer above the surfaces of the aircraft. Leakage drag, which is generally associated with fixed-wing aircraft and accounts for 1%–2% of total drag (Sadraey 2009), results from the change in momentum of the air flowing through the gaps between fixed and moveable surfaces (e.g., ailerons, flying wing flaperons, elevators, rudders, flaps, ruddervators). Based on the frontal area presented to the airstream, form drag (aka, pressure drag or flat plate drag) results from the unbalanced pressure distributions across the area of the projected shape presented to the viscous flow and the turbulence it creates. Thrust must overcome this pressure differential in order to move the aircraft forward. Interference drag results from the energy losses resulting from the interaction of airflows at the juncture of various components. Where stabilizers join empennage and wings mate to the fuselage, the flows converge and interact to produce turbulence, shearing, and separation of the boundary layers—all of which are wasteful of energy—creating interference drag. The existence of interference drag explains why total drag is actually greater than the sum of all the drag acting on all components of the aircraft. Finally, cooling drag results from the loss of momentum and total pressure in the air flowing over the powerplant to carry away heat (Sadraey 2009).

Increased drag translates into reduced payload, range, and endurance—all important design considerations. Features intended to reduce drag are apparent in the design of all

FIGURE 10.1
Skunk Works® 3 + Mach D-21 under development. (Image courtesy of Lockheed Martin Skunk Works®.)

FIGURE 10.2
Artist's illustration of the X-51 Waverider. (Image courtesy of NASA.)

UASs, but these are generally a more important consideration in fixed-wing aircraft where mission goals require the ability to stay aloft for extended periods. For example, consider the following. Elliptical wingtips and high aspect ratio (AR) wings reduce induced drag. (In its simplest form, for a symmetrical, Hershey-bar wing, the aspect ratio is simply a comparison of the length of the wing to its chord. For nonsymmetrical wings, where the chord varies from root to tip, AR is expressed as the square of the wingspan compared to the surface area of the wing.) High aspect ratio wings are long and narrow. Think of it this way: An infinitely long wing would have no tips, produce no vortices, and induce no drag. As the span of the wing goes to infinity so, too, does the aspect ratio. Just as high aspect ratio wings decrease induced drag, so, too, does a high fineness ratio fuselage reduce form drag. The fineness ratio is a comparison of the length of the fuselage to its width. A short wide fuselage has a low fineness ratio; a long narrow one exhibits a high fineness ratio. Where one carefully examines the design of the fixed-wing UAS, high aspect and fineness ratios are frequently very apparent (examples include the Aeromapper EV2, the Hi Aero Gabbiano, and the IDETEC Stardust). A high-wing aircraft suffers less interference drag than a mid-wing design. (High-wing UASs include several Israeli Aerospace Industry, or IAI, designs, e.g., Heron, Searcher, Mastiff, and Hunter, and the designs they influenced, e.g., the RQ-2 Pioneer and the RQ-7 Shadow.) Control surface hinges of sUAS are sometimes nothing more than an extension of the airframe material (e.g., carbon fiber cloth) or covering; these surfaces are attached by a thin skin of plastic film or composite and run the length of the surface, thus completely eliminating leakage drag. Additional characteristics will subsequently be discussed, but the necessity to minimize drag will always be among the factors to be given careful consideration in the design of any UAS, but most particularly those of fixed-wing configuration.

Heavier-than-air craft, whether manned or unmanned, whether fixed- or rotary-wing, must develop lift to fly. Lift is the force opposing the gravitational pull of the earth on the aircraft. All heavier-than-air craft, whether manned or unmanned, develop lift according to the same laws of physics. External aerodynamic forces acting on the surface of an airfoil are attributable to two forces: air pressure (or pressure differential) and air friction. Of these two forces, friction, which occurs in a very thin fluid layer existing at the wing surface, is much the lesser, and, for most considerations, can be ignored to produce an idealized condition for study (Hurt 1965). Thus, in this generalized, or ideal state, lift acting on an airfoil, whether wing, propeller, or rotor blade, can be said to develop as a function of pressure differentials acting across opposing airfoil surfaces.

A distinguishing characteristic of subsonic flow is that changes in velocity and pressure occur with relatively minor changes in density—for most purposes these can be considered negligible. This is another simplifying assumption, and is the reason subsonic airflow is said to be "incompressible"—if the density of the flow is held constant, then changes in pressure and velocity become the significant factors in producing lift. In its most simplified form, Bernoulli's equation states that Total Pressure equals the sum of Dynamic Pressure plus Static Pressure. According to the Bernoulli equation, or Bernoulli's principle (and Newtonian physics), because the mass flow remains unchanged, variations in velocity result in corresponding, but inverse changes in pressure, and vice versa. Thus, as the velocity (dynamic pressure) of the airflow over the cambered surface of an airfoil increases, static pressure above that surface decreases, while total pressure remains unchanged. The force of lift develops as the result of the pressure differential acting across airfoil surfaces. If we are describing a wing during flight with a more cambered upper surface, then lift is the result of the decreased static pressure (attributable to increased air velocity) acting above the wing's surface, and greater pressure (due to lesser airflow acceleration) on the lower surface (and possibly some amount of impact pressure resulting from wing angle of incidence). In the same way, the rotating blades of a helicopter main rotor produce lift that is controlled to produce thrust, the force opposing drag, for propulsion and, also, for maneuvering. Similarly, a propeller converts the shaft horsepower (shp) output from the UAS powerplant to a propulsive force by rotating in an air mass to produce a pressure differential across the propeller back (cambered surface) and the flat face (surface having the higher pressure). An alternative explanation is that the thrust resulting from the acceleration of a mass of air through the propeller (or rotor) disk develops according to Newtonian mechanics, that is, according to the laws of the conservation of momentum, energy, and mass (Seddon 1990). Thrust produced by any propeller or propulsor is represented by the letter "F," for force, and equals the product of the mass of air and the amount of acceleration imparted to that air mass (i.e., $F = m \times a$).

Unfortunately, a propeller is not terribly efficient at converting shaft horsepower to thrust. Propeller efficiencies may fall between 50% and 87%, though some newer NASA airfoils and advanced planform designs, as incorporated in the unducted fan (or propfan) powerplants developed for manned aircraft during the 1980s, were able to achieve around 90% conversion of input power to thrust. Geometric pitch is the *theoretical distance* a given point on the propeller (usually measured at the 75% radius or spanwise blade station) should advance in one revolution if no inefficiencies were present. Effective pitch is the *actual distance* a propeller moves through the air under specified conditions. The difference between geometric and effective pitch is referred to as slip. (More will be said about these terms in the section on UAS powerplants.) Propeller slip results from inefficiencies and represents losses in the conversion of input power to thrust. Factors that reduce the ability of the propeller to effectively propel (i.e., "pull" or "push") the aircraft through the air include, for example, aerodynamic drag, suboptimal angles of attack that produce stalled regions on the blade, and energy lost to vibration, noise, and tip flutter. These conditions result in energy conversions (or, losses) that diminish performance and account for the difference between the expected distance of travel (geometric pitch) and the actual distance covered (effective pitch) in a single revolution. Successively outboard blade stations travel increasingly greater distances and, consequently, at correspondingly higher speeds. The blade shank, the thick, noncambered section located just outboard of the hub, is traveling at a much slower rate than the propeller tip. Because the amount of lift produced by an airfoil section at a given blade station is a function of both airspeed (of the relative wind) and angle of attack, the propeller manufacturer designs the propeller so that its blade angle decreases from hub to tip. This is referred to as propeller twist. Twisting the blades in this way produces more uniform pressure distributions (lift) across the propeller disk, maintaining a relatively acceptable angle of attack while preventing blade sections from either stalling or turbining (being driven like a pinwheel). Twisting the blades improves propeller efficiency. Because accelerating a larger mass of air to a lower velocity for a given amount of thrust (recall that force, in this case, thrust is the product of mass times acceleration) requires exponentially less energy (energy consumption increases as the square of the increase in acceleration), larger propeller diameters have the potential to produce thrust more efficiently, but the design tradeoff is that increased tip speeds introduce higher energy (and power) losses. Similarly, adding blades using three or four instead of two, holds the potential to accelerate a greater amount of air through the propeller disk, but the increased losses due to drag and the additional weight and complexity (especially in a constant speed design) may offset any potential efficiency gains.

The preceding highlights several of the ways UAS designs are similar to those of manned aircraft. However, due to scale effects among other factors, certain differences exist, as well. On average, the size (mass) of both unmanned fixed- and rotary-wing aircraft are orders of magnitude less than their manned counterparts (Austin 2010). As we "scale-down" aircraft to smaller sizes, all attributes are not affected in the same way. For example, wing surface area, being two-dimensional, will vary inversely as the square of the scaling factor, whereas volume, a cubic, will vary inversely as the cube of the scaling factor. Thus, all other factors held constant, wing loading—that is, weight (mass in the gravitational field of the earth) divided by wing area—tends to increase with a decrease in UA size. Higher wing loading affects aircraft performance, decreasing the rate of climb and increasing that of descent. Decreasing the rate of climb may negatively affect endurance and, consequently, range. Increasing the rate of descent may result in excessively hard landings, necessitating the installation of a parachute recovery system. Higher wing loading may compel the designer to use a catapult launch system in lieu of the less equipment intensive hand-launch. Higher wing loading also reduces maneuverability. Conversely, any reduction in mass affects the ability of the UAS to resist the disturbing or upset force of wind gusts, an important consideration in many data gathering missions. Neither are rotorcraft immune to the effects of reduced scale. Narrower blade chords and smaller rotor disks produce low Reynolds numbers, which reduce the efficiency and lift of the rotor system (Seddon 1990) which, in turn, decreases range, endurance, and payload. Thus, in UAS design, size matters, affecting several operational characteristics which, may, in turn, affect the ability of the platform to meet mission requirements.

So, to bring the conversation full circle, size will frequently be dictated by the mission. A large platform may be necessary for stability, to carry enough fuel for the desired range and endurance and to mount the necessary sensor package or payload. Or, the agility of a smaller UAS package may be desirable for close-up work and for negotiating the tight spaces encountered during infrastructure inspections. On the other hand, an inappropriately sized UA may reduce system performance. Selection of appropriate design attributes and platform size and configuration or of an existing unmanned platform design that will optimize performance and effectively and efficiently fulfill the mission is best accomplished through the judicious selection of components and the careful consideration of the performance of the overall system. During the selection or design process, keep in mind that the overall UA system should be a synergistic construction of individual components with sometimes competing system demands and often diametric operational constraints. The selection or design process is one of reconciling these inconsistencies to maximize performance. Often, optimization of the entire system (to achieve the mission goal) may only be achieved by sacrificing performance of certain individual components and subsystems, which are integrated to form the UA system.

10.3 Airframe Designs

UAS designs may be broadly categorized according to how these aircraft develop lift. That is, whether the wings are fixed with relation to the airframe and carried forward with the aircraft through the atmosphere or whether, alternately, the lifting airfoil rotates around a fixed axis providing the capability of producing lift with zero airspeed and, theoretically, in an infinite number of directions. Whether fixed- or rotary-wing, unmanned aircraft in each class are designed in a myriad of configurations. This section of the chapter is devoted to an overview of UA airframe designs, corresponding attributes and considerations, as well as the advantages and disadvantages of types discussed.

10.3.1 Fixed-Wing Designs

For a variety of reasons, some of which have been discussed, unmanned aircraft are produced with much greater design diversity than those which are manned. For example, 70% of the latter are configured with a low wing and a T-tail or inverted-T empennage (Louge et al. 2004), whereas UASs are commonly designed with tail/empennage configurations that include cruciform (the Navy's Ion Tiger), T-tail (the Rustom-H and the Hi Aero Gabbiano), inverted-T (Northrop Falconer), V-tail (Northrop Grumman Global Hawk), inverted-V (General Atomics Predator), H-tail (AAI RQ-2 Pioneer), and Y-tail (General Atomics Reaper) stabilizer/control surface groups. Each has characteristics that may recommend them to a certain design solution. V-tails are an attempt to reduce the wetted area (i.e., surface areas in contact with air flows) of the tail surfaces over those of the T-tail/inverted-T designs, whereas, since air flowing off the vertical stabilizer is constrained by structure, induced drag may be diminished with the latter designs (though at the cost of some increase in parasite drag). Some claim the cruciform design combines the advantages of the T-tail and inverted-T designs (Sadraey 2009). Because the tailplane of an H-tail is bookended by the vertical stabilizers, wingtip vortices off the horizontal stabilizer are effectively eliminated, decreasing induced drag. In comparison to a V-tail, the Y-tail design increases yaw stability with a penalty in increased wetted area and, correspondingly, in drag. UAS wing configurations include conventional high-, mid-, low-wing aircraft, canards, and flying-wing designs. Certain of the flying-wing designs have a distinctive fuselage (Insitu ScanEagle and Gatewing X100) while others are of all-wing or blended wing-body construction (NASA's Helios, Boeing's X-48, and the Skywalker X8). (Because no humans are aboard, the fuselage may be entirely omitted, and the blended wing-body/flying-wing design is increasingly being used as an sUAS remote-sensing platform.) In general, the process of designing a UAS affords greater freedom in the selection of configuration (Gundlach 2012). Compared to manned aircraft, much greater design diversity exists in the global fleet of aircraft, which are flown unmanned. According to Gundlach (2012, 128), as a design consideration, UAS "[t]ails are the artistic pallet of the designer."

10.3.1.1 Factors in UAS Tail Designs

The purpose of the tail is to provide stability and the control moments needed to trim and maneuver the aircraft. Control (or maneuverability) and stability are antithetical design considerations—increasing one diminishes the other. To illustrate this point, it might be mentioned, parenthetically, that with the introduction of computer assisted fly-by-wire systems, it became possible to destabilize fighter designs (i.e., to introduce negative static stability into the aircraft) to make them more maneuverable. The General Dynamics F-16 was the first aircraft intentionally designed to incorporate a slight aerodynamic instability, known as relaxed static stability, or rss (Figure 10.3).

An aircraft maneuvers through three dimensions of space, and its lifting, controlling, and stabilizing surfaces must consequently provide control and stability about three axes: directional stability/control (yaw) about the vertical, or "z" axis; lateral stability/control (roll) about the longitudinal, or "x" axis; and longitudinal stability/control (pitch) about the lateral, or "y" axis. The tail is primarily responsible for controlling motion about the y- and z-axes. The stabilizing tail surfaces, the vertical and horizontal stabilizers in a conventionally configured design, are airfoils, and with the wing, these are termed "lifting surfaces" to differentiate them from the aircraft control surfaces (i.e., ailerons, rudder, and elevator). (On nonconventionally configured aircraft, the control surfaces might be, on a flying-wing UAS, elevons, or on an aircraft with an inverted-V- or V-tail aircraft, ruddervators.) In a conventionally designed aircraft, the empennage is the aft-most, skinned structure to which rudder and horizontal tailplane attach.

To provide pitch (longitudinal) stability, an aft-mounted tail assembly must provide a tail-down (nose-up) force. When properly designed, this surface produces lift in the direction opposite the wing. The aerodynamic center of the wing is the location on the chord of the wing about which pitching moments occur. Incompressible airfoil theory predicts that the aerodynamic center (AC) will fall at 25% chord regardless of wing camber, thickness, and angle of attack. In actuality, the unpredictable, chaotic nature of air flow causes the AC to fall between 23% and 27% (Hurt 1965). Think of the AC as the lateral axis (y-axis) about which pitch changes occur and the fulcrum about which the aircraft is balanced. With the center of gravity of the aircraft located forward of the ac, the tail-down force acts as a counterbalance. If airspeed decays, less down force is produced, the nose drops, airspeed increases and the equilibrium is restored. In this way, the horizontal stabilizer contributes

FIGURE 10.3
General Dynamics F-16 was the first aircraft designed with relaxed static stability (rss). (Image courtesy of NASA.)

to longitudinal (static) stability. This relationship between the aerodynamic center and the tail-down force produced by the tailplane is also why adverse loading of the aircraft to move the cg aft of the AC is too often disastrous: Both cg and tail-down force are acting in concert to cause the aircraft nose to pitch up and reduce, possibly catastrophically, control while simultaneously increasing the likelihood of a stall. Moreover, to provide pitch stability, especially at lower airspeeds, an aft tail must not stall before the wing—doing so would cause a loss of tail-down force and a severe pitch break while the aircraft wing continues to produce lift. This design requirement is generally accomplished using a tailplane with a lower aspect ratio than the wing, which increases the range of angle of attack before a stall occurs (Gundlach 2012).

In a conventionally designed aircraft, the vertical stabilizer is the primary source of directional control and stability. (Although a wing, especially one that is swept, will contribute to static directional stability, any effect is relatively slight.) Should the aircraft yaw, or sideslip, the change in angle of attack of the vertical stabilizer causes a side force (change in pressure differential acting on the major surfaces of the vertical stabilizer) to yaw the aircraft about the center of gravity in a restoring moment that will turn the nose of the aircraft into the relative wind. The size of the vertical stabilizer and the arm (distance between the stabilizer and the cg, or x-axis) will determine the effectiveness of the stabilizer in creating a restoring force following any displacement of the nose in yaw.

The tradeoff in any tail design is effectiveness in producing the desired stability and control moments versus weight and parasite drag. Consequently, as a general rule of thumb, the tail should be no larger than that which produces the desired level of stability and control, and no larger. A number of other factors affect the decision regarding the selection of configuration (whether T-tail, Y-tail, V-tail, etc.), and these will be subsequently discussed.

In closing this section, it might be worth noting that the design of all aspects of manned aircraft, from nose to tail and wingtip to wingtip, is guided and constrained by the Code of Federal Regulations (CFR) Title 14, entitled Aeronautics and Space (e.g., CFR 14, Parts 23, 25, 27, 33, and 35). In the emerging field of unmanned aircraft, UASs are, at this point, relatively unfettered by regulations and other external forces. (See the preceding section on exogenous design influences.) However, as unmanned aircraft are increasingly integrated into the National Airspace, this circumstance will certainly change as UASs become increasingly subject to regulation.

10.3.1.2 Conventional Wing, Inverted-T-Tail Aircraft

Gundlach (2012) offers a generalized definition of conventional aircraft describing these as constructions with tail surfaces located behind the wing. For the purpose of the ensuing discussion, I will narrow this definition by restricting the definition to those aircraft having tail surfaces installed in an inverted-T configuration (i.e., vertical and horizontal stabilizers attached to the empennage). During the time of UAS nascency, unmanned aircraft were manned aircraft modified to fly without an onboard operator (i.e., as a true drone, remotely piloted aircraft or autonomous UAS). Consequently, manned and unmanned aircraft were nearly identical and of conventional design. Examples include the Hewitt-Sperry Automatic Airplane (or Curtiss-Sperry Flying Bomb), Lawrence "Burst" Sperry's Verville-Sperry M-1 Messenger, DeHaviland's Tiger-Moth-based Queen Bee, the B-17s of Operation Aphrodite and the B-24/PB4Ys of Operation Anvil (Figure 10.4). During WWII, the U.S. developed purpose-built unmanned aircraft. These incorporated the common-for-the-time forward wing and inverted-T empennage, having a vertical stabilizer/rudder, forming the leg of the "T" and a horizontal tailplane/elevator located at the bottom

FIGURE 10.4
Upon completion of its 80th combat mission, the B-17 Flying Fortress, *The Careful Virgin*, was transferred to Operation Aphrodite and deployed, as an RPA, against V-1 emplacements in France. (Army Air Corps image.)

of the vertical member to create the characteristic inverted-T shape. All surfaces attached to a tapered empennage. Examples include the low-wing Interstate TDR-1, the high-wing Denny Radioplane OQ-1/2/TDD-1 (and its descendent, the Northrup MQM-57 Falconer), and the high-wing Naval Aircraft Factory TDN-1. UAS designs that resemble conventionally designed manned aircraft have become increasingly rare with the passage of time. Currently produced examples of UASs constructed with forward wings and inverted T-tails are much less common than similarly configured manned aircraft, and examples are generally confined to the least sized sUAS (e.g., the AeroVironment RQ-11 Raven, the PrecisionHawk Lancaster, and the Aeromapper EV2).

A primary advantage of the conventional design is that it is a proven configuration in which lifting surfaces can be conveniently attached to the empennage by conventional means. Sadraey (2009) claims that familiarity with this well-tested construction is the reason that 60% of manned aircraft are manufactured with forward wings and a conventional inverted-T tail. Another advantage inheres to the fact that this design is the simplest and easiest upon which to perform predictive computations and performance analyses. According to Sadraey (2009, 299), "The analysis and evaluation of the performance of a conventional tail is straight forward … If the designer has low experience, it is recommended to initially select the conventional tail configuration." Finally, a conventional tail configuration is lighter and less complex than some other designs (e.g., an H-tail design).

10.3.1.3 Twin-Boom, Pusher-Propeller Designs

In a twin-boom design, the booms, made of composite (often carbon fiber) cylinders or of monocoque construction, replace the empennage as primary structure and provide the attachment points for the stabilizers. Under the broad definition earlier ascribed to Grundlach, twin-boom designs are "conventional" configurations if defined in terms of wing loading and control functionality (i.e., in comparison to canard and flying-wing designs).

Twin-boom UASs have proliferated. These designs share a common heritage, the lineage of which can be traced back to the Israeli efforts to develop unmanned aircraft for military applications. Impressed by the performance and potential of the Firebee UA, the Israeli military placed an initial order with the Ryan Aeronautical Company for 12 of the aircraft which the Israelis used for engineering and test prototypes to develop modifications and improvements to support their military's specific operational requirements. The

Israelis renamed the modified Firebees Mabat (Hebrew for "Glance") and also purchased Northrop Chukar decoy drones (renamed Telem, meaning "Furrow"). These aircraft were deployed during the 1973 Yom Kipur War, and, although losses were heavy, the aircraft successfully completed their missions and proved their value. Israeli Aircraft Industries (IAI) manufactured the twin-boom, PT-6A-powered Arava STOL/utility/transport aircraft, and based on the engineering experience gained in producing this aircraft, one year later, in 1974, the Israelis began development of the Scout unmanned aircraft, a twin-boom, pusher propeller (as opposed to a conventional tractor propeller) design.

Thus began the proliferation of UAS twin-boom designs that have been produced in a variety of tail configurations. The Tadiran Mastiff, a contemporary of the Scout, was designed as a twin-boom T-tail. The upright T design removes the stabilizer and control surfaces from the turbulence of the propeller discharge and from the engine exhaust. The IAI Scout, the AAI/IAI Pioneer (RQ-2), and the IAI Hunter (RQ-5), Searcher and Heron are all twin-boom, H-tail designs. Although an advantage of the H-tail is decreased induced drag associated with the bookending of the tailplane that prevents the formation of vortices, the vertical stabilizers also act, in the same way as the wings of a biplane, to reduce the induced drag associated with the vertical surfaces, which can, consequently, be made shorter (with less wetted area, but increased aspect ratio). The two rudders can each be smaller while retaining control authority. H-tails also encourage directional stability over single stabilizer designs (inverted- and upright-T-tails) by reducing the yawing tendency induced by the propeller slipstream that impinges upon the vertical stabilizing surfaces. The tradeoff is that control inputs to two rudders make the system more complex, ground clearance is minimally reduced and the tail assembly is slightly heavier than some other configurations (e.g., inverted- and upright-T-tail designs).

The Aerosonde and RQ-7 Shadow, both produced by AAI, and the Penguin, manufactured by the UAV Factory, all incorporate inverted V-tails supported by a twin-boom structure (Figure 10.5). In comparison to a V-tail, an inverted-V design may produce slightly better yaw characteristics (compared to a noninverted tail) in a coordinated turn. Also, being in the slipstream of a pusher powerplant, an inverted-V twin boom may permit the use of smaller control surfaces and/or provide better low speed responsiveness,

FIGURE 10.5
UAV Factory Penguin BE. Note this electrically powered pusher design is constructed with an inverted-V tail supported by carbon fiber booms. (Image courtesy of UAV Factory.)

but at the cost of increased buffeting and parasite drag. The use of ruddervators complicates the control system—combining pitch and roll inputs using differential control surface movements can make electrical/electronic/mechanical control interfaces more complex and difficult to design and interface. Ruddervator flutter can be a problem as was discovered during early evaluation of the MQ-4C Global Hawk (adding counterbalancing weight to the control surfaces solved the problem) (Norris and Butler 2013). Due to increased complexity, the use of ruddervators can negatively affect both maintainability and reliability—a ruddervator (actuator) failure was blamed for the crash of a BAMS RQ-4A prototype at NAS Patuxent River on June 11, 2012. Greater complexity also increases the expense of manufacture.

Aside from the advantages and disadvantages ascribed in the preceding paragraphs to various characteristic tail constructions, certain positive and negative attributes inhere to the twin-boom design, itself. Twin-boom construction offers the designer an opportunity to create an aircraft having less "wetted" area and drag, although the drag produced by using two stabilizers and additional control surfaces may negate this advantage. The use of twin-boom construction also affords flexibility in meeting design goals. During design, the arm of the tail may be increased or decreased with relative ease to counterbalance anticipated payload weights and provide the necessary stabilizing and control moments. For example, control surface arm and stabilizer moment may be increased by moving the entire tail aft, to improve stability and controllability (at the expense of reduced maneuverability). Gundlach (2012) notes that a twin-boom pusher design "… yields large, close-coupled tails and elongated noses. A minimum weight in the nose is needed to counter the engine moment, which is generally satisfied by avionics, communications system, or payload. The General Atomics Predator A, Warrior and Reaper systems take this approach" (p. 135).

Twin-boom construction is analogous to a box, which surrounds a pusher propeller to enhance safety. The "boxing" of the structure also stiffens the assembly and, on V-tails, decreases surface side-loading to reduce torsion (twisting). On the negative side, the tail surface located in the discharge of a pusher prop will tend to increase interference drag and minimally reduce propulsive force, while exposing the tail assembly to greater buffeting and vibration. Finally, with the engine mounted in the rear of a pusher design, greater fuselage volume is available for the payload, while sensor packages will be provided a clear view without the distorting influences of engine exhaust and heat.

Another option available to design UASs with a field of view undistorted by exhaust gasses is that of a twin-boom, twin-engine aircraft, such as the InView Unmanned Aircraft System, having wing-mounted powerplants (Figure 10.6). Despite this advantage, multi-engine, fixed-wing UASs are not common. Center-line thrust (pusher/tractor design) twin-engine configurations represent an attempt to overcome the undesirable yaw characteristics that occur when operating on a single wing-mounted engine (e.g., during an engine-out incident), but this option suffers from the combined disadvantages of both pusher and tractor designs plus the complexity of a twin-engine installation (Gundlach 2012). Nonetheless, one prominent example of a twin-boom, center-line thrust UAS is the RQ-5 Hunter developed jointly by IAI and TRW, Inc. (TRW was acquired by Northrop Grumman in 2002).

10.3.1.4 Flying Wings

As the name implies, the distinguishing characteristic of a flying wing is that this design features a single horizontal lifting surface, although vertical members in the form of

FIGURE 10.6
Twin-boom, twin-engine InView UAS. (Image courtesy of Barnard Microsystems.)

winglets, wingtip endplates, and vertical stabilizers may be present. UAS Winglets (Insitu ScanEagle) function as they do on manned aircraft to reduce tip vortices and induced drag. Extending below the wing, wingtip endplates are often provided at the tips of sUAS flying wings (e.g., the Gatewing X100 and the IAI Malat Mosquito) to decrease induced drag (parasite drag will be increased) and to protect the underside of the aircraft during a landing. True vertical stabilizers installed on UASs are an attempt to overcome, to some degree, the inherent directional instability of a flying wing design. Although vertical stabilizing airfoils are relatively uncommon on a UAS, one example is that of the WASP III Battlefield Air Targeting Micro Air Vehicle (BATMAV). (Due to directional instability and a lack of computerized stability augmentation systems, vertical stabilizers were also incorporated in the design of early manned flying wings, such as the Northrop YB-49. Figure 10.7.) Control and possibly stabilizing inputs are accomplished through differential movement of elevons (a portmanteau of aileron and elevator) installed on the trailing edge of the wing. The fuselage may be clearly discernable (Institu ScanEagle) or nonexistent (Skywalker X8). According to Gundlach (2012, 120), when conventional aircraft are defined under the broadened definition in which the categories of twin-boom and conventional forward-wing aircraft having an empennage/tail assembly are conflated, "[f]lying wings are the second most prevalent UAS configuration."

The flying-wing design affords certain advantages over other designs. Its construction is straightforward with few parts, qualities that simplify assembly and manufacture. The flying wing is attractive to designers because it offers the theoretical potential for high aerodynamic efficiency (low drag for the amount of lift produced) and greater efficiency, which leads to lower energy consumption and increased range. These characteristics also imbue aircraft of this design with excellent gliding ratios. Flying wings are robust, and sUAS are often constructed without landing gear, saving weight and reducing parasite drag. This design affords considerable flexibility in payload positioning, although aircraft cg is critical and the load must be judiciously distributed. Also, because UAS flying wings are most frequently of pusher design, the lack of a forward-mounted powerplant affords the same advantages as a twin boom—that is, greater fuselage volume for sensor packages and a clear view without the distorting influences of engine exhaust and heat. For military UASs, an added advantage of the flying wing is a low radar cross section, or

FIGURE 10.7
Northrop YB-49 flying wing (circa 1947-48). Note the vertical stabilizers necessitated by the lack of a flight dynamics computer. (U.S. Air Force image.)

RCS, which is a measure of the detectability of the aircraft. The characteristic of minimal RCS is a significant reason why unmanned combat aerial vehicles, such as the Northrop UCAV X-47B, among other entries in the Navy's Unmanned Carrier-Launched Airborne Surveillance and Strike (UCLASS) competition, and stealthy reconnaissance UASs, such as the Lockheed Martin RQ-3 DarkStar and the RQ-170 Sentinel, are of flying-wing design. Again, in these instances of military UAS applications, it can be seen that mission goals dictate design characteristics.

Of course, as with most things in life, there's no such thing as a "free lunch," and certain negative aspects inhere to the flying-wing design. Due to the lack of stabilizing surfaces, flying wings exhibit the inherent characteristic of directional instability, although the autopilot or stabilization augmentation systems are able to mitigate this instability to varying degrees. Tradeoffs exist between wing loading and speed, maneuverability, and stability. A flying wing is sensitive to displacement by wind gusts and air turbulence. Designing a flying wing requires the use of "non-standard," blended airfoils—the aeronautical engineer cannot just consult a catalog of wing designs (e.g., NACA Report No. 824) to select the airfoil for an appropriately shaped wing. Finally, flying wings have a narrow cg range and balancing the payload, autopilot, receiver, avionics, antennae, etc., can be challenging.

10.3.1.5 Canard UASs

Canard UASs incorporate a single, horizontal airfoil, smaller than the wing and located ahead of the main lifting surface. Unlike an aft horizontal stabilizer, which produces a downward force to raise the nose, the forward canard produces lift acting in the same direction as the main wing. Based on function, two types of canards may be identified: control and lifting canards. The former acts as a longitudinal control (pitch) surface and is often designed with zero angle of attack. Control canards are not intended to support the weight of the aircraft in flight and, consequently, are not heavily loaded. Lifting canards share the weight of the aircraft with the main wing. This type of canard is highly loaded and the main wing acts as a horizontal tail surface. One would expect that this configuration, in which a secondary lifting airfoil shares the load, would make possible a smaller main wing, but that is not necessarily the case. The foreplane produces a downwash, which interferes with the flow to the main wing reducing lift and increasing induced drag, thus reducing the combined effect of the two lifting surfaces. The use of a canard may permit the designer to move the center of gravity relatively far aft, which may be beneficial in pusher designs. Regardless of whether the foreplane is control or lifting canard, to maintain aircraft controllability the canard must stall before the main lifting surface. Consequently, the main wing cannot reach full lift, which reduces performance and increases the criticality of observing the cg range—adverse aft loading (i.e., rearward of the aft cg limit) can produce deep, unrecoverable stalls.

Unmanned aircraft incorporating a canard wing are much more rare than conventional, twin-boom or flying wing designs. The reason so few canard UASs exist, according to Gundlach (2012, 113), is "because there are no aerodynamic advantages over the conventional configuration ... [because a canard design] has more parts than a flying wing." I would add that the advantages of other designs (greater flexibility in payload location, unobstructed view for sensors, more robust construction, etc.) enhance the ability to achieve mission goals and influence designers to select an alternate configuration. Although not common, examples of canard UASs exist and include the Highly Maneuverable Aircraft Technology (HiMAT) remotely piloted aircraft (RPA) developed by NASA in the mid- to late-1970s to demonstrate the feasibility of an unmanned fighter for aerial combat; the ADCOM YABHON-R Medium Altitude Long Endurance (MALE) UAS currently under development in the United Arab Emirates; and the L-3 Mobius, an optionally piloted vehicle, or OPV, capable of autonomous flight. Despite relatively few examples of canard UASs, this design has been evident since the dawn of powered, heavier-than-air flight—the Wright Flyer was a canard design.

10.3.2 Rotating-Wing or Rotary-Wing Designs

In comparison to the fixed-wing UAS, rotary-wing aircraft embody characteristics that may be beneficial in certain applications and disadvantageous in others. Rotary-wing UASs are highly maneuverable and capable of vertical takeoffs and landings. The tradeoff is a general reduction in payload carrying capability, range, and endurance—the reduced performance being predominantly the result of inefficiencies associated with the rotor system (attributable to the complexities of rotor aerodynamics and the low Reynolds numbers associated with scale effect). The aerodynamics of rotary-wing flight is well beyond the scope of this chapter. (For example, Boeing, supported by NASA, has produced a two-volume tome, *Rotary-wing Aerodynamics*, NASA Contractor Report 3082, on the topic.) Nonetheless, the reader will find a few of the more salient aerodynamic characteristics and their ramifications described in each of the sections that follow. This brief overview will suffice for an introduction to the flight and design of rotary-wing UASs. Those readers interested in a more detailed description will have no difficulty locating additional resources.

10.3.2.1 Helicopter UAS

The aerodynamics and flight physics associated with a helicopter are very complex. A helicopter simultaneously produces both lift, to oppose the weight of the aircraft, and thrust, represented by a vector acting in the direction opposite that of flight, by accelerating a mass of air through the rotor disk. The energy imparted to the air mass is provided by the fuel or battery, converted to mechanical power by the aircraft powerplant, and transmitted through gear reduction (e.g., in a transmission) to the mast which supports the main rotor blades. The helicopter gear reduction (transmission) is necessary to convert the high rpm, low torque output of the motor to low rpm, to maintain a low rotor rpm for aerodynamic efficiency (generally, the transonic regime should not be entered), and high torque to move the blades through the viscous fluid of the atmosphere. In a hover, 60%–70% of this power is consumed in producing lift (referred to as induced power), while the remainder (referred to as profile drag power) is expended in overcoming parasite drag (Gessow and Myers 1985). Because low Reynolds numbers are characteristic of narrow chord/short span blades (Schafroth 1980) and the energy required to accelerate an air mass increases as the square of the acceleration, increasingly larger diameter rotors become increasingly efficient due to the ability to process greater amounts of air through rotor disk (Gessow and Myers 1985). Another contributing factor is that increasing the length, and therefore the aspect ratio, of the rotor blades reduces the drag induced by producing lift. The tradeoffs are increased parasitic drag and the reduction in rotor rpm, which will reduce dynamic pressure and tend to diminish lift, necessary to keep tip speeds subsonic.

In a no-wind hover, lift is produced symmetrically across the rotor disk. Things change dramatically with any relative motion of the air mass with respect to the rotor disk (as would occur as the result of a gust of wind or from directional flight). Assuming a two-bladed rotor system, the blade moving into the relative wind, termed the advancing blade, produces greater lift, while the airfoil traveling in the same direction as the relative motion of the air mass, the retreating blade, produces less lift. This destabilizing condition, known as dissymmetry of lift, is mitigated through effective rotor system design. Because the relative airspeed of the advancing blade produces more lift, this blade tends to rise while the retreating blade produces less lift and will descend. This is a fortunate circumstance, because in ascending, the angle of attack of the advancing blade decreases to reduce lift, while, as the retreating blade descends, its angle of attack increases to increase lift. Thus, rotor systems are designed to allow a predetermined amount of blade flapping, to mitigate the effects of dissymmetry of lift. Blade flapping may be designed in a helicopter rotor by using a flapping hinge installed in the hub (fully articulated rotor), by using a teetering (aka teeter-totter) hinge in a semi-rigid rotor head supporting two rotor blades (common on Bell helicopters), or by allowing flexing in the blades and hub (modern rotor hubs, or flextures, of manned helicopters may use composite materials and rubber-like elastomeric bearings to permit adequate blade flexing). Enter the law of conservation of angular momentum which states that angular momentum, calculated as the product of mass times angular velocity times distance (of the center of gravity from the spin axis) does not change (under idealized conditions, i.e., no friction). When a helicopter blade flaps in response to dissymmetry of lift, its center of gravity shifts: the advancing blade flaps upward driving the cg inward, while the retreating blade flaps downward to cause the cg to move outward. If mass does not change but the distance of the cg to the spin axis does, then velocity must react inversely for angular momentum to be conserved. The commonly used analogy is that of figure skaters withdrawing arms inward to increase their rate of spin—this is another example of the effects of the law. Because flapping causes the blades to accelerate (lead) or decelerate (lag), provision for this movement must also be accommodated and is accomplished through hub flexing and blade bending or installation of a lead-lag (or drag) hinge incorporated in the rotor hub. (Tail rotor blades also encounter dissymmetry of lift and may also be provided with hinges to accommodate blade flapping.) Although sUAS helicopters have been fitted with hinges to allow blade flapping (e.g., Zeal produced aluminum teetering hinges for high-performance applications), the moments and forces acting on smaller rotor systems are generally low enough that flexing and bending can accommodate most blade movement.

Large unmanned helicopters (e.g., the Northrop Grumman Fire Scout and the Boeing A160 Hummingbird) accomplish variations in lift and thrust in the same way that manned helicopters do, that is, by maintaining a desired rpm while effecting blade pitch changes about a feathering hinge installed in the hub. To increase lift (and altitude), a collective pitch change causes all blades to simultaneously increase angle of attack by the same amount. Thus, lift across the disk is increased. The rotor heads of smaller unmanned helicopters (e.g., the AutoCopter and the Yamaha RMAX) often incorporate feathering hinges, as well. Moreover, rather than directing inputs straight from the servos through the swash plate to the pitch change mechanism, still smaller helicopters (e.g., the ZALA 421-02 and the T-Rex) may also use a flybar (aka paddles or stabilizer bar) system, similar to that found on two-bladed Bell and Hiller manned rotary-wing aircraft. The flybar system consists of a bar, fitted with paddles or weights at the end, installed at 90° to the main rotor blades and perpendicular to the main rotor mast. Typically, the flybar accepts control inputs (e.g., from servos in a UAS) and transmits them, through mechanical linkages, to the swashplate to be sent, in turn, to the pitch change mechanism. The purpose of the flybar assembly is to provide additional cyclic inputs in response to wind and turbulence, which, in turn, results in greater stability and control. The smallest, electric unmanned helicopters (e.g., the Proxdynamics Black Hornet Nano), which are sometimes classed as micro-aerial vehicles or MAVs, generally have fixed-pitch main rotor blades and increase or decrease lift by varying rotor rpm.

All helicopters, manned and unmanned, large and small, develop directional flight in a similar way. To produce directional thrust, the pitch of the blades must change differentially to tilt the rotor disk in the direction of desired travel. Tilting the rotor disk is accomplished through differentially changing the pitch of the blades, through cyclic inputs, as they orbit around the mast axis. Due to gyroscopic precession, these inputs must occur 90° ahead of maximum tilt. To illustrate the effects of gyroscopic precession, assume that the direction of blade rotation is counterclockwise as viewed from above the disk (common for manned helicopters produced in the United States). If the longitudinal axis of the helicopter is the 0°/360° reference line extended forward of the helicopter, then, to fly forward, the input to increase pitch must occur at the point 90° in the direction of rotation, while that to reduce pitch must occur at the point 270° in the direction of blade rotation. Due to gyroscopic precession, maximum blade deflection will occur 90° later and the rotor disk will tilt forward, directing thrust more aft and moving the helicopter forward.

Another design feature of rotor blades is that, like propellers, they often incorporate twist, increasing blade angle from tip to hub. (Because the angular velocity of the inboard blade stations is lower, without an increase in blade angle less lift would be produced nearer the hub.) As with propellers, this is done to reduce stress and increase efficiency by more evenly distributing lift across the entire rotor disk. This twist is less apparent in helicopter rotors, largely because it is distributed across a much greater span.

Torque effect is present in both fixed-wing and rotary-wing designs, but much more dominant in the latter. Torque effect develops as the result of Newton's Third Law, "for every action there is an equal and opposite reaction." Power delivered to the rotor causes the fuselage to display the propensity to rotate in the opposite direction. This tendency varies linearly—increase power, increase torque effect. Torque effect is countered on single rotor helicopters, both manned and unmanned, by the tail rotor (aka, anti-torque rotor). If the thrust produced by the tail rotor is less than torque effect, the fuselage will yaw in the direction opposite main rotor blade rotation; if thrust produced by the anti-torque rotor is greater than the torque effect force, the fuselage will yaw in the direction of main rotor rotation. Using two rotors affords the advantage of torque cancellation, whereby the two rotor systems rotate in opposite directions and thus the torque of one cancels that of the other and no other anti-torque system (e.g., a tail rotor) is necessary. Examples of UASs with two main rotor systems include both coaxial, contra-rotating, and tandem, counter-rotating designs. The difference is that in the former example, the blades rotate around a common axis, while in the latter, rotation occurs about two independent axes. The first rotary-wing UAS, the QH-50 DASH, developed by Gyrodyne for the Navy in the late 1950s and early 1960s, was a coaxial, contra-rotating design powered by a Boeing turboshaft producing 300 shaft horsepower. During the late 1980s and early 1990s, Sikorsky experimented with a coaxial unmanned design known as the Cipher and Cipher II, which subsequently evolved into the USMC's Dragon Warrior UAS. Currently produced examples of a tandem, counter-rotating UAS rotor system may be found on the unmanned version of the Kaman K-MAX, a helicopter capable of lifting an external load equal to its own weight, and the IAI Ghost, which resembles a Chinook manned helicopter in configuration. The K-MAX rotor system, wherein the blades mesh, much like those of an old-fashioned, mechanical egg beater, is also referred to as a synchropter or intermeshing rotor design.

10.3.2.2 Multi-Rotors

Multi-rotors exist mainly in quad-, hexa-, and octocopter configurations. Notice that all have an even number of rotors for torque cancellation. (Though much less common, three rotor, "Y-frame" multi-rotor designs also exist, but these must use a tail rotor or rotor tilting, which greatly complicates flight dynamics and aircraft control, to counteract the torque effect induced by an odd number of rotors.) Current manufacturers of sUAS multi-rotors include DJI, Aerobot and Aeryon. Multi-rotors are almost exclusively small UASs, powered by electric motors. They afford the advantages of VTOL flight, the ability to hover and loiter on station, agility, and a relative freedom from vibration, but at the expense of limited range, altitude, and endurance.

Recall that the earlier discussion of drag principally focused on fixed-wing aircraft. The reason that the points made in that section were more relevant to those aircraft is attributable, in part, to the fact that drag increases with airspeed. The missions for which VTOL aircraft (e.g., helicopters and multi-rotors) are designed generally do not dictate high airspeeds. Moreover, where endurance and range may be sacrificed to obtain other desirable attributes, such as low vibration or maneuverability, a rotary-wing platform may be selected over a fixed-wing configuration. The ability to fly in any direction, rather than in only a straight line, negates the advantages of streamlining, and multi-rotor designs, consequently, do not incorporate fairings or fillets or any other strategy to reduce form or interference drag. Aerodynamically, multi-rotors are very dirty—antennae, wires, controllers, and motors protrude into the airstream. In comparison to the design of fixed-wing UASs, the low-speed operational envelope and directional agility of rotary-wing aircraft render the consideration of drag less important in the design process.

10.3.2.3 Other Rotating-Wing UASs

Examples of UAS rotary-wing designs that are neither helicopter nor multi-rotor exist. One is the Eagle Eye tiltrotor, powered by a single Pratt & Whitney Canada (P&WC) turboshaft producing more than 600 shp. Designed and constructed by Bell, the Eagle Eye closely

FIGURE 10.8
Honeywell Tarantula Hawk landing. (Image courtesy of Honeywell Aerospace.)

resembles the larger manned versions of the company's tiltorotor designs. To transition from helicopter mode to airplane mode, the tiltrotor is capable of rotating the nacelles, located at the end of its wings, from the vertical position to a horizontal orientation and then back to vertical for a landing. Although the Eagle Eye is capable of flying like an aircraft, it is strictly a VTOL aircraft—the blades of the rotor-props are too long to accommodate a conventional, run-on landing. In 2002, the U.S. Coast Guard placed an order for Bell Eagle Eyes to bolster the USCG's deep water surveillance program, but the order was subsequently frozen pending restoration of funding.

A second example of a novel UAS rotary-wing design is that of the single rotor, ducted-fan Honeywell RQ-16 Tarantula Hawk (T-Hawk) powered by a four HP boxer twin (Figure 10.8). The ducted fan is extremely efficient (negligible induced drag) allowing for a rotor many times smaller than would otherwise be required. The T-Hawk overcomes torque effect by deflecting the thrust of the fan with fixed turning vanes and maneuvers by using louvers to vector thrust (direct or deflect fan discharge) in the direction opposite the desired direction of travel.

10.4 Powerplant Designs

As is the case with UAS airframes, great diversity is found in the availability powerplants for unmanned aircraft propulsion. An example of every type of manned aircraft powerplant, categorized according to operating cycle and design, has been installed on UASs—and more. The main objective of this section is to introduce the reader to this diversity and provide a brief justification for why or reasons why these are not used in certain UAS designs. Please note that the list of powerplants discussed later is not exhaustive, but includes those powerplants that are most commonly installed on unmanned aircraft.

(For our discussion, the pulsejets installed on unmanned aircraft that have included the V-1, Republic Loon or the Enics T90-11 UAV will not be covered; similarly, neither will the scramjet that powers the Boeing X-51 be discussed.)

10.4.1 Four-Cycle Engines

Four-cycle, or four–stroke engines operate according to the Otto cycle (after Nikolaus August Otto), which is an idealized thermodynamic cycle describing the operation of a spark ignition, reciprocating engine consisting of five events and requiring four strokes of the piston to complete the cycle. The piston is linked to a throw on the crankshaft through a connecting rod. As the throw rotates around the crank axis, it will pull the piston downward in the cylinder drawing in air, which has been mixed with fuel in the proper proportion, through the open intake valve. This is the intake stroke. During the next 180° of rotation, the throw drives the connecting rod/piston assembly upward to compress the fuel/air mixture in the cylinder. Compression is necessary in a heat engine to provide sufficient expansion of the air during combustion and the power stroke to extract a useful amount of power. During the compression stroke, both intake and exhaust valves are closed. Due to the inertia stored in the rotating/reciprocating engine components, the ignition event generally occurs before the piston has reached its maximum upward travel on the compression stroke allowing the stored energy to carry the piston through top dead center as the compression process is completed and the burning fuel/air mixture is attempting to expand. The combined effect of compression and combustion provides maximum downward push on the piston during the power stroke. During this 180° of crank throw rotation, the force acting on the top of the piston is transmitted to the crankshaft through the connecting rod. The engine generates power for the production of thrust only during the power event. The next 180° of rotation of the crankshaft pushes the piston upward in the cylinder to expel the exhaust gases through the exhaust valve, and overboard through the exhaust system. Thus, 4 strokes and 720° of crankshaft rotation are necessary to complete the five events of the Otto cycle.

Four-stroke motors can afford certain design benefits: less vibration (in comparison to a two-stroke engine), quieter operation, relatively high torque, and a broader power band. Where range and endurance are important factors in the anticipated mission of the UAS, four-stroke engines offer the major advantage of providing the highest fuel efficiency of any internal combustion engine. On the other hand, they are also generally heavier, more complex, and expensive and do not, in comparison to other powerplants (e.g., two-cycle engines), commonly power unmanned aircraft. One example of a UAS four-stroke powerplant would be the 120 horsepower (hp) Teledyne Continental IO-240-B7B installed on the Eagle ARV UAS. Another is the Austrian-produced, (opposed) four cylinder, 115 hp, turbocharged Rotax 914F engine that powers the RQ/MQ-1 Predator A. The IAI RQ-5 Hunter was also powered by a Moto Guzzi two cylinder, four-stroke powerplant. Small four stroke engines, available in sizes of 7.5 cc's and above, may be installed on sUAS platforms. An example would be the Barnard Microsystems InView UAS powered by a pair of 29.1 cc (1.8 cubic inch) displacement powerplants that burn 100 low lead aviation fuel to which is added synthetic oil in a 20:1 ratio to provide lubrication.

10.4.2 Two-Cycle Powerplants

Two-cycle engines complete the same five events (i.e., intake, compression, ignition, power, and exhaust) required of any practical heat engine, but in just two piston strokes and 360°

of crankshaft rotation. Although slightly more complex two-cycle designs exist (e.g., those using rotary and reed valves), the simplest construction requires no valves and very few parts. As the piston begins upward on the compression stroke, a low pressure area is created below the piston. Occupied by the piston connecting rod and the crankshaft, this area is enclosed and termed the crankcase. Air, mixed with fuel in the proper mixture ratio and a small amount of oil for lubrication, is drawn into the crankcase and stored. At the same time, above the piston is a charge of air and fuel, previously stored in the crankcase during the preceding compression stroke, which is being compressed by the upward motion of the piston. The ignition event is timed at the proper point around the maximum upward travel of the piston (top dead center, or TDC) to provide maximum expansion of the gases and the greatest downward force on the piston through the connecting rod to the crankshaft. This is the power stroke. With the downward travel of the piston, an exhaust port is uncovered to begin the scavenging of the cylinder. As the piston continues downward, an intake port is uncovered and the decreasing crankcase volume forces the fuel/air mixture stored there through the intake port and into the cylinder. The fresh charge pushes the remaining spent gases out of the cylinder through the exhaust port. The intake port and then the exhaust port are covered by the piston as the fresh charge is compressed and the process is repeated so long as the engine continues to operate. Thus the two-stroke engine is capable of completing all five events in one complete crankshaft rotation.

Some fairly obvious advantages of the two-stroke design are simplicity, with few moving parts, compactness, and low weight. This last attribute contributes to a high specific power (power-to-weight ratio), at least within a narrow power band. Compared to other powerplants, two-stroke designs are relatively inexpensive. One disadvantage is a lack of smoothness in operation that creates airframe vibration. Where a vibration-free UAS platform is necessary for susceptible payload sensors, the shaking induced by a two-stroke powerplant can prove difficult and costly to isolate and dampen. Two-strokes are also notoriously noisy with a relatively high rate of fuel consumption. This inefficiency has a negative impact on range and endurance. Another serious disadvantage associated with two strokes is the poor performance that occurs during off-load power (Fahlstrom and Gleason 2012). At low rpms, the exhaust gases do not scavenge well in a two-cycle engine and, consequently, dilute the fuel/air charge drawn into the combustion chamber. The resulting lean mixture produces intermittent firing, popping, rough operation and may even "kill" the engine. This characteristic can be fatal to the UAS if it occurs during an attempted recovery at low engine rpm.

In the earliest days of UAS design, most aircraft were manned aircraft modified to respond to exogenously originated or preprogrammed commands, and purpose-built UAS powerplants did not exist. As late as the 1960s, the selection of a UAS two-stroke engine was limited to those that were commercially available and included generator, lawn mower, and chain saw engines, but these quickly proved to be "… unsatisfactory in terms of life at high output, level of vibration, fuel economy and reliability" (Catchpole and Parmington 1990, 1). Research, driven largely by military requirements and funding, developed new designs and evolved these to the point where, today, two-stroke engines power a multitude of diverse unmanned platforms. Manufacturers, including Cosworth, Sachs, Desert Air, Graupner, Husqvarna, Hirth, 3W Modellmotoren, and Ricardo, among others, currently produce two-stroke engines, explicitly for UASs, capable of operating on gasoline, kerosene fractions, and glow fuel (a mixture of methanol, nitromethane, and synthetic or castor oil to provide lubrication). Examples of UASs on which these are (or have been) installed include the RQ-2 Pioneer, UAV Factory Penguin, the IAI Searcher, the

General Atomics Gnat 750, and the SVU-200 unmanned helicopter designed by American, Dennis Fetters, for export to China.

10.4.3 Electric Motors

The operation of UAS electric motors will be more fully discussed in the next chapter on UAS electrical systems, so the treatment here will consequently be brief. UAS electric powerplants are generally brushless DC motors (BLDC or BL motors). Advantages of brushless motors (compared to brushed electric motors) include greater torque and specific power, more torque per amp (higher electrical efficiency), lower maintenance, and a longer life (no brushes, slip rings or commutator to pit, burn, or wear) and less electromagnetic interference (EMI), which is often an extremely important consideration for UA systems. (EMI can induce spurious signals in autopilot, avionics, and communications systems to the extent that the vehicle may be uncontrollable and subsequently lost.) BLDC motors may be of inrunner or outrunner design.

An inrunner BLDC motor is constructed by mounting permanent magnets on the rotor (connected to the output shaft), which are simultaneously attracted and repelled by the magnetic fields produced by the stationary windings (or stators) positioned around the magnets. In an outrunner design, the outer shell, in which the permanent magnets are installed, rotate around the central stator. Again, the permanent magnets are attracted/repelled by the magnetic fields established in the stator windings. The output shaft, to which the propeller or VTOL rotor drive system connects, is integral to this outer shell. Brushless motors often use neodymium permanent magnets (also known as NdFeB, NIB, or Neo magnets), which are made of the rare earth metal, neodymium, alloyed with iron and boron. These are the strongest permanent magnets yet developed. Neodymium magnets are available in several grades (strengths)—the stronger the magnet, the more it costs. The use of strong magnets makes a BLDC motor very efficient.

Due to inherent advantages, sUAS electric powerplants are most commonly, though not exclusively, of outrunner design. An outrunner will rotate much more slowly than an inrunner, and, yet, produce more torque. Because the outrunner stator is interior to the motor, the windings and internals can be sealed to protect them from the environment. Outrunner windings are not subjected to centrifugal loading which may extend motor life. Air cooling of the windings is not necessary, since heat can be dissipated by conduction. This is significant because one factor limiting the maximum sustained power output by a brushless motor is the amount of heat produced by the BLDC—too much heat may permanently damage stator winding insulation and weaken rotor magnets.

Electric motors are installed on a large proportion of sUAS fixed-wing aircraft and helicopters and nearly exclusively on multi-rotor VTOLs. Examples include the UAV Factory Penguin BE, the UTC Aerospace Vireo, and the Aeryon Labs Scout and SkyRanger (Figure 10.9). Although electric motors are very quiet and provide extremely smooth, vibration-free operation, the primary design tradeoff is that of reduced endurance and range. Endurance for the Scout quadcopter is approximately 25 minutes, while the published flight time of the fixed-wing Vireo is one hour, though this depends a great deal on winds and experience has demonstrated that, on a blustery day, this is an extremely optimistic expectation. Published endurance for the Penguin BE is 110 minutes. Compare this to the published endurance of 20+ hours for the Penguin B powered by a two-stroke 3W 28i two-stroke engine.

FIGURE 10.9
Aeryon Labs Scout quadcopter with outrunner BLDC motors. (Image courtesy of Aeryon Labs, Inc.)

10.4.4 Gas Turbines

Gas turbines that power manned and unmanned aircraft are referred to as Brayton cycle (or Joule cycle) powerplants. Named after the American engineer, George Brayton, other descriptors alluding to Brayton cycle characteristics are constant pressure (referring to an idealized condition where the combustion process is assumed to occur without changes in combustor pressure) and open cycle, a reference to the fact that all processes are constantly ongoing as air (aka, working fluid) is processed through the powerplant (as opposed to the discrete cycles of two- and four-stroke engines). The main constituent components of an aircraft gas turbine are the inlet, compressor, diffuser, combustor, and turbine section. These are assembled to create what is known as the core engine (of a turbofan) or the gas generator (of a turboprop or turboshaft engine). The gas generator or core engine is the basic assembly upon which aircraft gas turbines rely to produce the gases or working fluid necessary for the engine to function as a propulsor. Although greater detail will subsequently be introduced, for now, we can differentiate aircraft gas turbines as follows: turbojets were installed on older fixed-wing designs that relied strictly on the reaction thrust generated by the engine; turbofans are more efficient propulsors, consisting of a core engine and fan module, found on newer fixed-wing airplanes and UASs; turboprops, consisting of a gas generator, gearbox, and propeller, are installed on fixed-wing aircraft, both manned and unmanned; and turboshaft powerplants are installed on helicopters and tiltrotors. An aircraft gas turbine is mechanically simple, but the physics and thermodynamic principles associated with its operation are complex. Gas turbines are considered to be the most reliable and smoothest operating of all powerplants available for installation on unmanned aircraft (Fahlstrom and Gleason 2012). They generally burn heavy fuels (e.g., kerosene hydrocarbon fractions) and are, by far, the most expensive of UAS powerplants. Gas turbines are generally installed on larger, high-performance UASs.

The turbojet, turboprop, turboshaft, and turbofan all consist of the same basic component assemblies in which the five events characteristic of an internal combustion engine occur. Air enters the inlet and is ducted or routed to the compressor where the air undergoes an increase in static pressure and total pressure. Compressor design may be axial flow, centrifugal, or a combination of the two. A stage of axial compression consists of set of rotating blades, or rotors, installed around the circumference of the compressor drum, and a set of downstream stationary blades, or stators, installed in the compressor housing. The compressor drum and, consequently, the rotors are driven by the turbine(s) to accelerate air into the stators. The stators slow the airflow and increase working fluid static pressure, before directing it into the next stage of compression. An axial-flow compressor is extremely efficient, but exhibits a very low per-stage compression ratio, approximately 1.1:1 to 1.25:1. This high efficiency is the reason that the majority of large aircraft gas turbines are constructed with axial-flow compressors, which may achieve an overall compressor pressure ratio of 40:1, or more. Axial-flow compressors are found in the gas turbines powering large UASs (e.g., the Global Hawk) and UCAVs (e.g., the X-47B). Centrifugal compressors achieve a much greater per-stage pressure increase, but at a much lower efficiency. The use of centrifugal compressors can create a compact powerplant, and these are integral to the TPE-331 installed on the General Atomics Reaper. Due to losses that increase substantially with each boost in pressure, no more than two stages of centrifugal compression will be incorporated into an aircraft gas turbine. The compound compressors, integral to some turboprop and turboshaft gas turbines (e.g., the Pratt & Whitney Canada, or P&WC, PT6, and the Rolls Royce 250, and M250), are configured with multiple stages of axial compression discharging into a single stage of centrifugal compression. Passing out of the final compressor stage, the air enters a divergent duct, known as the diffuser, where the working fluid slows and static pressure increases. Taken together, these components, the inlet, compressor, and diffuser are sometimes collectively referred to as the cold section of the engine.

The hot section consists of the combustor, turbine section, and exhaust duct. Often the diffuser duct will be constructed with vanes that reduce airflow turbulence and direct the working fluid into the combustion section (aka, combustor) where fuel, continuously sprayed from the fuel nozzles, mixes with the air just before being ignited to rapidly oxidize in the combustion process. Working fluid temperature and energy reach maximum values in the combustor. Upon exiting the combustion section, the working fluid enters the turbine section to pass through multiple stages of turbine. Unlike a stage of compression, a stage of turbine consists of a set of upstream stators (aka, the diaphragm or nozzle) and a downstream set of the turbine blades installed in a rotating disk. This rotating assembly, or rotor, is also known as the turbine wheel or T-wheel. The nozzle assembly directs the working fluid, at the proper angle and velocity, into the cupped turbine blades causing the rotor to spin like a pinwheel. Because energy is incrementally extracted by each successive T-wheel, less is available to each downstream stage, and the diameter of each rotor is successively larger to uniformly capture the appropriate amount of power in each subsequent downstream stage. The rotor disks are fastened to the turbine shaft and the power extracted by the rotors is used to drive the compressor and any propulsor airfoils (i.e., the fan, propeller, or rotor blades of a helicopter). From the turbine section, the working fluid enters the exhaust duct to be directed overboard.

Aircraft gas turbines often incorporate two (or, less commonly, three) sets of turbines, the compressor, or gas generator turbines and the power, or free turbines. This distinction is a key feature differentiating various classes of gas turbine designs. Although other configurations exist (e.g., the PT6), most often the power or free turbine shaft rotates within the compressor turbine shaft. Because the two assemblies are independently supported by separate bearing systems, they can rotate at different rpms, with the more upstream turbine assembly, the compressor turbine, rotating faster. The gas generator turbine constantly drives the compressor to process air in the continuously ongoing process of the Brayton cycle. No turbine system driving the compressor will extract all of the energy available in the working fluid. In a turboprop, turbofan, or turboshaft powerplant, some power must remain in the working fluid to drive the propeller, fan, or rotor blades. Only a turbojet does not have a power turbine assembly. Nonetheless, some energy must remain in the working fluid to provide thrust during acceleration of the working fluid through the propelling nozzle of the exhaust duct. The output shaft of a turboprop or turboshaft will drive a gear reduction system. If the propeller or rotor blades reach transonic speeds, aerodynamic performance dramatically deteriorates and the output rpm of the power turbine is too high to drive the propulsor airfoils at an rpm acceptably low to operate efficiently. Consequently, the gear reduction system (the gearbox, gear reduction unit or the transmission of a helicopter) converts high rpm, low torque to low rpm, high torque to drive the prop or rotor blades. Due to the heavy loads encountered by gearboxes integral to turbofan installations, an aircraft fan is rarely driven through gear reduction.

Turbojets, developed during WWII are very noisy and, in terms of fuel consumption, extremely inefficient, particularly at lower airspeeds. On the other hand, they are without the additional complexity and weight of an integral gearbox, prop or fan, and so exhibit a high specific thrust. Turbojets have not been recently installed on UASs but have provided the propulsive force for the Ryan Firefly (General Electric YJ97-GE-3 turbojet producing 1814 kg, or 4000 lb of thrust), the Ryan Firebee (Continental J69-T-29A turbojet generating 771 kg, or 1700 lb of thrust), and Firebee II (Teledyne CAE J69-T-6 turbojet producing 835 kg or 1840 lb of thrust), a UAS frequently flown on reconnaissance missions in Vietnam and as recently as 2003 in Iraq.

As previously discussed, accelerating a larger mass of air to a lower velocity for a given amount of thrust (thrust is the product of mass times acceleration) requires exponentially more energy (energy consumption increases as the square of the increase in acceleration). Because, in comparison to the older turbojet design, a turbofan imparts a relatively low level of acceleration to a larger mass of air (for a given level of thrust), turbofans will always be the more fuel-efficient design. Additionally, the air discharged from the fan contains lower levels of energy per unit volume, which is one reason that fanjets are inherently quieter than turbojets. (Another is the higher frequency of the noise energy in the fan discharge.) Turbofans are installed on some of the most sophisticated UASs yet produced. These include the Northrop Grumman Global Hawk, powered by a Rolls Royce F137 turbofan, developing approximately 3400 kg (7600 lb) of thrust; the Northrop Grumman X-47B UCAV, powered by the Pratt & Whitney F100-PW-220U, producing 7257 kg (16,000 lb) of thrust; and the Lockheed Martin Polecat (Figure 10.10), powered by two Williams FJ44-3E turbofans (Figure 10.11), each developing 1365 kg (3010 lb) of thrust. Because maintaining the lowest possible RCS is an important design characteristic of many military UASs (e.g., the X-47B and Lockheed Martin's DarkStar and Polecat), the fans installed on these aircraft are generally small, producing a medium to low bypass ratio (a comparison of the amount of air discharged by the fan relative to the amount entering the core engine). Not infrequently, the bypass ratio of UAS turbofans is very low, possibly less than 1:1.

Turboprops and turboshaft gas turbines are so similar as to be essentially the same. The free or power turbines of both designs drive a gear reduction unit to reduce the output rpm and increase the torque delivered to the propulsor airfoils. In fact, aircraft gas turbines (e.g., P&WC PT6, Lycoming T-53, and the Rolls Royce 250) have often been installed on both helicopters and fixed wings, and the primary factor determining whether these are considered turboprops or turboshafts is the type of aircraft on which they are installed, that is, a turboshaft if installed on a rotary-wing aircraft and a turboprop if installed on a fixed-wing aircraft. One example of a reconnaissance UAS powered by two turboprops is the centerline thrust IAI Eitan ("Steadfast") powered by two PT6-A's developing 900 kW

FIGURE 10.10
Polecat developed "in-house" at Lockheed Martin's Skunk Works®. (Image courtesy of Lockheed Martin Skunk Works®.)

(1200 shp) each. Another turboprop-powered UAS is the General Atomics Reaper (previously known as the Predator B), powered by the Honeywell TPE-33-10 producing 671 kW (900 shp). Examples of turboshaft powerplants installed on UASs include the Honeywell T53-17A-1 installed on the Kaman (K-1200) K-MAX. This powerplant is capable of producing 1341 kW (1800 shp) of power, but is flat-rated to 1118 kW (1500 shp) for takeoff and 1006 kW (1350 shp) in flight. A powerplant may be flat-rated for other reasons (e.g., to provide sea level performance to a certain pressure altitude), but in this case it is done to increase powerplant and transmission life. Another example of a UAS turboshaft installation is that of

FIGURE 10.11
Two Williams F44 turbofans (above) powered the Lockheed Martin Polecat. Williams International produces a family of small, lightweight civil-certified FJ33/FJ44 turbofan engines (roughly in the 1000–4000 lbf range). (Image courtesy of Williams International.)

the Pratt & Whitney Canada PW207D turboshaft, which develops 478 kW (641 shp) and powers the Bell Eagle Eye tiltrotor.

10.4.5 Wankel or Rotary Powerplants

The Wankel engine is named after its inventor, the German engineer, Felix Wankel who received a patent for his design in 1929. The engine consists of triangular rotor, having slightly curved, equilateral sides, rotating within an epitrochoidal chamber, sometimes described as a two-lobe stator (Falstrom and Gleason 2012). The triangular shape of the rotor creates three volumes in the chamber, each of which will produce, in a single revolution, the five necessary internal combustion engine events—that is, the equivalent of three complete Otto cycles. Thus, because three power pulses occur in one revolution of the rotor, the specific power of the engine is high. This rotating assembly is the only moving part in the engine. Because Wankels convert the force of the pressure created in the combustion chamber into rotation without reciprocating motion, these are extremely smooth-running engines—a significant consideration where a steady, stable, vibration-free UAS sensor platform is desired. Falstrom and Gleason (2012, 81) emphasize the importance of this characteristic saying, "Vibration is the deadly enemy of electronics and sensitive electro-optic payload systems and is much of the reason for the lack of system reliability of UAV systems." Wankels are also very compact with a high specific power. Rotary engines can be made to operate quietly when fitted with an effective muffler system. Although Wankels are reasonably fuel efficient and capable of high rpms, torque output is somewhat low, a problem that can be solved using a gear reduction system. Rotary engines are installed on a number of UASs, including the RQ-7 Shadow and the RQ-2C Pioneer both powered by UEL AR-741 rotary engines producing 28.3 kW (38 hp) and the Schiebel S-100 Camcopter, powered by an Austro AE50R Wankel producing 41 kW (55 hp).

10.4.6 Heavy Fuel Engines

Heavy fuel engines (HFEs) burn heavy fuels, that is, more dense than gasoline. If mineral based (i.e., not biofuels), heavy fuels (HFs) are obtained from kerosene petroleum fractions. UAS heavy fuels include Diesel, JP-5 (kerosene-based fuel with a relatively high flash point developed for carrier use), JP-8 (similar to Jet A-1), and biodiesel hydrocarbons. Jet B and JP-4 (aka, avtag or by the NATO code, "F-40") are "wide-cut" (containing 50%–70% gasoline fractions) and may not be recommended for use in certain UAS heavy fuel engines. Other HFEs are capable of burning either gasoline or kerosene fractions. Although Diesel engines (which may be either two- or four-stroke configurations) burn the same fuels as other HFE engines, the difference is that non-Diesel HFE engines rely on spark ignition, rather than the heat of compression, to ignite the fuel–air mixture. Heavy fuel UAS engines may be of Diesel, Wankel, two- or four-stroke design. To ensure more efficient combustion, two-stroke HFEs frequently have two spark plugs installed in each cylinder.

The push by the military and governmental agencies to encourage adoption of heavy fuel UAS powerplants (aka, the single fuel initiative) can be explained by the many advantages resulting from burning heavy fuels. In comparison to gasoline, HF is energy dense, containing more energy per volume (about 15% more)—consequently, the potential for improved endurance inheres to the use of heavy fuels. In comparison to gasoline, HFs generally store longer without degradation. Some HFEs are capable of operating on a 2-stroke oil/gasoline mix (e.g., 3W-International HFEs). Because HF has a higher flashpoint, the operation of an HFE UAS may be considered safer. HF may be more readily available in remote areas and less developed countries. HFEs are generally very fuel efficient (consuming ≈20% less by volume than Otto cycle engines), which improves endurance. Examples of UAS HFE installations include the General Atomics Gray Eagle in which is installed a Thielert Centurion 1.7 turbo-diesel producing 101 kW (135 hp); the USMC Dragon Warrior Vantage incorporating a heavy-fuel Wankel engine powering the main rotor (the tail rotor is driven electrically) generating 31 kW of power (42 hp); and the new IAI Super Heron powered by an Fiat DieselJet powerplant (of Italian manufacture) developing 147 kW (200 hp).

10.4.7 Propellers on UASs

The propeller is generally considered to be an integral component of a powerplant installation. On a fixed-wing aircraft (rotary-wing aircraft and multi-rotors were covered in previous sections), the combination of powerplant and propeller create the propulsor, the unit responsible for generating a propulsive force. With the exception of those examples powered by a turbojet, turboshaft, or fan, all fixed-wing UASs generate a propulsive force by accelerating a mass of air through a propeller disk. All of the preceding UAS powerplants, excluding jet, fan, and turboshaft gas turbines, drive propellers. Thus, a propeller is not an insignificant component of the majority of fixed-wing unmanned aircraft. UA propellers may be of fixed or variable pitch (e.g., ground-adjustable) or of constant speed design. Propellers for UAs are made from a variety of materials: wood, carbon composite, fiberglass, aramid, aluminum, nylon, fiberglass-reinforced nylon, steel, etc. Manufacturers include Sensenich, McCauley, Aerovate, Northwest UAV (NWUAV), among a large number of others. Sensenich has been producing propellers for target drones and reconnaissance UASs since the 1950s (e.g., for Dennyplane aircraft and the Northrop Falconer).

Small UASs most often use fixed-pitch propellers. These small propellers often have their diameter and pitch information displayed on the propeller back (cambered surface). The first number indicates the diameter of the prop disk (the distance, in inches, measured from tip to tip on a two-bladed propeller) and the second represents the geometric pitch, or the distance the prop should theoretically travel in one revolution (see the preceding discussion under UAS Flight Dynamics and Physics). The pitch of a fixed-pitch propeller should be chosen to align with the mission and flight characteristics of the UAS: Selecting a propeller of lesser pitch will reduce drag, and all other factors constant result in higher rpms, a characteristic favoring lower speed operation and the takeoff/landing segments of the flight. Choosing a higher-blade angle (or greater pitch) would be warranted for higher airspeed applications such as efficient operation at cruising speeds.

Large UASs, such as the turboprop Reaper, will use constant speed propellers in an attempt to maintain an efficient angle of attack (around two to four degrees) where the lift to drag ratio is maximized. Such a system is much like those found on manned aircraft: a single-acting propeller governor will port high pressure oil, diverted from the powerplant lubrication system, to a piston in the propeller hub to effect pitch change or release oil to allow some combination of spring force, aerodynamic forces, and the force exerted by counterweights (if installed) to effect a pitch change in the opposite direction. The term, single-acting, refers to the fact that the governor will direct high pressure oil to cause the blade angle to change in only one direction. A single-acting governor will direct oil to the propeller to either increase pitch or decrease pitch—depending upon manufacturer design; it can do either—but not both. A double-acting governor will route high pressure oil to either side of the propeller piston (a component of the pitch change mechanism) to both increase and decrease blade angle. The Reaper uses a four-bladed

McCauley propeller, which is very similar to those found on manned aircraft. Aerovate and Northwest UAV both produce constant speed propellers for UASs smaller than the Reaper; propellers manufactured by the former use aerodynamic forces counteracted by spring force to accomplish changes in pitch, whereas the designs of the latter may be either electrically or hydraulically actuated.

10.5 Launch and Recovery Systems

All unmanned aircraft must be launched and, although some are intended to be expendable, most will (hopefully) be recovered. Just as the nature of the mission touches upon nearly every aspect of UAS design, so, too, does the choice of airframe and powerplant influence, even dictate, the method of launching and recovering many UAS configurations. The author has identified a few examples at points throughout this chapter. Recall that a design with less wing area will exhibit less drag, but the associated increase in wing loading may compel the designer to use a catapult launch system in lieu of the less equipment intensive hand launch. Moreover, higher wing loading will increase the rate of descent to possibly produce excessively hard landings, necessitating the installation of a parachute recovery system. The relationship works in the other way, too. That is, the desire for a particular means of launching and recovering an aircraft may predetermine airframe and propulsor configuration, something that largely drove the design and development of the UTC Vireo. The Vireo was developed to compete for a government contract emphasizing compactness, portability, and the ability to be hand-launched and recovered without benefit of a capture device (subsequently discussed). Consequently, the Vireo sUAS was developed around a flying wing, mounted high on the fuselage, to facilitate hand launches and with belly skid (integral to the battery pack) and folding propeller to increase survivability during landings in rugged, remote areas. The launch and recovery elements of any UAS are components of a subsystem existing as integrated or discrete entities comprising a portion of the entire unmanned aircraft system. In concluding this chapter, the author will discuss the relationship of the launch and recovery subsystem to the entire unmanned aircraft system.

Perhaps the oldest methods of initially applying the dynamic energy necessary to achieve lift for unmanned aircraft is the abrupt release of potential energy via catapult and the much more gradual and gentler acceleration of the hand launch. Both means were used, from the earliest days of flight, to launch models of gliders and powered aircraft before committing to the more consequential attempts at manned flight. Early examples of unmanned, hand-launched aircraft include the contra-rotating feathered rotor devices developed by Lomonosov (1754) and Launoy and Bienvenu (1783), the scale models of the patented powered monoplane designed and flown by the du Temple brothers (c. 1857–1870) and the research models constructed by Otto Lilienthal, Samuel Pierpont Langley, and the "father of aerodynamics," Sir George Cayley, all flown around the turn of the nineteenth century; John Stringfellow purportedly employed a catapult to launch models of the Ariel, the first patented, powered aircraft design (1848), and Elmer and Lawrence "Burst" Sperry used a catapult engineered by the future designer of the famed Norden bomb sight to launch the Hewitt-Sperry Automatic Airplane (aka, the Sperry Flying Bomb) in 1918 (Jarnot 2012).

Probably the earliest and certainly the simplest method of launching UAs, still very common today for lofting sUAS, is the hand launch. The primary advantage is obvious—no additional launch equipment is needed in the field. The time and logistics associated with transporting, setting-up, and tearing-down the extra equipment is completely eliminated. Consequently, support costs are also reduced. This type of launch is very gentle and (unless the attempted launch fails) the concern over additional forces placed on the UA during acceleration is also eliminated. Problems may arise where insufficient energy can be imparted to the aircraft to achieve liftoff—particularly in no-wind or light-wind conditions. Maintaining altitude and imparting the proper trajectory during launch may also be challenging. Aircraft which are most likely suitable for hand launching are those with a low wing loading and high specific static thrust. "Most hand-launched unmanned aircraft weigh less than 20 lb with wing spans under 10 feet" (Gundlach 2012, 441). Examples of hand-launched sUAS include the Aerovironment RQ-11 Raven, the Wasp and the Puma, the Lockheed Martin Stalker, and the Crow developed by Rich Brown and Nathan Maresch at the Salina campus of Kansas State University.

Catapult launchers (including what are sometimes referred to as rail launchers) and "tensioned line" systems (aka High Start or Hi Start systems) use one of several methods to store the initial energy subsequently imparted to the unmanned aircraft to provide the dynamic energy necessary to produce lift. The application of this energy cannot be too abruptly applied. (Langley's Great Aerodrome experienced structural failures when the spring-powered catapult accelerated the ungainly aircraft to 27 m/s [60 mph] in a mere 21 m [70 feet]. The Wright Brothers also employed rail launch systems, and catapult/rail launch systems thus predate the use of landing gear.) Small, relatively light UASs may be launched by a catapult that stores energy in a bungee. Bungee cord has improved over the years, and currently produced bungee has much greater elasticity, and consequently greater capacity to store energy, than older materials could provide. The British company, Universal Target Systems, currently produces a bungee catapult capable of launching a 105 kg (approximately 231 lb) UAV to a launching speed of 24 m/s (about 54 mph) (Novaković and Medar 2013). Larger fixed-wing unmanned aircraft may be launched from a pneumatic catapult. For example, both UAV Factory and Arcturus UAV manufacture catapult systems capable of storing compressed air to launch their aircraft. In comparison to pneumatic launch systems, advantages generally afforded by bungee launchers include compactness, lighter weight, less complexity, and a smaller footprint once assembled in the field. Although bungee launchers afford several advantages, pneumatic launchers are more common (Gundlach 2012). Carrier deployed unmanned aircraft, such as the Northrop Grumman X-47B UCAV, are launched from the same steam-powered catapult as their manned counterparts.

Other commonly used launch techniques include rocket-assisted takeoff (RATO or, sometimes, JATO for jet-assisted takeoff) as used to launch the German V1, the Ryan FireFly and FireBee, and the Northrop BQM-74 Chukar, air-drop launches (Lockheed Skunkworks D-21 and the Boeing X-51 Waverider) and even car-top launch systems (UAV Factory Penguin and the Sperry Flying Bomb) where the unmanned aircraft is carried to takeoff speeds by an automobile and then released to lift from the carrier cradle. Larger UASs, generally weighing more than approximately 455 kg or 1000 lb (e.g., the Global Hawk and aircraft in the Predator/Reaper class) operate exclusively from paved (or, at least, improved) runways and often rely on retractable landing gear for launches. Of course, a variety of configurations exist, and, for example, the UAV Factory Penguin incorporates fixed tricycle landing gear to operate from grass fields and pavement, but, as previously mentioned, may alternatively be launched from a car-top cradle or pneumatic catapult.

RATO/JATO launch systems are simple in construction and afford high specific thrust (thrust to weight ratio) and modularity. Rocket-assisted takeoffs are particularly beneficial when the lack of a runway and the complexity, logistics and footprint of a catapult launcher precludes its use. An early example of an RPA incorporating two RATO bottles integrated into its fuselage is the Northrop MQM-57 Falconer, a UA whose lineage is traceable through the 15,000 OQ-2/TDD-1 drones and their variants produced during WWII back to Reginald Denny's original Dennyplane, powered by the Dennymite engine. The MQM-57, which carried a high resolution camera, is an early example of a successful tactical reconnaissance UA. The RATO bottles could abruptly loft the Falconer, which was powered by 72 hp McCulloch O-100, from a zero length launch ramp. (The aircraft relied on a parachute recovery system.) Heavier UAs, requiring a relatively high rate of acceleration to lift-off, may also benefit from a RATO/JATO launch system. The German V-1, with its short, stubby wings and relatively high wing loading, required a RATO, steam catapult, or airdrop launch. More recently, RATO systems have been used to launch the AAI RQ-2 Pioneer and the TRW/IAI-developed RQ/MQ 5A/B/C Hunter in areas where space available for takeoff was limited. In designing RATO/JATO systems, alignment with the aircraft's center of gravity is critical to maintaining control during launch.

An airdrop launch affords several advantages. When combined with a parachute recovery system, which also eliminates the drag associated with a landing gear carriage, an airdrop launch allows designers to optimize the wing for high-speed flight. (This is also true of a RATO/JATO launch system where a parachute recovery is employed.) Since the UA is usually at a relatively higher altitude than other launch systems provide, aircraft takeoff weight is reduced while range and endurance are increased. Most UA are airdropped from under the aircraft, an approach generally considered safer than carrying the aircraft above the mothership. The Lockheed M-21/D-21 air launch, in which the D-21 UA was to lift from a modified SR-71 mothership, failed to provide sufficient separation resulting in a fatal accident (described in the next paragraph), after which the D-21 was dropped from beneath a B-52.

Regardless of the type of launch mechanism, certain desirable characteristics generally inhere to devices used to impart the dynamic energy necessary to loft the UAS. The requisite length of the launch varies directly with the square of change in UAS velocity and inversely with acceleration, whereas, "[t]he launch energy imparted to the UA is proportional to the takeoff gross weight [and the imparted] … kinetic energy … is proportional to the relative launch velocity squared. This implies that launch speed can be a more significant launch energy driver than UA weight" (Gundlach, 2012, 424). Although transferring a certain level of energy to the vehicle is necessary, it is, at the same time, desirable to minimize forces during launch (and recovery) to prevent the overstressing of structures, to reduce wear and tear on the launch mechanism, and to minimize the field footprint of any associated apparatus and supporting devices (e.g., a compressor and accumulator for pneumatic launchers). Thus, the rate of UA acceleration and the amount of generated energy converted to force to be absorbed by the launching device is significant. Also, launch mechanisms must provide sufficient clearances during the entire operation. This seems obvious, but failure to provide adequate clearance between Lockheed's D-21 supersonic UA and the M-21, a modified SR-71 Blackbird, during an air launch resulted in a fatal accident in which the Launch Control Officer, Ray Torick, drowned after ditching over the ocean. (The pilot, Bill Park, who had previously ejected from a Blackbird at 500 feet and 200 knots, survived, but both the D-21 UA and the M-21 mothership were destroyed.)

Gundlach (2012, 425) defines the recovery (or capture) operation as the successful activity of "… transitioning the UA from a flying state to a nonflying state." Recovery systems include the use of capture nets, parachute deployment, arresting devices, and proprietary apparatus such as the Insitu SkyHook® (Figure 10.12). The recovery phase can provide

FIGURE 10.12
A Boeing Insitu RQ-21A Blackjack captured by the company's proprietary SkyHook® recovery system. (U.S. Navy image.)

greater challenges than the launch operation. The approach requires adherence to an airspeed controlled within narrow limits, near stall, where the aircraft is susceptible to wind gusts, and a precise ground track to intersect the location where the aircraft is to be captured or recovered. Moreover, upon capture, the dynamic energy carried by the UA must be dissipated in a controlled way to minimize stresses induced in the platform and recovery device. Even a belly landing can produce very large forces on the fuselage as it compresses to absorb energy, particularly where descent rates are high.

The necessity of using a launch and recovery system may be dictated by the design and mission of the aircraft, but the choice of recovery may also afford certain inherent advantages. A belly skid recovery, which relies on the friction between the ground and the UA airframe to arrest forward motion, eliminates the drag associated with landing gear to enhance range and endurance and obviates the need for a runway. Larger, often tactical, UASs which may not well withstand the forces and abrasion experienced during a belly landing, often rely on a net recovery. Nets can also be used as an arresting barrier in conjunction with conventional landing to more rapidly slow an unmanned aircraft while reducing the aircraft's rollout. Similarly, arresting cables may be employed in much the same way. The SkyHook® proprietary recovery system employs a unique configuration wherein a vertical arresting cable is clutched by a grasping claw located at either wing-tip of certain Insitu UASs (e.g., ScanEagle and Integrator). This configuration provides an extremely small footprint and is, consequently, often found installed on a ship deck or trailer. Parachute recovery systems, used on a wide range of aircraft (e.g., sUAS, tactical and air-launched UA) provide low descent rates and relatively soft landings. Parachutes can also be employed as emergency flight termination and drag inducing (drogue 'chute) devices. Unmanned aircraft suspended under a parachute canopy may also be snagged for recovery during descent, a common means of retrieving the Ryan Firebee during the Vietnam War.

The largest UASs (e.g., the Global Hawk) commonly employ the most complex command and control system for launch and recovery. The Global Hawk's launch and recovery element (LRE, aka, the satellite control station) is intended to be rapidly deployable and capable of controlling the aircraft in the field to a distance of approximately 370 km (200 nautical miles) (Austin 2010). Once the UAS is aloft, control of the aircraft is "handed-off" to the mission control element (MCE) for BLOS operation during the majority of the mission. Upon completion of the operation, the Global Hawk is flown to within range of the LRE and control relinquished so that the aircraft may be landed at its base of origin. The most advanced launch and recovery systems are fully integrated with the navigation avionics, autopilot, and aircraft subsystems to provide the capability of fully automated takeoffs and landings. One example is the Advanced Launch and Recovery System (ALRS) available on the IAI Heron II. The ALRS can not only perform preprogrammed autonomous takeoffs and landings, but also "hand-off" aircraft control to the ground station of a forward base of operations. According to Israel Aerospace (2002 under Advanced Ground Control Stations), the ALRS "… minimizes human error during [critical] … phases of flight and reduces the cost of operator training."

10.6 Conclusion

As can be seen from the great diversity in airframe construction, available powerplants and combinations thereof, design options for UASs are numerous affording the designer and engineer tremendous flexibility and control over the final aircraft configuration. Many factors ultimately affect the decisions that culminate in the final design, but the most significant among these is the mission goal. After all, the UAS exists solely to efficiently and successfully achieve some objective, most often in the civilian world, the acquisition of remotely sensed data. The selection of airframe configuration, the type of powerplant, and the integration of these into the entire unmanned aircraft system should be accomplished with full consideration of how the UAS is to be used and of the ultimate goal of the mission for which the system is to be flown.

DISCUSSION QUESTIONS

10.1 Why does the author state that, in designing any UAS, form follows function?

10.2 List factors in the design of UASs. Why are these important? Of these, which two are likely the most significant? Why?

10.2.1–10.2.3 Describe economic and exogenous design factors and tell how these may influence UAS design.

10.2.4 What are the four main forces acting on any heavier-than-aircraft during flight? Describe how lift and thrust develop. What are the sources that sum to produce total airframe drag? Describe each. Explain why total airframe drag is greater than the sum of the drag acting on each individual component. Explain how an increase or reduction in drag affects other UAS design parameters.

10.2.4 Define aspect ratio and fineness ratio and tell how these influence UAS performance and design.

10.2.4 Describe how a propeller develops thrust. Define propeller blade face, blade back, geometric pitch, effective pitch, and slip. What are sources of propeller inefficiencies? Describe these. Why are propeller and rotor blades "twisted" so that the blade angle changes from blade station to station?

10.2.4 What is meant by "scale effect?" How does scale effect affect UAS, particularly sUAS, design?

10.3 When classified according to the method of lift production, what are the two broad categories of UAS?

10.3.1 Gundlach (2012) states that the great diversity in tail designs is analogous to an artist's pallet. List and describe the various tail configurations available to the designer of UASs.

10.3.1.1 Describe the X, Y, and Z axes as these relate to aircraft stability and control. What is meant by "tail-down" or "nose-up" force? Why is this important? How is it designed into unmanned aircraft?

10.3.1.2–10.3.1.5 Compare and contrast the various configurations of control/lifting surfaces employed in UAS design to impart stability and control about the X, Y, and Z axes, providing advantages and disadvantages of each. Identify aspects or characteristics unique to each (e.g., the use of ruddervators or elevons).

10.3.1.3 Briefly discuss the origins and history of twin-boom, pusher unmanned aircraft designs that are so common today.

10.3.1.5 Describe the two types of canard wings and discuss their use on UASs.

10.3.2 Discuss the benefits and drawbacks of selecting a rotary-wing design over that of a fixed-wing.

10.3.2.1 Explain the flight dynamics and their control in a helicopter UAS. How may torque cancellation be achieved? What is the difference between counter-rotating and contra-rotating propellers and helicopter rotor systems?

10.3.2.2 Why are the majority of multi-rotors configured with an even number of rotors? What is a Y-frame multi-copter? How is torque cancellation achieved on Y-frame sUAS?

10.3.2.3 The Honeywell Aerospace Tarantula Hawk is a novel approach to sUAS rotary-wing design. Describe the construction of the

T-Hawk and its rotor system. How does Honeywell achieve torque cancellation on this sUAS?

10.4–10.4.6 List and describe the various powerplant configurations available to the designer of UAS, identifying aspects or characteristics unique to each and the advantages and disadvantages of each design when installed on UASs. Explain the operating cycle of each. Give examples of UASs on which each is installed.

10.4.3 Describe the construction of inrunner and outrunner BLDC motors, giving the advantages and disadvantages of each. Why are neodymium magnets commonly used in the rotors of these motors? What will be the effect of heat on these magnets?

10.4.4 Explain the differences between turbojet, turboprop, turboshaft, and turbofan gas turbines? Give examples of UASs on which each type of gas turbine has been installed?

10.4.7 Identify those UASs most likely to use a fixed-pitch propeller and those where a constant-speed propeller would be used. A constant-speed system is heavier and much more complex. Explain why, in the face of the preceding statement, these would be used. Identify those flight regimes where higher or lower blade angles would be desirable and justify your answer. When compared to a "cruise" fixed-pitch prop, would a "climb" prop have a greater or lesser blade angle?

10.5 Describe the various launch and recovery systems that are a part of any unmanned aircraft system, providing characteristics, advantages, and disadvantages of each. Provide examples of UASs using each. Provide the phrases represented by the acronyms LRE, MCE, and ALRS and tell how each are used in the command and control of UASs.

References

Austin, R. 2010. *Unmanned Aircraft Systems: UAVS Design, Development and Deployment*. Chichester, UK: John Wiley and Sons.

Catchpole, B.G. and B. Parmington. 1990. *Design and Preliminary Development of an Engine for Small Unmanned Air Vehicles*. Propulsion report 184. Melbourne: Department of Defense, Aeronautical Research Laboratory.

Fahlstrom, P.G. and T.J. Gleason. 2012. *Introduction to UAV Systems*. Chichester, UK: John Wiley and Sons.

Gessow, A. and G.C. Myers, Jr. 1985. *Aerodynamics of the Helicopter*. New York: Frederick Unger Publishing.

Gundlach, J. 2012. *Designing Unmanned Aircraft Systems: A Comprehensive Approach*. Reston: American Institute of Aeronautics and Astronautics.

Hurt, H.H., Jr. 1965. *Aerodynamics for Naval Aviators*. Washington, DC: Naval Air Systems Command.

Israel Aerospace Industries. 2002. Advanced Ground Control Stations. http://www.iai.co.il/2013/34404-37386-en/IAI.aspx (accessed April 12, 2015).

Jarnot, C. 2012. History. In *Introduction to Unmanned Aircraft Systems*, eds. R. K. Barnhart, Hottman, S. B., Marshall, D. M. & Shappee, E., 1–16. Boca Raton, FL: CRC Press/Taylor & Francis Group.

Louge, M., A. Halterman, A. Billington, E. Franjul, J. Gong, Y. Lee, J. Nersasian, C. Ozkaynak, J. Pei, and M. Ujihara. 2004. Mars unmanned aircraft 2003–2004. Working paper, Odysseus Team, Cornell University. http://www.southampton.ac.uk (accessed June 6, 2014).

Norris, G. and A. Butler. Triton's turn. *Aviation Week & Space Technology*, May 27, 2013.
Novaković, Z. and N. Medar. 2013. Analysis of a UAV bungee cord launching device. *Scientific Technical Review* 63–3:41–47.
Sadraey, M. 2009. *Aircraft Performance Analysis*. Saarbrücken, Germany: VDM Verlag Dr. Müller.
Schafroth, D.M. 1980. Aerodynamics, Modeling and Control of an Autonomous Micro Helicopter. PhD diss., Eidgenössische Technische Hochschule Zürich, Schweiz (Swiss Federal Institute of Technology in Zurich).
Seddon, J. 1990. *Basic Helicopter Aerodynamics: An Account of First Principles in Fluid Mechanics and Flight Dynamics of the Single Rotor Helicopter*. Oxford, UK: BSP Professional Books.

11

UAS Subsystem Nexus: The Electrical System

Michael T. Most

CONTENTS

11.1 Introduction

An electrical system is essential to the operation of all controllable unmanned aircraft (UA). Even the least complex, remotely piloted aircraft (RPA) rely upon the electrical power to receive, process, and distribute input signals to achieve command and control and often for propulsion. Among the most diminutive of the commercially available small unmanned aircraft systems (sUAS) is the electrically powered Proxdynamics PD-100 (aka, Black Hornet Nano) micro-aerial vehicle (MAV) which measures approximately 10 × 2.5 cm (4 × 1 in.) and weighs just 16 g (2.1 ounces) including the battery which provides 20 min, or more, of flight time. The PD-100 is capable of remote operation as an RPA or of autonomous flight enabled by an onboard autopilot with integrated avionics (i.e., GPS receiver, inertial measurement unit, and three-axis gyro system). Because the MAV's payload consists of an electrically positioned, electro-optical sensor capable of streaming (downlinking) either a live video feed or still images, the PD-100 is well suited to various 3-D (dirty, dull, and dangerous) missions including close-quarters search and rescue following a natural disaster, inspection of nuclear and chemical manufacturing facilities, and hostage rescue and military applications. As unmanned aircraft increase in size, platforms generally become more capable of carrying larger, heavier payloads, which, in turn, provide greater mission capabilities. Consequently, as a generalized observation, it could be stated that the complexity of the associated onboard systems become greater as UA platforms become larger. The intricacy and sophistication of the onboard systems of large unmanned aircraft (e.g., the Northrop Grumman Global Hawk, the Lockheed Martin Sentinel, or the NASA-modified General Atomics Reapers of the Ikhana and Altair programs) become comparable to those of large, turbine-powered manned aircraft.

The electrical system of any of the gamut of an unmanned aircraft is the singular attribute that interconnects all other components and subsystems comprising the UAS. The electrical system provides the power for controlled flight, subsystem intercommunication, onboard intelligence, telemetry, navigation, and payload operation. The electrical system is the subsystem nexus common to all UASs.

11.2 UAS Electrical Systems: General Characteristics

On the macro-level, the electrical systems of unmanned aircraft may be characterized in broad terms, generally descriptive of those found in the majority of UASs. What follows is intended to provide a basic introduction and overview of electrical systems and to lay the groundwork for the more detailed descriptions that follow in the subsequent sections.

To function, all aircraft electrical systems must have a source of power—in UASs, electrical sources may take the form of a battery, generator, or, in some cases, solar panels, fuel cells, or some combination of these. The power source provides energy, known variously as electromotive force, EMF, or voltage, to move electrons, which have mass, through a conductor. Measured in units known as amps, this flow of electrons constitutes the electrical current. Assuming a given level of opposition to current flow, known as resistance and measured in ohms, the rate of electron movement is a function of the amount of voltage, or energy, applied. This relationship is described in Ohm's law (named after the German mathematician and physicist Georg Simon Ohm, who first suggested the effects of varying voltage on current). Ohm's law is mathematically expressed in the following way: Current flow (represented by the letter I or A) equals the applied voltage (represented by the letter E or V) divided by resistance (R), that is, $I = E/R$ or $A = V/R$. The unit of resistance, the ohm, is often represented by the upper case Greek letter omega, Ω. Larger values of resistance are measured in Kilohms (K-ohms), Megohms (M-ohms), or even Gigaohms or Gigohms (G-ohms). Resistance is present in all conductors and components—even power sources themselves have a certain amount of internal resistance. (Although for most purposes, e.g., when calculating circuit values or performing generator load analyses, power source internal resistance is generally considered negligible; battery internal resistance becomes an important consideration as batteries discharge. This characteristic and the responsible chemical processes are subsequently discussed.) As mentioned earlier, the electrical current is measured in amps, or in smaller subunits, such as milliamps or microamps, and less often in larger units, such as Kiloamps or K-amps. In physics, mechanical power, measured in watts (SI unit) or horsepower, is a measure of the time rate of doing work, which can be defined as the distance an object moves as the result of the application of energy per unit of time. Because electrical power is derived by measuring the rate at which electrons, which have mass, move through the circuit (this rate is equivalent to 1 coulomb, or the charge associated with approximately 6.241×10^{18} electrons per second) as the result of the application of energy (voltage or EMF), the definitions of electrical and mechanical power are similar. Electrical power, which is the product of applied voltage and current flow ($P = E * I$), is expressed in watts, kilowatts (K-watts), or megawatts (M-watts).

Aviation quality wire (i.e., manufactured according to FAA standards and specifications) is sized according to the American Wire Gage (AWG) system in which larger numbers indicate a smaller diameter wire. For example, a fairly small diameter wire stamped with AWG 24 is larger than an AWG 26 conductor. Although a manufacturer could order wire in odd AWG sizes, aircraft wire manufactured according to FAA specifications is generally available only in even numbered AWG diameters. Aviation quality wire is stranded for flexibility and manufactured only in aluminum (not commonly used due to lower conductivity, higher resistance, and susceptibility to corrosion) or copper. The resistance of a conductor, measured in ohms, is a function of the material of which it is made (given by a constant of material resistivity and represented by a lower case Greek rho, ρ). Given a particular conductive metal (e.g., copper or aluminum), resistance then varies directly (increases) with length and inversely with cross-sectional area (larger diameter, less resistance)—the longer the wire and smaller the diameter, the greater the resistance. Expressed mathematically, the resistance, in ohms (Ω), of a wire can be calculated using the formula, $R = (\rho L)/A$. All the electrical circuits will have resistance associated with the components (e.g., motors, autopilots, avionics, servos, receivers, etc.) and the interconnecting wire, but unnecessary resistance consumes voltage that is then unavailable to the load to do useful work. Resistance generates heat as it converts one form of energy, voltage, to another, thermal energy. The energy thus consumed in propelling electrons through circuit resistances and lost to the component to be energized is known as a voltage drop. An excessively large voltage drop in the external circuit reduces component performance and service life. Consequently, when designing any UAS electrical system or installing a new component, selecting a wire of suitable diameter is critical to satisfactory system operation and longevity. For a given UAS circuit, anticipated peak current flow, the maximum voltage drop acceptable in the circuit, and conductor length, from bus to component ground, are likely the most important factors in selecting the correct AWG wire size (diameter). Guidance on correct wire selection can be found in a variety of sources, including manufacturer specifications and Advisory Circular AC43.13-1B published by the FAA.

Another important characteristic of wire is the dielectric quality of the insulation. It is not flippant to state that the purpose of the insulation surrounding the conductor is to keep circuit current in the wire. A measure of the ability of the insulation to perform this paramount function has been historically referred to as its dielectric strength, though preferred terms, depending upon the context, might be relative permittivity or dielectric permittivity. If the wire abrades or is cut or damaged by solvents or fuel, its dielectric characteristics are diminished, weakening the insulation at a point where, if voltage is sufficient to overcome the remaining resistance at that location, current will leak from the circuit in a short circuit generating heat and impairing or interrupting system operation. If the system does not incorporate circuit protection in the form of a fuse or circuit breaker, the short circuit may produce enough heat to cause an in-flight fire and loss of the aircraft. Where installed, such circuit protection devices should be installed as near the electrical power source as possible to protect the maximum length of circuit wiring.

Because most sUAS airframe structures are made of nonconductive composites (e.g., carbon fiber and/or fiberglass), the likelihood of a fire induced by a short circuit is likely greater in larger, higher performance unmanned aircraft of traditional semi-monocoque construction, in which the aluminum fuselage/empennage can be used as a return path for a single-wire electrical system. Examples of such unmanned aircraft would include the Northrop Grumman Global Hawk/Euro Hawk and the Lockheed/Boeing Tier Three-minus DarkStar (Figure 11.1), both of which have been designed around a fuselage constructed of aluminum. Because all the electrical systems require a closed-loop circuit through which to conduct the current, a nonconductive structure presents designers of composite UASs with a challenge not facing those tasked with creating structures of metal. An aluminum aircraft structure can provide the return path to the power source. In such a configuration, the positive leads of direct current (DC) power sources, typically a battery,

FIGURE 11.1
Lockheed Martin/Boeing Tier III- (minus) RQ-3 DarkStar on the ramp following delivery to Dryden Flight Research Center, Edwards, CA. (Image courtesy of NASA.)

often a generator, are connected to a distribution bus to which the positive side of the component circuits also connects. When the negative leads of the power source(s) and components are connected (grounded) to the airframe, a complete circuit is provided. Current flow to circuits is then controlled by semiconductor switching devices and, on larger UASs in which current flow may be higher, by switch-controlled relays or solenoids. Because only one conductor carries current from the bus to the component with the airframe acting as the conductive return path to complete the circuit to the source, this arrangement of wiring is referred to as a single-wire system—and, it is something that is not possible where the UAS structure is constructed completely of dielectric (i.e., composite) materials. Composite UAS manufacturers must either embed conductive strips in the structure to be used as a common distribution bus or shared ground or double the amount of wire used, which costs more, adds weight, consumes internal volume available for other components, complexifies wire routing, decreases useful load and, therefore, the weight of any potential payload, while reducing both range and endurance.

Another design consideration associated with circuit wiring is the creation of electromagnetic interference or EMI (aka, radio-magnetic interference, or RMI) in the circuits. EMI results in a voltage being impressed upon a conductor creating a spurious signal, which can corrupt or confuse the information carrying signal in sensitive autopilot, communication, and navigation circuits to produce a system malfunction. EMI can even cause a critical UAS system to fail completely (FAA 2012). At least one Aerosonde UAS has been destroyed as the result of EMI produced by an unshielded magneto ignition circuit interfering with the autopilot and avionics systems. Potential sources of such spurious signals are many and include ignition systems, switch-mode power supplies, electronic speed controllers (ESCs), electric motors, servos, and even conductors themselves. Any change in the intensity of current flowing in a conductor results in an expansion or contraction of the associated electromagnetic field that surrounds the wire. Should the resulting relative motion cause this varying field to cut across an adjacent conductor, a secondary voltage will be impressed in the wire that is electrically insulated from the one carrying the initially varying current flow. If this secondary circuit is carrying data to a critical component, the impressed (spurious) signal, or EMI, can corrupt the original data signal, with potentially undesirable or catastrophic results (e.g., a crash and loss of the aircraft). Because the strength of the EMI-producing magnetic field decreases as the inverse of the square of its distance from the receiving conductor, maximizing separation between conductors is sometimes an effective strategy for reducing or eliminating electronic noise or interference. Twisting wires is also a simple way to cause the interacting fields to cancel and, thus, reduce EMI. A more positive EMI-suppression strategy is the use of shielded wire or cable (FAA 2012). (An electrical cable consists of multiple insulated conductors, sometimes twisted, and routed together in a common sheath.) Another commonly employed method used to reduce or eliminate EMI in UAS electrical systems, particularly those found in sUAS, is the use of chokes made of sintered soft iron or ferrite (ceramic materials compounded with iron (III) oxide), which are available in a variety of sizes, dimensions, and shapes, such as beads or tubes. Where used in sUAS installations, the wires associated with EMI-producing circuits are often passed through the centers of toroid- or donut-shaped soft iron chokes and coiled in turns around their surfaces. The chokes are passive and are not connected to the circuit. Absorbing and dissipating energy while being frequency sensitive and tunable (by varying characteristics such as size, shape, and material of construction), these chokes are somewhat analogous to the inductors used in electrical circuits as electronic noise (EMI) filters.

Electronic filters are created from reactors, that is, inductors and capacitors. Although they produce an opposition to current flow in an alternating current circuit, reactors differ from resistors in several important ways: (1) the amount of opposition, termed reactance where attributable to reactors, varies with frequency and, where both inductance and capacitance are present in a circuit, it completely disappears at the resonant frequency; (2) reactors create a phase shift between voltage and current, whereas resistors do not, and; (3) because reactors alternately absorb and return energy to the circuit, they do not convert voltage to heat in the same way as a resistor. A final EMI-suppression strategy relies on the use of a capacitor (or capacitors) in parallel with the load and/or an inductor (or inductors) in series to create a low-pass (LP) electronic filter. An LP filter permits the passage of low-frequency voltages (and associated currents) while higher frequencies are filtered or blocked. Direct current has zero frequency and is freely passed. A high-pass filter, which will block lower frequencies (longer wavelengths) while passing those of higher frequencies (shorter wavelengths), can be constructed from some combination of inductors in parallel and capacitors in series with the load. Because both low- and high-pass filters are frequency sensitive, both types can be tuned to pass or block frequencies or bands by varying the value of the reactors used in the construction of the filter. Consider the following as an example of the use of an electronic filter to suppress the noise. Direct current aircraft fuel pump motors are constructed of an armature, made of many turns of insulated wire windings, repulsed by the magnetic field produced by the stators. The relative motion created by the rotation of the armature windings in the magnetic field of the stators induces a high-frequency alternating current that, were it not filtered out by the capacitive noise filter integrated into the motor, would generate a high-frequency EMI. Consequently, one potential application of an electronic low-pass filter would be incorporation into the design of fuel pumps installed in the tanks of UASs powered by gasoline or heavy fuel engines.

A point to be made, in closing this section, is that UAS electrical systems share commonalities though the components and overall system design will vary with platform size and mission. Nearly all UASs, from the smallest RPA to the largest, most sophisticated high-altitude, long-endurance (HALE), Tier II +, and Tier III- aircraft, carry a battery. All UAS electrical systems are powered from a primary source, most often a generator or battery, and provide the available power to the electrical loads through a distribution system. Most have some form of voltage regulation to provide proper voltages to the various components. These similarities, as well as the differences, will be discussed in the remainder of this chapter.

11.3 sUAS Electrical Systems

Not very long ago—a matter of a few years—the small unmanned aircraft flying today would not have been possible. Improvements in certain enabling technologies that include power storage (batteries), artificial intelligence (power plant controllers and onboard computing), electronics miniaturization, and electric motor design have enabled the rapid evolution of small, electrically powered unmanned aircraft. For example, in 2001, a conventional-for-the-time "brushed" motor, suitable for powering a sUAS, required a gearbox and weighed approximately 269 g (Logan et al. 2007). As small electric motors evolved, brushless inrunner motors, comparable in power to brushed motors, became available. (Inrunner and outrunner brushless DC motors or BLDCs will be subsequently described.) Although, at about 209 g, a comparable brushless inrunner weighed about 22.3% less than the brushed motor it replaced, it still required a gear reduction system (Logan et al. 2007). Today, a brushless DC outrunner motor, not requiring a gearbox but of comparable power output, would weigh approximately 181 g, a savings of 32.7% over the brushed motor of 2001. Electrically powered sUAS currently use either brushless inrunner or outrunner electric motors to drive the propeller or rotor system.

The smallest among sUAS are generally powered by electric motors. The use of an electric motor affords certain advantages in the design and operation of UASs, providing quiet, smooth, and reliable, low-vibration propulsion. Exhaust gas and noise are eliminated. Larger sUAS may use an internal combustion engine for propulsion and to drive a generator as the primary source of electrical power. The addition of an electrical generator will require more sophistication in power distribution and control and generally complexifies the entire electrical system. Thus, the choice of power plant will affect the design of the entire electrical system. An unmanned aircraft system is a holistic integration of components, the entirety designed to successfully achieve intended mission goals. Factors such as payload, endurance, range, vibration levels, operational altitude, and maneuverability, among others, may factor into the choice of a power plant, and thus, indirectly influence the design of the electrical system.

11.3.1 All-Electric sUAS

Arguably, the first all-electric unmanned aircraft flew in 1957 (Noth 2008). An RPA, the Radio Queen, was powered by a brushed motor having a permanent magnet stator. Electricity was supplied by a silver-zinc battery. Contemporary examples of all-electric, fixed-wing sUAS powered by an electric propulsor include the AeroVironment RQ-11 Raven, the UTC Aerospace Systems Vireo, the UAV Factory Penguin BE (Figure 11.2), and the PrecisionHawk Lancaster. Almost all multi-rotor sUAS are all-electric, and even some large unmanned rotorcraft, such as the IAI Panther tiltrotor, which, at 65 kg (~143.3 lbs.), approaches three times the FAA maximum weight for sUAS (<55 lbs.), rely exclusively on electrical energy to operate the propulsor and all onboard systems. The use of an electric powerplant dictates much in the design and construction of the sUAS electrical system, including the requirement to not only control motor rpm, regulate voltage, and provide

FIGURE 11.2
View of all-electric Penguin BE payload bay. Visible in the bay, just forward of the wing, is the lithium–polymer (Li–Po) battery cartridge and just aft of the wing, atop the powerplant cowling, is the air scoop to direct cooling air around the BLDC inrunner electric motor. (Image courtesy of UAV Factory.)

usable DC power for the propulsor, but also to power the receiver, servos, avionics, onboard computers, and payload. What generally will not be found on an all-electric sUAS are airborne charging systems and backup power sources (i.e., a dedicated emergency battery) to provide power for starting and to energize the electrical system in the event of a loss of the electrical generator.

11.3.1.1 Power Sources for All-Electric sUAS

The most common source of power in all-electric sUAS is one or more batteries. Power sources for larger, all-electric unmanned aircraft include solar panels (NASA Pathfinder and Centurion), fuel cells (U.S. Naval Research Labs Ion Tiger), or a combination of both (NASA Helios HP03) (Figure 11.3). (The NASA aircraft also carried batteries to supplement the primary sources of electrical power.) Batteries afford several advantages as sUAS power sources. They exhibit relatively high energy density, and unlike liquid fueled aircraft, which become increasingly lighter during flight, all-electric UASs powered by batteries do not experience a shift in their center of gravity as power is consumed. On the other hand, batteries are typically the heaviest component onboard an sUAS, and battery weight may be a factor determining the method of launch and the most important factor limiting the range and endurance of the aircraft. Moreover, batteries consume large volumes of space, and "[f]or a given energy storage, even Li–S batteries occupy four times the volume of that of fossil fuels, presenting a problem for all other than short-range [unmanned aircraft]" (Austin 2010, 291).

The basic unit of a battery is an electrochemical cell in which energy conversions occur. Without an external circuit connected to the cell terminals, chemical reactions cause ionization within the cell electrolyte and the active material on the positive plates or electrodes (the cathode) and the negative plates or electrodes (the anode) (Gundlach 2012). During this process, the active plate material is converted to the inactive material as cations form at the positive plates of the cell. Simultaneously, anions, with extra electrons available to become electricity, migrate to those plates, which are negative. Once the active

FIGURE 11.3
NASA Helios flying at a speed of approximately 25 mph and 10,000 feet off the coast of Kauai, Hawaii. (Image courtesy of NASA.)

material is inactivated through the ionizing reactions, it is unavailable to produce additional free electrons. The ongoing conversion of active material (to that which is inactive) explains why all cells and batteries eventually self-discharge over time. Aside from taking part in the chemical reactions occurring within the cell, another function of the liquid or paste electrolyte existing between the positive and negative plates is to allow migration of the ions within the cell and, thereby, conduct current internally through the cell. During discharge (aka, galvanic action), when an external load is connected across the terminals of the cell (or battery), the rate of ionization increases as electrons are drawn from the negative electrode under the attraction of the cations on the positive terminal. In this way, the rate of chemical reactions occurring within the cell increases in response to the demand for the power necessary to operate circuit loads. If active material is converted and consumed during the chemical reactions of the galvanic discharge, the cell cannot be restored to its pre-discharge condition. Nonrechargeable cells and batteries are said to be the primary sources of power—for example, a conventional carbon-zinc Leclanché dry cell is often referred to as a primary cell. In a rechargeable cell or battery, the active material is not consumed and lost, but is only converted, and the chemical reactions that produced the electrical current to power the external circuitry can be reversed through the application of electricity from an external source (e.g., a generator or solid state battery charger) to restore the electrolyte and active plate material to their original states. The process of producing chemical reactions through the application of electrical energy (DC) is known as electrolysis or electrolytic action. Rechargeable power sources in which the battery constituents can be electrolytically restored are referred to as secondary cells or batteries.

Nominal cell voltages are a function of electrolyte and active material chemistries, which determine the rate at which ionization can occur to drive free electrons through the external circuit. (A more technically stated description would be that the available output voltage is a function of the net reduction potentials of the positive and negative plates or electrodes.) The term "battery chemistry" refers to the materials from which the cell components (positive and negative plates and the electrolyte) are produced. For example, lithium polymer and lithium-ion polymer batteries are constructed of similar materials and develop voltage using comparable chemistries: graphite for the anode, lithium cobalt or, rarely, lithium magnesium for the cathode, and a lithium salt electrolyte. In nickel–cadmium batteries, the electrolyte is potassium hydroxide (KOH), while the negative plate is cadmium and the positive plate is nickelous material (e.g., nickel (III) oxide-hydroxide). Examples of nominal voltages associated with cells used in UAS applications, as either main or supplementary power sources, include hydride cells (nickel–cadmium, or Ni–Cd, and nickel metal-hydride, or NiMH) producing 1.2 V per cell, lithium cells (lithium-ion, or Li-ion, and lithium–polymer, or Li–Po) generating 4.2 V, and lithium–sulfide cells (Li–S) producing 2.1 V per cell.

Unlike voltage, the capacity of a cell is not a function of chemistry, but rather of the amount of active plate or electrode material exposed to the electrolyte. Capacity may be defined as the ability of a cell (or battery) to deliver a specified constant current flow over a given period of time, often standardized at an hour, to a given voltage level. The specified or expected rate of discharge is often referred to as the C-rate, for capacity rate. The C-rate is also sometimes said to be shorthand for charging/discharging rate (Fahlstrom and Gleason 2012). For example, a fully charged lead-acid (LA) cell will have a nominal voltage of approximately 2.0 V (or possibly a tenth or two higher). The same LA cell is considered discharged at 1.75 V. If the cell has a published ampere-hour rating (AH rating) of 1 amp, the C-rate discharge would be 1 amp for 1 h. At the end of that 60 min, if the cell voltage is 1.75 V or higher, the cell is capable of delivering 100% of its rated capacity. If a "topped-off" or fully charged LA cell is discharged at the C-rate to reach 1.75 V in one-half hour, then the cell is at 50% of its capacity. A battery can discharge at much higher rates than the specified C-rate, but for correspondingly shorter periods. At some point, the rate of discharge will become too great, and the battery will be irreversibly damaged by the heat. Another commonly used term, related to capacity, is energy density, which refers to the cell (or battery) capacity per weight. Similarly, power density refers to maximum deliverable power, expressed in watts per unit weight (Fahlstrom and Gleason 2012).

Cells of similar chemistry can be connected either in series or parallel (or both) to create a battery. Connecting cells in series additively increases the output voltage of the battery while placing cells in parallel similarly increases capacity. As an example, connecting two Li–Po cells in series produces a battery, commonly referred to as a 2S, having a nominal voltage of 8.4 V, whereas placing the cells in parallel would create a battery having double the capacity but the same EMF (4.2 V) as the individual cells. Thus, battery voltage is a function of both cell chemistry and the number of cells connected in series, and capacity of the battery is, keeping all other factors constant, the result of battery size (determined by plate size) and, possibly, the number of cells connected in parallel. Sufficient voltage is necessary to overcome circuit resistance and provide adequate power to operate the power plant, servos, and other circuit loads. Where necessary, the voltage will be regulated to a lower level appropriate to components requiring less energy to operate. Capacity affects endurance and range. However, because capacity is also directly related to battery size, the ability of a battery to deliver current may be a limiting factor in sUAS attributes related to weight, such as the ability to hand launch the aircraft or in the size of the payload that may be carried. In a "from-scratch" build, battery packs can be created with cells in both series and parallel configurations to meet the design needs and mission goals of the UAS platform (Gundlach 2012). For example, the UAV Factory designed the Penguin BE with a 48-cell, 640-W Li–Po battery pack that is constructed as a quick-change cartridge capable of delivering regulated onboard voltages of 6 and 12 V (Figure 11.4).

11.3.1.2 Electric sUAS Propulsors

Electric motors are energy converters, using electricity to produce a mechanical output in the form of torque force. All DC electric motors operate through the magnetic attraction/

FIGURE 11.4
Penguin BE 48-cell, 640-W lithium–polymer battery pack constructed as a quick-change cartridge capable of delivering regulated onboard voltages of 6 and 12 V. (Image courtesy of UAV Factory.)

repulsion of the interacting magnetic fields associated with a rotating component (rotor, shaft, or armature) and the stationary magnetic field (often referred to as the stator). The dissimilar polarity of the fields—north (armature) and south (stator), and vice versa—create an attractive force that pulls on the armature or shaft to produce rotation. Similarly, the rotor fields are repelled by the stator fields having similar polarities. This interaction causes the armature (or output shaft) to rotate producing a torque force that can be used to drive the propulsor rotor or propeller (or other electrical components such as landing gear, flaps, and control surfaces). The magnetic fields of the armature and stator can be associated with an electric current flowing through the multiple turns of wire to create an electromagnet or generated by permanent magnets. Although stator and armature fields may be generated by permanent magnets or an electrical current, permanent magnets cannot be used, in the same motor, for both. The reason is that the polarity of one field must be oriented, timed, and synchronized to attract (opposite field polarities) and simultaneously repel (fields of the same polarity) the fields associated with the stationary and rotating motor components. Using permanent magnets to generate both fields would cause the rotor to "lock-in" at a neutral position and rotate no further. Controlling the orientation and timing of the fields is critical in producing rotation and the resulting torque force. The means by which field orientation and timing are achieved varies according to the design of the motor, that is, whether the motor is of "brushed" or "brushless" construction.

In a brushed motor design, the (stationary) stator field may be produced using either permanent magnets or multiple turns of current carrying wire wound the field poles or field pieces. The stator field remains stationary. The timing of the switching operation that controls the orientation of the magnetic field associated with the rotor is accomplished by a mechanical device known as the commutator. The commutator is divided into discrete segments, electrically insulated by a dielectric material such as mica. Opposing segments (i.e., located 180° apart on the commutator) are the terminating points of the ends of one winding. Stationary carbon brushes convey current from the external power source to the armature windings by contacting commutator segments. As current flows into one segment and out of the opposing one, a magnetic field is created in the rotor (armature) winding and oriented in such a way that the interacting attractive and repulsive magnetic forces of the armature and stator fields cause the rotor to develop torque. As the armature rotates, the brushes contact another pair of commutator segments to energize the windings. The direction of the current flow and polarity and orientation of the associated magnetic fields are controlled by the commutation process in such a way that a torque force develops to rotate the armature on its bearings.

Brushed motors can operate directly on source DC, and motor speed can effectively be controlled by increasing or decreasing the strength of the magnetic field by using a variable resistance to vary the intensity of the current flowing in the windings. In comparison to brushless designs, the control of brushed motor rpm is much more easily achieved. However, brushes produce friction that increase heat and reduce output power, constantly wear and will eventually require replacement. Significantly, the interface of the brushes and commutator is a source of potential arcing that may cause EMI and corrosion. In addition, the brushes are made of carbon, a semiconductor material having a measure of inherent resistance that reduces the overall efficiency of the motor. As will be subsequently described, brushless motors control the timing and orientation of the associated magnetic fields by what might be referred to as electronic commutation—no brushes, no segments, and no arcing. Brushless motors offer greater reliability, operate more quietly, and afford greater specific torque and thrust. BLDC propulsors also provide more torque per ampere of consumed current. In the nineteenth century, remotely piloted unmanned aircraft (e.g., the Radio Queen and the Magicfly) were powered by brushed motors. Not until improvements in the transistor design and microprocessor technology permitted, around the onset of the millennium, the economic miniaturization of the necessary controlling micro-circuitry (in the form of an electronic speed control, ESC) did the widespread use of brushless motors in sUAS propulsors became possible (Büchi 2012). Although small, brushed motors (e.g., the RS 540 and Speed 600) continue to be available for hobby RPA, the advantages of brushless motors dictate their pervasive use in electrically powered UASs.

Brushless motors are constructed as either "inrunners" or "outrunners" using permanent and electromagnets. Both designs are used to develop sUAS thrust. One commonality among inrunner and outrunner BLDC motors is that the permanent magnets are always installed on (or in) the rotor, while the stator conducts the electric current that establishes the electromagnetic field. Another is the use of neodymium magnets (also known as NdFeB, NIB, or Neo magnets), the strongest type of permanent magnet yet developed, to increase efficiency and power output. The previously mentioned motor controller, also commonly known as an electronic speed control or ESC, is a microprocessor that acts as an inverter/switching device to convert DC to a bidirectional pulsing current (as opposed to a sinusoidal waveform) to provide the desired voltage frequency and voltage amplitude to the motor. The purpose of the motor controller, or ESC, is to time, sequence, and orient the polarity of the electrically induced magnetic field to properly attract/repel the magnetic fields associated with the permanent magnets. The interaction of flux fields creates an angular force causing rotation of the output shaft and the development of torque to drive the rotor or propeller—all without brushes. A sensor (e.g., a counter-EMF (CEMF) or back-voltage sensing circuit, optical or Hall effect device) detects rotor position and speed to send rpm information to the ESC through a feedback loop to provide the means by which the controller can maintain the desired level of power output (rpm is one component in formulae used to calculate engine horsepower). The ESCs that control small RPA and sUAS most commonly rely on a feedback system that senses CEMF in the de-energized (or "floating") BLDC motor stator winding to control rpm. This strategy is effective in governing both inrunner and outrunner motors. Due to BLDC stator design (which are wound in either delta or wye configurations), an ESC does not recognize the type of motor and any given controller is "blind" to whether it is controlling the speed of an inrunner or outrunner aircraft powerplant.

An electronic speed controller becomes warm during operation. The MOSFET transistor found in many ESCs dissipates the greatest amount of power during the switching operation that controls the voltage pulses used to maintain the desired engine rpm. Consequently, the greatest amount of heat is generated in the ESC while the motor is operated at less than full rpm—reduced motor speed requires more frequent switching to control the stator field currents. In general, the higher the current rating of the ESC, the less is its internal resistance, and a larger, heavier, more expensive controller (i.e., having a higher current capacity rating) will afford the advantages of generating less heat, providing a longer service life, and, possibly, yielding very slightly improved range and endurance. Because the ESC generates heat, potentially considerable amounts of it, providing adequate cooling is important, especially where the controller incorporates the additional functionality of an integral BEC (described in the following section). Consequently, additional ducting may be necessary to provide cooling air to the motor controller, and possibly the motor itself (Figure 11.5). The use of heat sinks may also be effective in reducing ESC temperatures. Cooling the ESC translates into more efficient operation, less resistance (which increases with temperature), and lower voltage drop at the controller (the voltage drop increases with resistance), which, in turn, provides greater power to the motor. Moreover, the ESC, itself, can be destroyed by excessive amounts of heat.

Often, an all-electric sUAS will carry a BEC (battery eliminator circuit or battery elimination circuit), which, because a battery is still required, is a bit of a misnomer. The BEC may be a discrete component or incorporated into the electronic speed controller, especially likely in those ESCs having lower power ratings. Obviously, the BEC cannot eliminate the need for a battery in an all-electric UAS. However, by providing generally lower voltage (usually around 5 V) to the receiver, servos, avionics, and payload gimbal/sensor(s) independent of the electric energy provided to power the motor, the use of a BEC eliminates the need for an additional battery (or batteries) dedicated to non-powerplant components.

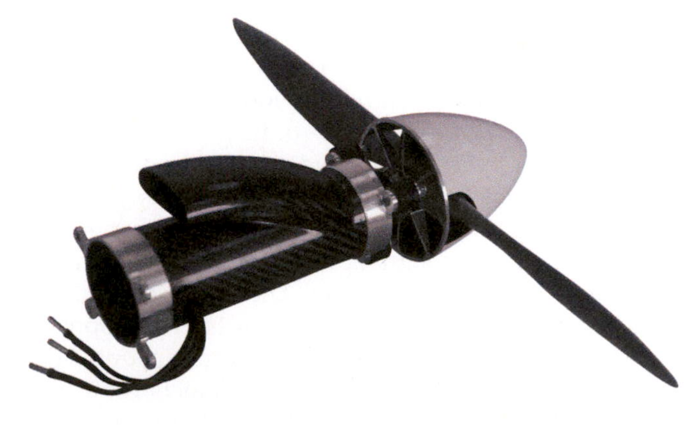

FIGURE 11.5
Carbon fiber forward-facing air scoop and carbon fiber cooling shroud installed around Penguin BE inrunner BLDC motor and gear reduction. (Image courtesy of UAV Factory.)

Lower capacity BECs, for example, those rated at approximately 5 amps, or less, generally use a less expensive, linear voltage regulator, whereas those having higher current ratings use a more efficient switch-mode regulator. (Linear and switch-mode regulators are subsequently discussed in detail.) Regardless of whether the BEC is a discrete component or integral to the ESC, the two components are electrically connected in parallel, drawing power for the entire electrical system from a single source battery.

Because the permanent magnets of an inrunner BLDC are located on the output shaft of the motor, the rotating mass is located close to the axis of rotation and the rotor, in comparison to that of an outrunner design, and tends to spin considerably faster. (For an explanation, the reader can review the discussion in the section "The law of conservation of angular momentum in the helicopter aerodynamics" provided in Chapter 10.) Consequently, these may be suitable to directly drive a small rotor in a ducted fan propulsor, or will require gear reduction to reduce rpm and increase torque to rotate a larger propeller. Inrunners may be slightly more efficient than outrunners, which can contribute to improved endurance and range. Another advantage of the inrunner design is the compactness associated with a housing of smaller diameter, which can result in reduced fuselage diameter and less drag. On the other hand, outrunners also afford advantages to the UAS designer. Because the permanent magnets of outrunner motors are located in the housing, the propeller or rotor is mounted directly to the rotating motor case. In comparison to an inrunner, this characteristic creates a much greater rotating mass located further from the center of rotation, which means that outrunners spin more slowly while producing greater torque. No gear reduction is generally necessary. Because the outrunner stator is interior to the motor, the windings and internals can be sealed to protect them from the environment. Moreover, outrunner windings are not subjected to centrifugal loading that may extend motor life, and air cooling of the windings is often not necessary, since heat can be dissipated by conduction. Both inrunners and outrunners provide the propulsive force for sUAS. For example, both the AeroVironment RQ-11 Raven (Figure 11.6) and UAV Factory Penguin BE use integral planetary gear reduction systems incorporated into inrunner propulsors, whereas the Aeryon Labs Scout quadcopter uses four outrunners to produce both lift and thrust.

11.3.2 Non-Electrically Powered sUAS

Small unmanned aircraft are also powered by propulsors consuming hydrocarbon or alcohol-based liquid fuels. Although the onboard electrical source for the autopilot, avionics, and payload may be a Ni–Cd, Li–Po, or Li–ion battery and as sUAS become larger with more powerful internal combustion (IC) engines, the installation of a generator as a power

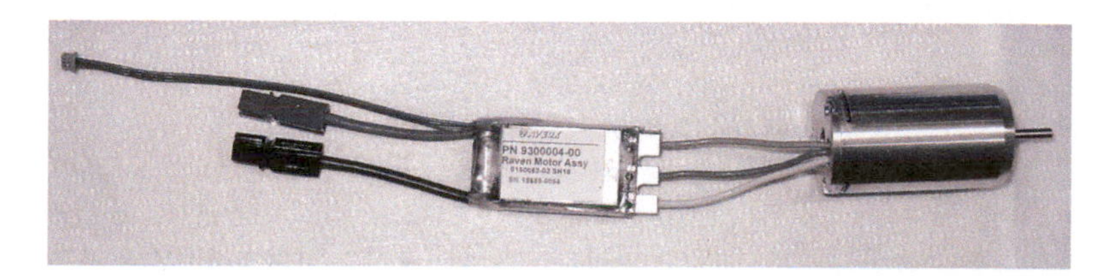

FIGURE 11.6
Aveox 27/26/7-AV inrunner BLDC motor with integral gear reduction and ESC as installed on the RQ-11 Raven. (Image courtesy of Aveox, Inc.)

source becomes more attractive. At the same time, the complexity and sophistication of the electrical system will also likely increase as the aircraft size becomes larger.

Powering UAS electrical systems from an electrical generator becomes attractive, where an IC engine is powerful enough to provide maximum propulsive force while having the additional capacity to drive a generator capable of delivering sufficient electrical power to energize all electrical components under the most adverse circumstances (with a margin of safety in the form of additional reserve capacity). A generator converts some of the horsepower developed by the engine to power for the electrical system by providing relative motion between a conductor (a winding consisting of multiple loops or turns of wire) and a magnetic, usually electromagnetic, field. All generators produce an alternating current output that must be either mechanically or electronically rectified to DC. A conventional DC generator is much like the brushed motor previously described. The stationary field windings (collectively known as the stator) are energized with a DC to produce the magnetic field in which the armature rotates. Brushes riding on commutator segments convert the output to a DC. (Most often, enough residual magnetism exists in the field poles, about which the stator is wound, to excite the armature when the generator shaft is first rotated. If the magnetism is insufficient, the field must be "flashed" from a suitable DC source such as an automotive or aircraft battery for output to be achieved.) In fact, the construction of a DC motor and generator is so similar that a motor can be used as a generator simply by energizing only the field windings and taking electrical power from the brushes. The familiar alternating current (AC) generator, commonly known as an alternator, creates a magnetic field in the windings of the rotor. (Electrical energy is transferred to the rotating armature via brushes and slip rings, so the direction of current flow through the windings is consistent in polarity and intensity.) This field (typically) sweeps across three discrete stator windings that produce a three-phase AC output which is rectified (electronically commutated) by a diode bridge circuit coupled to a low-pass filter to produce a very clean DC output. Though more complex, alternators are generally lighter and more efficient than the older DC generator design.

Regardless of the power source, whether battery or generator, voltage regulation is necessary. As previously mentioned, where current demands are not great (~5 amps or less), the smallest of the UASs may use linear voltage regulators. Although generally an integrated circuit, linear voltage regulators function in a way analogous to a voltage divider network wherein the desired voltage is taken from the appropriate divider reference point and any excess is dissipated as heat. This is very inefficient, and the electrical efficiency of linear regulators is typically only around 40% and may fall to as little as 14% (Dimension Engineering 2014). Linear regulators generate a lot of heat, particularly where current flows are elevated, and may require a large heat sink, which will increase the overall size of the installation. On the other hand, linear regulators are simple, inexpensive, and, where current flows are low, acceptably inefficient. For higher current flows, switch-mode regulation, which maintains the desired voltage by switching on and off, is a better option. Power is mainly consumed during the switching operation, but since the regulating section (e.g., pass transistor) is mostly energized or de-energized, very little power is wasted. In comparison to a linear regulator, a switch-mode regulator may be 85% efficient, delivering more than twice as much useful power for given voltage (Dimension Engineering 2014). The generator output of larger sUAS is controlled by a solid state device known variously as a generator control unit (GCU) or generator power unit (GPU), among other terms. Because the acronym GPU could be confused with that representing ground power unit, I prefer GCU and will use generator control unit in the text of this chapter. The GCU regulates voltage and generally performs other functions, which may include reverse current protection and a means to prevent excessive current flows. Where some of the generator output is used to charge an onboard battery, the GCU output must be greater than battery voltage to overcome the battery's internal resistance and provide the energy for the electrolytic chemical reactions that will cause the battery's active material (and possibly electrolyte) to be restored (i.e., charge the battery). (This was discussed earlier in the chapter.) The GCU also provides power to a distribution bus or individual components (e.g., avionics and payload), depending on how the electrical system is designed, at the proper voltage. The reverse current function of the GCU is to protect the generator when battery voltage may be higher than generator output. This can happen, for example, at low engine rpms. Because the battery is connected to the generator through the control unit, a high battery voltage will pass through the generator rotor and stator windings and tend to drive the armature like that of a motor. This condition, known as "motoring" or "generator motoring," may cause the generator which is, at the same time, being driven by the power plant to become damaged.

As the sUAS electrical system becomes more complex, the use of an electrical bus, or common connection point for wires and multiple circuits, becomes more desirable. All circuits must convey electrical energy to power the component and provide a return path to the source. The smallest of sUAS use wire for the entire circuit, from the power source to component and back to the source. This is referred to as dual-wire system to differentiate this type of circuitry from the single-wire system used to save space and weight and to simplify the installation on those aircraft of conventional aluminum design. However, because the electrical system is the nexus or interconnecting subsystem for all UAS electrical components, as the airframe becomes larger and the number of subsystems becomes greater, undesirable attributes tend to develop in the overall electrical system which must be resolved. The amount of wire required in a dual-wire system adds increasingly to the weight of the aircraft and tends to morph into an inscrutable tangle of unidentifiable conductors—a circumstance that renders troubleshooting exponentially more difficult. In a conventional, semi-monocoque design, the structure can be used to eliminate 50% of the wiring required for the electrical system. This is one advantage afforded in the construction of the structure of very large UASs, such as the Global Hawk (Figure 11.7), being either partially or entirely made of metal. The use of electrical buses can also reduce the

FIGURE 11.7
The Northrop Grumman RQ-4/MQ-4 fuselage is of semi-monocoque aluminum construction. (U.S. Air Force image.)

amount of wire required for the circuit and, at the same time, provide a safer (e.g., reduce the likelihood of EMI and the chafing or abrading of insulation that produces a short to ground or other conductor) and better organized method of routing wire. In smaller sUAS, electrical buses may be a simple terminal strip, identical to those found on small manned aircraft, with copper or aluminum bus bars installed across the terminal studs to interconnect the circuits. Where current flow is greater, a bar of copper or aluminum can provide a common connection point. Connecting circuit protection devices (e.g., fuses, circuit breakers, and fusible links) to the bus will limit current flow in the event of a short or ground fault—a condition that will generate heat and possibly melt wire insulation to cause a fire. To protect the maximum length of the conductor, this circuit protection should be connected as nearly as possible to the source of power, which is often at the bus. A small distribution bus may be connected across one side of multiple circuit breakers to act as a common connection and reduce the amount of wire necessary in the circuitry. The purpose of including circuit protective devices in the electrical system is not to prevent damage to the electrical devices that are installed in the UAS. Where circuit components must be protected from excessive current flow, an internal fuse or circuit breaker is used to protect the component (e.g., the UAV factory installs an internally accessed slow-blow fuse in series with the power lead to the heater controller of the Penguin's heated pitot/static probe). A well-designed electrical system, particularly appropriate to the larger UAS, isolates flight-critical and noncritical circuits from each other. This is accomplished by powering two or more subsystem buses, often termed essential and nonessential buses, from the main distribution bus. A large, slow-blow fuse, known as a current limiter, may be used to connect nonessential buses to the main bus. In this way, a fault in the noncritical-to-flight circuitry causing excessive current flow will "blow" (i.e., melt) the fuse to open the circuit. Current flow in the damaged circuit will cease preventing an electrical fire while allowing the critical circuits required for flight to remain energized.

11.4 Electrical Systems for Large UASs

The payloads, avionics, and communication systems of large UASs can consume substantial amounts of power, and the electrical systems of these aircraft share much in common with their manned counterparts. For example, as do many manned aircraft, the General Atomics Predator B (aka MQ/RQ-9 Reaper) is powered by a Honeywell TPE-331 which drives a starter/generator to power the ship's electrical systems. Operating as a DC motor while starting, both armature and field windings of a conventional starter/generator are energized to rotate the engine compressor through the accessory gearbox. As the hot gasses of the working fluid are generated, the starter is increasingly unloaded as the turbines provide greater torque to take the engine beyond its self-acceleration, or self-sustaining, speed. At this point, the starter can be de-energized. As the fuel metering system provides sufficient fuel to the hot section to complete acceleration of the engine to idle, the field windings of the starter generator can be re-energized to create the electromagnetic field in which the de-energized armature windings rotate. Armature output is mechanically rectified through a commutator and brushes to provide a DC electrical output to the main power bus. If the operation of payload, communications, or avionics devices requires AC, solid state inverters will convert the DC and deliver the alternating current at the desired voltage and frequency to one or more AC buses. The use of a starter/generator saves considerable weight (and space on the accessory gearbox), while reducing the overall complexity of the accessory drive system. Further, because only one component is constantly driven (instead of both starter and generator) by the power plant, greater power is available to drive other accessories or for propulsion. Recently, several companies have developed brushless permanent magnet generators and starter/generators specifically for UAS applications, for example, Northwest UAV for sUAS applications and Innovative Solutions for large unmanned aircraft. Under an Air Force SBIR/STTR (Small Business Innovation Research/Small Business Technology Transfer) contract, Innovative Power Solutions developed a brushless starter/generator (BS/G) for the Northrop Grumman RQ-4 Global Hawk/EuroHawk, powered by the Rolls-Royce AE3007 (aka F137). These newly developed BS/G designs require less maintenance and are even lighter than the conventional starter/generators still found on the majority of smaller gas turbine power plants.

In comparison to sUAS, a greater number of factors influence the determination of wire size for larger unmanned aircraft. As with sUAS, the selection of the proper wire diameter is guided by the length of the wire run, anticipated peak current flow, and acceptable voltage drop from bus to ground. However, other factors, including aircraft service ceiling, number of wires installed in a bundle or conduit, and intermittent operation may also affect the choice of wire selected for use in the electrical systems found on large UASs. The selection of wire with sufficient diameter is essential to ensuring proper circuit operation, providing adequate component service life and precluding the creation of excessive heat, which can lead to in-flight fires. On the other hand, using the smallest possible wire with these desirable attributes would save weight and space, while increasing range, endurance, and payload carrying capacity. Thus, determination of the correct wire size is best achieved through the Goldilocks method of wire selection—not too small and not too large. Again, the FAA publication AC43.13-1B, among other sources, provides information, formulas, and guidance on selecting wire, circuit protection and appropriate insulating materials, and on proper routing and EMI protection of aircraft electrical circuits while considering all of the foregoing factors. Although this publication is intended for use in designing and maintaining manned aircraft, its contents are also appropriate for a UAS. The electrical systems and components of a UAS may even be installed in a pressure vessel, similar to that designed into commercial transport aircraft, pressurized by bleed air from the gas turbine power plant. This design not only provides a more benign operational environment, but also reduces the risk of arcing which can damage electrical components, and possibly under certain conditions, create an electrical fire resulting in the loss of the ship.

As UASs become larger, more capable, and more powerful and the complexity of unmanned aircraft increases, their onboard electrical systems will become increasingly similar to those of manned aircraft. Just as the pneumatic and hydraulic pressure/actuator systems of the Boeing 787 Dreamliner have been replaced by electrical systems, the same revolution is also occurring in the design of unmanned aircraft. According to Ralph Livingston, Chief Engineer of Abbott Technologies Inc., built-in test equipment for use in troubleshooting will soon be commonplace on UASs and "[p]neumatic and hydraulic actuators and controls [will become] … the dinosaurs of the unmanned aerial vehicle (UAV) world" (Livingston 2013, 1). The higher energy efficiencies afforded by replacement with electrical components of pneumatic/hydraulic system pumps, lines, and actuators result in reduced heat loads, cooling requirements, and increased UAS range and endurance (Livingston 2013). On the other hand, greater demands are correspondingly placed on the electrical system in terms of increased complexity and generating capacity that translates into larger, heavier generators and more circuitry—again, no free lunch.

Aside from technological advances, two other drivers of this homogenization will be the influences of regulation and industry standardization that are expected to increasingly come to bear on the design process. (Exogenous design factors were discussed in the previous chapter on UAS design.) It is likely that FAA regulations for manned aircraft, covering design, maintenance, operation, and manufacture will largely be applied to UASs. As an indication of this likelihood, consider the following: In September 2009, the FAA published the results of the Unmanned Aircraft System Regulation Review (DOT/FAA/AR-09/7), intended to examine the feasibility of integrating UASs into the NAS. The study findings were that 30% of existing manned regulations would apply directly to UASs without changes, 42% could be interpreted to apply, and 12% would apply once revised. That is, UAS maintenance, operation, construction, and design could be governed by 84% of existing regulations. Moreover, 44% of advisory circulars would be directly applicable, and the remaining 56% could be applicable through modification or interpretation. In other words, much of the regulatory structure covering the design, construction, and maintenance of electrical systems (and every other aspect of UAS operations) already exists. Moreover, trade groups and industry representatives are also encouraging the standardization of large (and small) UAS electrical systems to reflect those of manned aircraft. For example, one requirement of CFR 14, Part 23, applicable to manned aircraft, is that a load analysis be performed to ensure that the generating system is capable of producing adequate power to supply the needs of the electrical system with an extra design margin to provide for expansion and a level of additional safety. A similar requirement has been recommended by members of a prominent standards organization making recommendations to the FAA's UAS Aviation Rulemaking Advisory Committee (aka ARAC). Groups advising the FAA have made similar recommendations that UAS be held to the same standards as manned aircraft in multiple areas, including continued airworthiness, flight manual contents, risk management, certification, communications, weight and balance, markings, maintenance and structural integrity, among others. That manned aircraft requirements will similarly be mandated for those which are unmanned, particularly those among the largest of UASs, is likely. In fact, groups proposing industry-wide standards and those representing the UAS industry in advising the FAA frequently make recommendations regarding regulations and standardization, using terms such as "established," "appropriate," and "accepted" in reference to engineering and verification techniques. The use of such terminology and phrasing undoubtedly encourages similarities in UAS and manned aircraft systems designs.

11.5 Conclusion

The electrical system is the essential nexus of the unmanned aircraft, interconnecting avionics, payload, command and control, onboard computers, and receivers, among other components, into a synergistically holistic, integrated system. It forms the backbone of the UAS, providing the means of autonomous and remotely piloted flight, data acquisition, telemetry, and, in some cases, propulsion. The electrical systems of sUAS are sophisticated, but relatively straightforward in design. As the unmanned aircraft platform becomes larger and more complex, so, too, does the electrical system that becomes increasingly similar to those found on large, turbine-powered manned aircraft.

DISCUSSION QUESTIONS

11.1 Why can it be said that the electrical system is the ubiquitous nexus of unmanned aircraft systems?

11.2 List and define/describe the basic units of electrical measurement. Use these to write the formulae for Ohm's law and electrical power. Mention how the definitions of mechanical and electrical power are similar or related. Why is aircraft wire, used in both manned and unmanned aircraft, stranded? List other important characteristics associated with aviation wire and tell why each is significant. Describe the AWG system. Give the formula for determining wire resistance and elaborate the definitions of the unit and explain their meaning and position (i.e., whether a direct or inverse relationship) in the formula. What is meant by the term voltage drop? How does a voltage drop in the external circuit affect component operation and service life? Give examples of circuit protection devices. What is the purpose of these? With respect to location, where are circuit protection devices best installed. What is the difference between single-wire and dual- (or double-) wire circuits? Why are composite airframes considered to be dielectric (i.e., define dielectric and relate this to UAS airframe construction), and how does this type of construction affect UAS electrical system design? Define EMI/RMI and tell why this is significant to dependable and safe UAS operation. Give strategies for mitigating or eliminating EMI, and tell why these are effective. How are high-pass and low-pass filters constructed? Which would most likely find application in UAS airframe circuit construction and tell why you believe your response is correct.

11.3 What has allowed sUAS to proliferate in the current technological environment? Identify some of the enabling technologies. Discuss the interaction of design decisions, based on mission goals, which affect UAS power plant and, correspondingly, electrical systems.

11.3.1 Identify and describe the all-electric unmanned aircraft that flew in 1957. What is the significance of this flight?

11.3.1.1 List the types of batteries commonly providing electrical energy for UAS ground and flight operations and describe cell chemistry and construction. Give nominal cell voltages for each type of battery. Define battery capacity and tell how this relates to the "C-rate." What factors determine total battery voltage and capacity? What is a 2S battery? What is the nominal voltage of a 3S Li–Po battery? What is meant by "energy density" or "power density?"

11.3.1.2 Describe, in a general way, the operation of an electric motor and the operation of brushed and brushless motors. What is commutation? What is the difference between electronic and mechanical commutation and on what types of motors would these be found? How is rpm differently controlled on brushed and brushless DC motors? Describe inrunner and outrunner BLDC motors. Compare and contrast these, giving advantages and disadvantages of each. What is the purpose of a BLDC motor controller, an ESC, and a BEC? Provide examples of sUAS using inrunner and outrunner BLDC motors and give ways in which these installations would differ.

11.4 List types of non-electric powerplants available to the UAS designer, giving advantages and disadvantages of each choice (a review of Chapter 10 may be helpful). How will the choice affect the electrical system design? Describe, compare, and contrast linear and switch-mode voltage regulators. What is a GCU and what functions does it perform? Define the term "motoring."

11.5 Describe the starter/generator as installed in the TPE-331 turboprop powering the RQ/MQ-9 Reaper. What recent innovation developed for the Global Hawk/EuroHawk is likely to supplant the more conventional brushed starter/generator? What advantages does the Goldilocks method of wire selection offer?

References

Austin, R. 2010. *Unmanned Aircraft Systems: UAVS Design, Development and Deployment.* Chichester, UK: John Wiley & Sons.

Büchi, R. 2012. *Brushless Motors and Controllers.* Norderstedt, Germany: Herstellung und Verlag.

Dimension Engineering. 2014. *A Beginner's Guide to Switching Regulators.* Under "What's wrong with a linear regulator?" Accessed July 27, 2014, https://www.dimensionengineering.com/info/switching-regulators.

Fahlstrom, P.G. and T.J. Gleason. 2012. *Introduction to UAV Systems.* Chichester, UK: John Wiley & Sons.

Federal Aviation Administration (FAA). 2012. *FAA Aviation Maintenance Technician Handbook—Airframe.* Newcastle, WA: Aviation Supplies and Academics.

Gundlach, J. 2012. *Designing Unmanned Aircraft Systems: A Comprehensive Approach.* Reston: American Institute of Aeronautics and Astronautics.

Livingston, R. 2013. *Intelligent Aerospace.* Important considerations in designing electrical power systems for unmanned aircraft. Last modified January 13, 2014. http://www.intelligent-aerospace.com/articles/2014/01/uav-electrical-power.html.

Logan, M.J., J. Chu, M.A. Motter, D.L. Carter, M. Ol, C. Zeune. 2007. Small UAV research and evolution in long endurance electric powered vehicles. In *Proceedings of AIAA Infotech @ Aerospace 2007 Conference and Exhibit,* Rohnert Park, California, AIAA Paper 2007–2730, 7–10 May, 2007.

Noth, A. 2008. *History of Solar Flight.* Autonomous Systems Lab. Zurich: Swiss Federal Institute of Technology. Accessed July 15, 2014. http://www.asl.ethz.ch/research/asl/skysailor/History_of_Solar_Flight.pdf.

12

Communication Systems

Saeed M. Khan

CONTENTS

12.1 Introduction

A simple and practical way to introduce UAS communications to the newcomer is through the concept of data links. Data links convey vital information to and from the unmanned aerial vehicle (UAV) and the ground control station (GCS), wirelessly in most cases. This information or data are used for controlling the UAV manually or through automation by manipulating control surfaces and throttle. Figure 12.1 illustrates UAV–GCS communication links for a small commercial remotely piloted vehicle. The link is also used for downloading captured images and telemetry data among other things. Enhanced commercial systems may employ the use of satellite links such as the Inmarsat Network which allows UAV and GCS to take advantage of a global IP network (Wagenen 2015, p. 1). For military UAS systems additional complexity and functionality is required in the design of the data link systems stemming from need to access the World Wide Web;

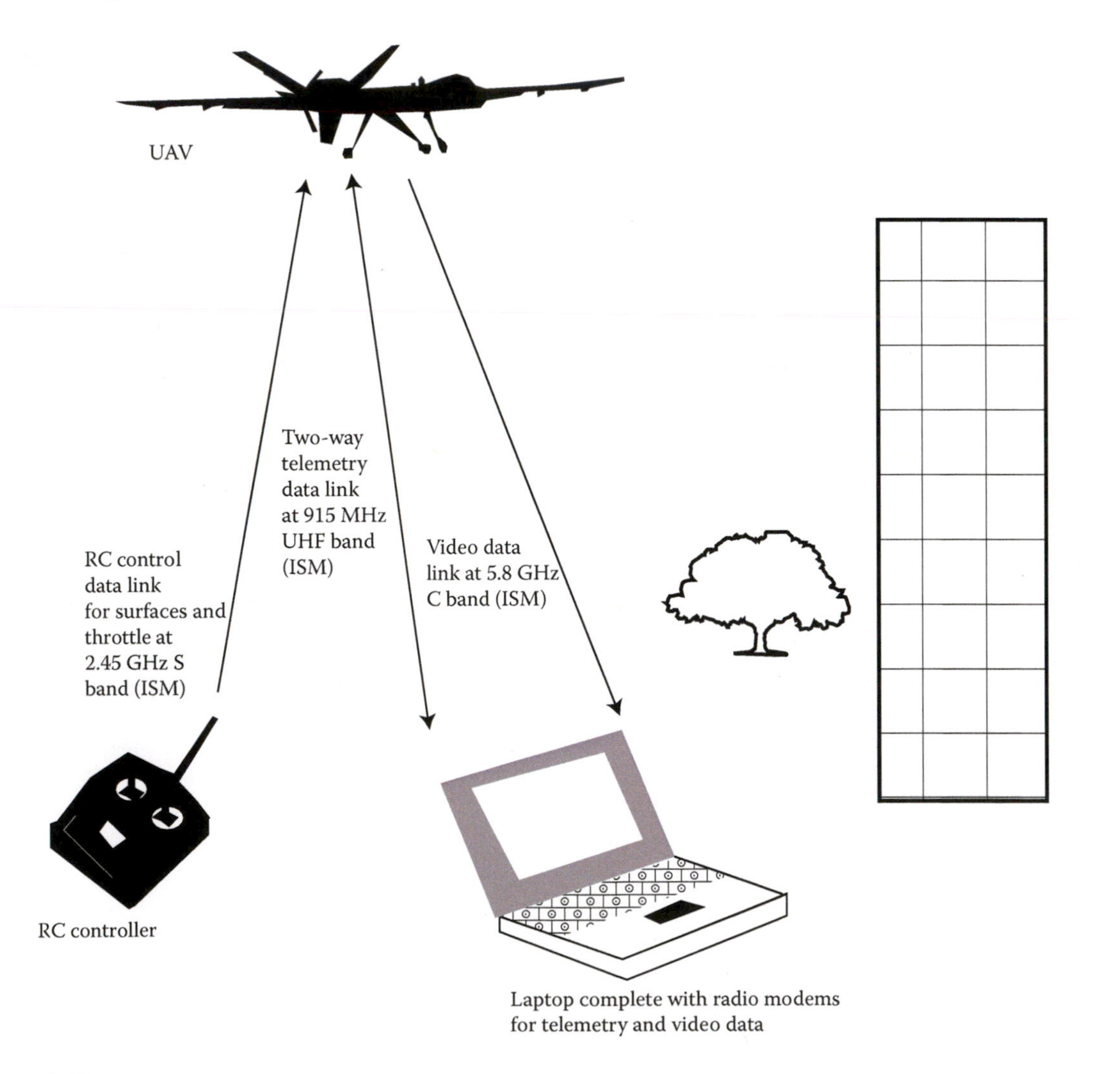

FIGURE 12.1
A UAS data link system for commercial applications.

resist unintentional interference; lower the probability of interception; security considerations; resistance to deception; and anti-jam capabilities (Fahlstrom and Gleason 2012, pp. 191–204). In other words, the functionality of these links makes them the lifeblood of any UAS system.

A commonly used classification for data links is line-of-sight (LOS) and beyond-line-of-sight (BLOS). In LOS systems the UAV is always in line of sight with the GCS. BLOS systems may involve other links, such as a satellite that is in line of sight to one of the parties (UAV/GCS) but not both. Communication links can also be characterized as being point-to-point or point-to-multipoint (Figure 12.2). An interesting link is one that forms between a swarm of unmanned vehicles where they can talk to each other while maintaining connection to a ground station. While the link architecture can take many more complex forms, they are usually derivatives of the ones that have been discussed.

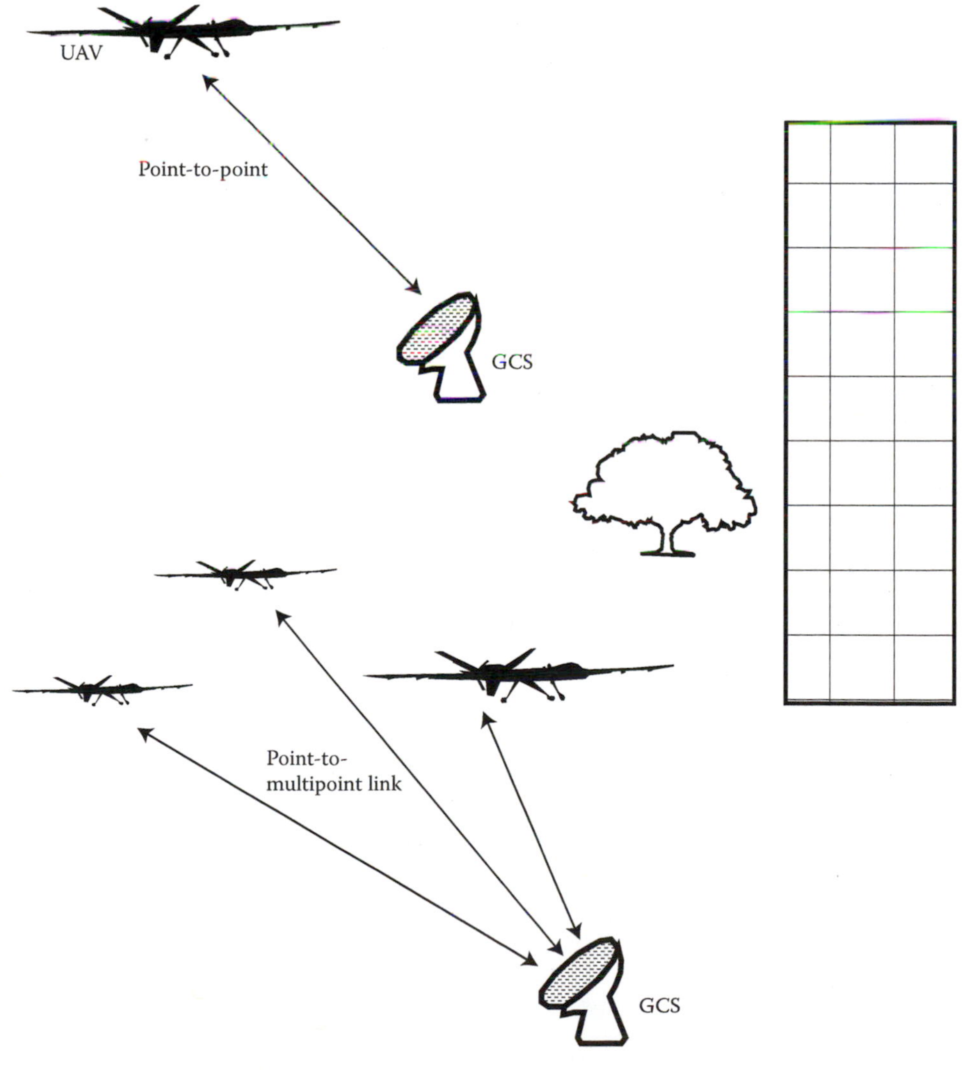

FIGURE 12.2
Point-to-point and point-to-multipoint links.

12.2 Electromagnetic Wave Propagation in Free Space

How is wireless communication even possible in free space, given that it excludes the possibility of a guiding system such as a two-wire line or a coaxial line? The answer of course is through *radiation* of electromagnetic waves. Radiation can come from many sources caused by time-varying electromagnetic fields. Undesired radiation can come from electronic circuits and appliances of all sorts, but properly designed *antennas* serve as highly efficient sources for controlled radiation.

An antenna radiates efficiently when electromagnetic energy is guided through a transmission line under matched conditions (meaning most of the incoming energy is transmitted to the antenna with little reflection taking place at the antenna-transmission line interface). Time-varying voltages and currents in a transmission line carry set up electric fields and magnetic fields. The electrical fields start with positive charges and end in negative charges, and magnetic fields are formed around current-carrying conductors. These *fields do not end at the antenna but have been observed to be radiating into free space outside.* The question then arises as to what holds these fields in place without the presence of charges or how these guided waves are detached from the antenna. One can conclude that while the *charges are required to excite these fields they are not required to sustain them.* As an analogy to this phenomenon it has been observed that a pebble when tossed into a pool of water will create ripples long after it has settled in the bottom (Balanis 2005, pp. 7–16). Some commonly used types of antennas include the horn, dipole, monopole, spiral, and patch antennas (Figure 12.3).

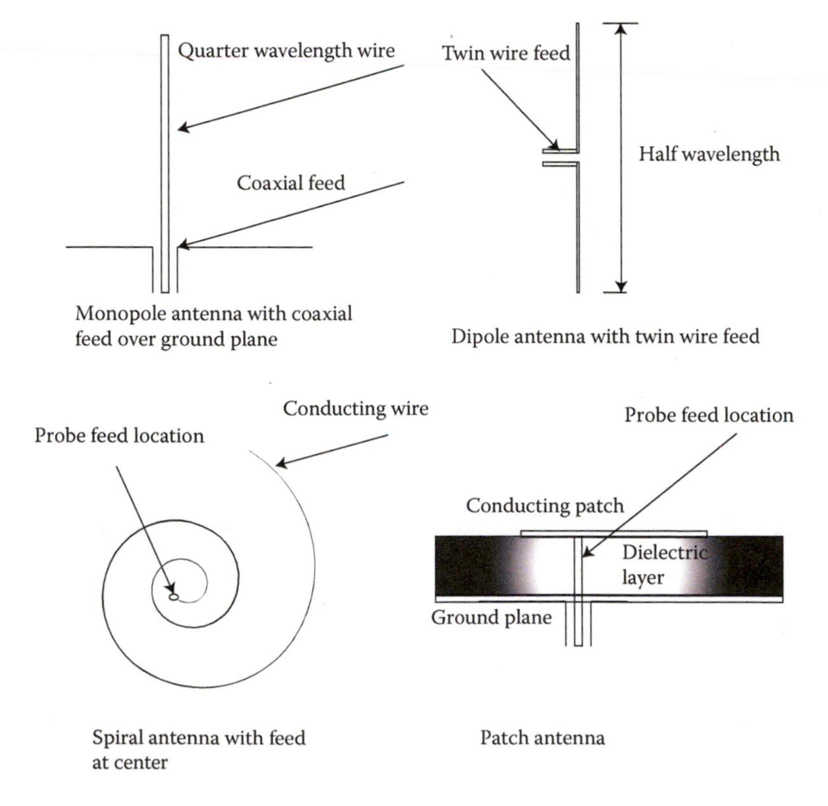

FIGURE 12.3
Common antenna structures.

Another important point about the radiated wave is that it may arrive at the receiver via not only a direct path but also other paths (*multipaths*) due to reflection, refraction, and diffraction. Reflection waves are caused when a signal bounces off objects before reaching the receiver while refracted waves are caused when the waves pass through different media on their way to the receiver. Refracted waves can cause loss in signal amplitude while traveling through lossy media. Diffraction is a phenomenon that occurs when a wave bends around a corner. Figure 12.4 is an illustration of these different wave-propagation phenomena. Multipath signals can lead to both destructive and constructive interferences with the direct signal. Generally speaking multipath signals that arrive almost at the same time as the direct signal are more harmful and can cause more errors in the information received.

For appreciable radiation to take place effective lengths of the devices/circuits/antennas should be comparable to a wavelength. Remembering that the relationship between wavelength (λ) and the frequency (f) in free space is simply ($\lambda = c/f$, where c is the speed of light in free space or 3×10^8 m/s) at lower frequencies, both the wavelength and the radiator get to be very large, making it unsuitable for wireless communication; for example, at a frequency of 1 kHz, the free space wavelength is 300,000 m and at that size it is very difficult to construct an efficient radiator of appropriate size (~λ/4 and greater).

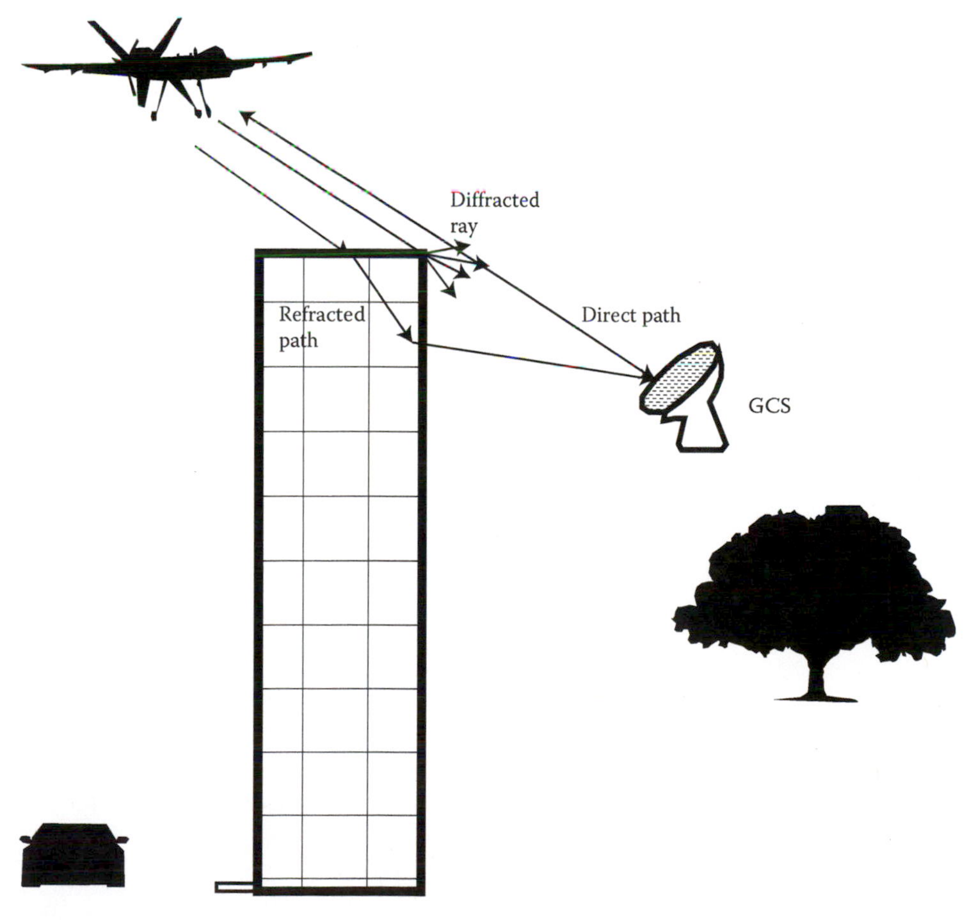

FIGURE 12.4
Obstructed direct path, refracted path, and edge diffraction of electromagnetic wave.

12.3 Basic Communication System and Its Elements

While there are many different types of data links there are certain elements that they all need to possess. All these elements play an important role in the design of the system whether this system is civilian or military; LOS or BLOS; point-to-point or point-to-multipoint; they all share these commonalities. We will begin by taking a look of at the basic elements of any electronic communication system and expand our discussions on each of these elements and for the special case of UAV–GCS communication.

Figure 12.5 shows a typical wireless communication system where communication takes place in only one direction from transmitter to receiver (aka *simplex system*). While the transmitter and receiver are separate entities in Figure 12.5, it is quite common to have systems where each unit is capable of both transmitting and receiving (aka *full duplex system*) at the same time with a device known as a duplexer which isolates a receiver from the transmitter and allows the use of a common antenna. Since transceiver architecture is basically composed of a receiving path and a transmitting path that contains the same basic elements with the exception of the duplexer, a separate discussion is not required at this stage.

12.3.1 Modulation

Modulation combines information with a carrier signal. The first section on the transmitting side holds the information source which can be analog or digital. Analog is converted

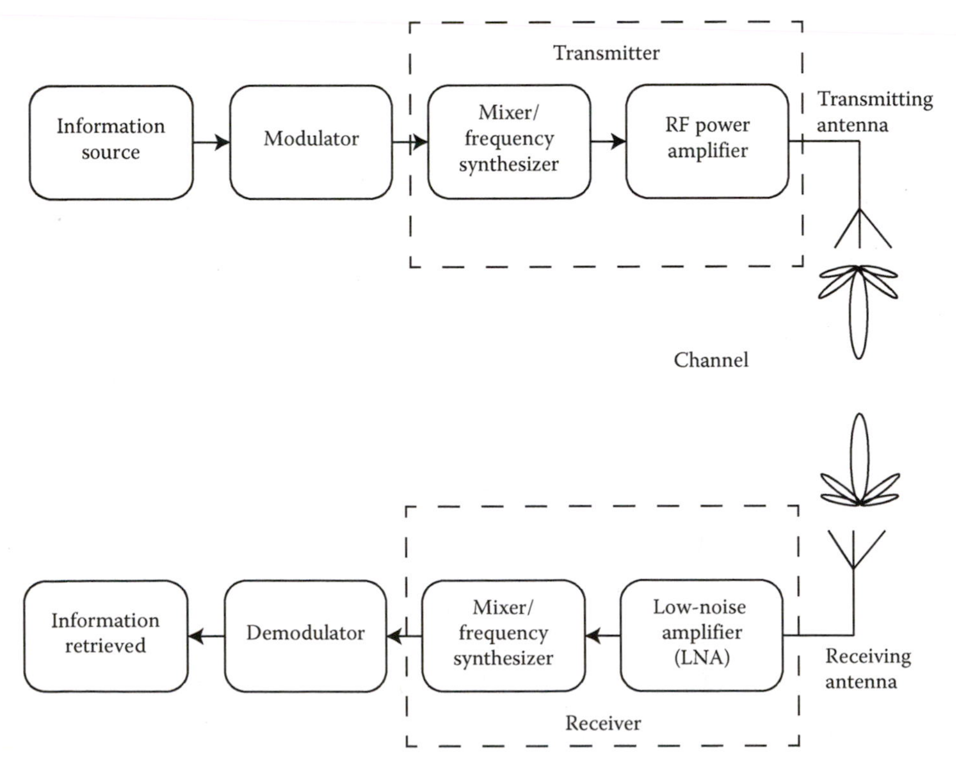

FIGURE 12.5
A simplified block diagram of a communication system.

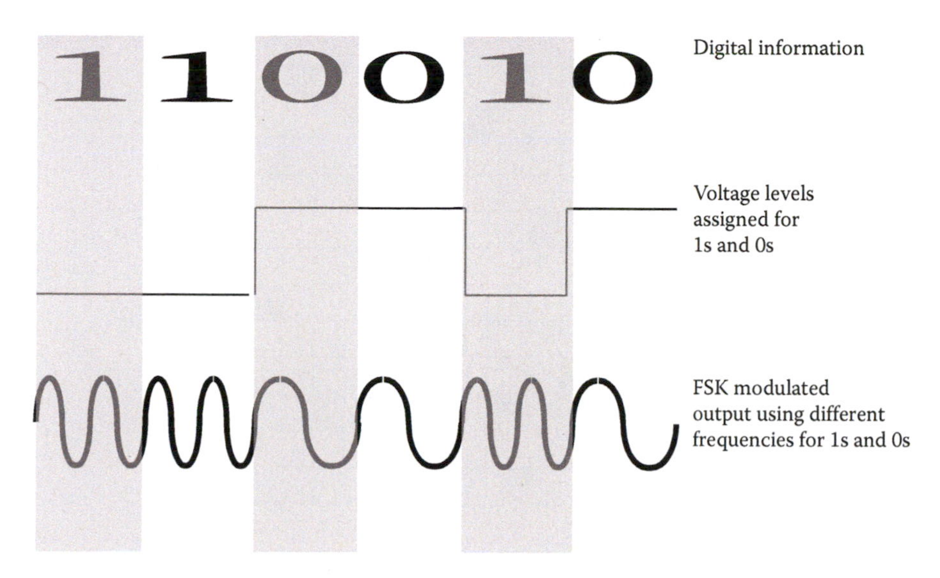

FIGURE 12.6
Frequency shift keying (FSK) modulation of digital data.

to digital data if digital data are required by the *modulator* which is the next section following the information source. Commonly, transceivers use the term *modem* to identify a system block that is able to serve both as a modulator and demodulator in such systems.

For the special case of frequency shift modulation (FSK), modulation that is widely used in UAV–GCS communication, all data need to be in digital form. Figure 12.6 shows how this particular type of modulation can be accomplished. First the digital data are converted to two different voltage levels. In this case, a low voltage level indicates a digital "1" and a high voltage level indicates a digital "0" (although the opposite is also possible). The FSK modulation uses two different frequencies to modulate the input digital data stream with the higher frequency being used to modulate 1s and a lower frequency to modulate 0s.

12.3.2 Transmitter

The transmitter takes in the modulated signal and outputs an amplified radio frequency (RF) signal transmission, for example, transmitter takes the FSK modulated signal and outputs an RF waveform. An RF power amplifier in the transmitter section delivers the signal to the antenna for transmission. In the special case of UAV–GCS using FSK modulation an added complexity is thrown in, that is, the output center frequency is not kept constant but hops around. This is discussed in the following section.

12.3.2.1 Frequency Hopping Technique for Transmission

The frequency hopping technique is what is known as a spread spectrum technique because it changes frequency of transmission in a pseudorandom fashion as if to spread the energy over a larger bandwidth. The key idea behind any spread spectrum technique comes about from trying to make it practically impossible for others to listen in. In the frequency hopping technique, an unauthorized person, who is trying to hear a conversation, would pick up unintelligible blips since the frequency changes in a manner that only allows people with knowledge of the order of the frequencies of transmission (Figure 12.7)

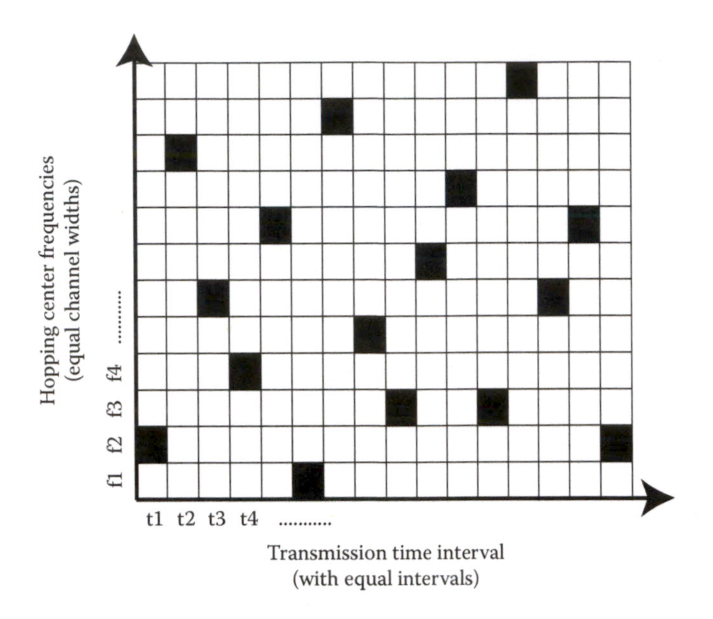

FIGURE 12.7
Frequency hopping scheme using pseudorandom frequencies.

to listen in. Blackened squares in the figure indicate how the center frequency varies over time. Someone with an intention to figure out this pseudorandom pattern may have to wait a long time. In fact, sometimes these patterns can be designed to go on for years without repeating. Another important advantage is that the noise from the channel (air for wireless) is also spread out over the entire band making a better distinction between the signal and noise possible. Frequency hopping is accomplished in the transmitter by mixing the modulated FSK signal with a frequency synthesizer that changes randomly.

12.3.3 Channel

The connection between the receiving and transmitting antennas is made through the channel for the wireless link. Assuming good atmospheric conditions, the amount of power transferred to the receiver using a LOS link in air depends on the gain of the two antennas, the polarization of these antennas, and the distance between them. Important topics of directivity, gain, and polarization of an antenna are discussed next.

12.3.3.1 Antenna Directivity

Before we can discuss the directivity of an antenna, we need to define an isotropic antenna. Basically an isotropic radiator is one that radiates equally in all directions. It is a hypothetical concept, one that does not exist in the real world since it requires the existence of a point-source antenna that is impossible to build. Figure 12.8 is a graphical representation of the radiation properties (aka radiation pattern) for a directive antenna and the hypothetical isotropic radiator.

One also needs to understand the concept of radiation intensity of antenna to define directivity. From the IEEE standard definition of terms for Antennas (IEEE 1983), in a given direction, the power radiated from an antenna per unit solid angle is its radiation intensity,

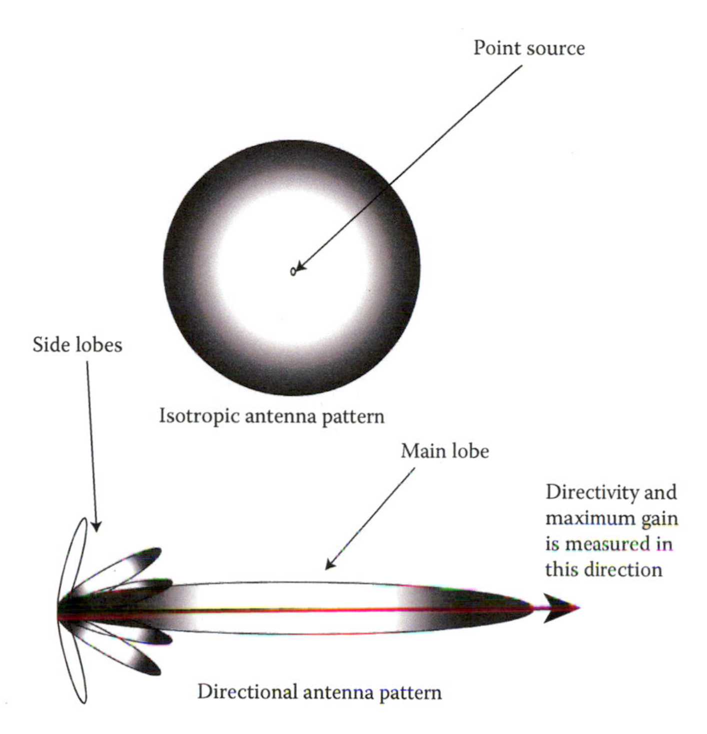

FIGURE 12.8
Radiation pattern of isotropic and directional radiator.

or in simpler terms, the amount of power radiated by the antenna in a certain direction far enough away from the antenna so as not to impact its electromagnetic characteristics (far field). The radiation intensity remains independent of distance in the so-called far field of the antenna (MTI, p. 1).

All antennas have some directional properties and they radiate better in some directions than others. If it is assumed that both the antenna and the isotropic radiator radiate the same amount of total power, then the directivity of the antenna is the ratio of the maximum radiation intensity to the radiation intensity of reference antenna such as an isotropic radiator. Moving forward D_0 (dimensionless) will be used to refer to the directivity of an antenna.

12.3.3.2 Antenna Gain

The antenna gain is *simply, the product of the directivity with efficiency which accounts for cable, connector losses, and dielectric losses* (Sevgi 2007, p. 212). Gain and directivity refer to maximum gain and maximum directivity unless stated otherwise. *Moving forward G_0 (dimensionless) will be used to refer to the maximum gain of an antenna. The terms conduction efficiency (e_c) and dielectric efficiency (e_d) are difficult but they can be experimentally determined. Based on our definition the following relationship exists between directivity and gain.*

$$G_0 = e_c e_d D_0 \tag{12.1}$$

It is very common to provide the gain in decibels,

$$G_0(\text{dB}) = 10\log_{10}(G_0) \tag{12.2}$$

When the reference antenna is an isotropic one, dBi is sometimes used in the place of dB. Thus, Equation 12.2 may be rewritten as

$$G_0(\text{dBi}) = 10\log_{10}(G_{0i}) \tag{12.3}$$

12.3.3.3 Antenna Polarization

Electromagnetic (EM) waves produced by antennas can have linear, elliptical, or circular polarization associated with them (Figure 12.9). The *amount of power transferred in the antenna link through the channel depends on how well matched the polarization of the receiving and transmitting antenna are*; in that case when they are properly matched, optimal power transfer can occur.

All EM waves possess an electric field and a magnetic field direction. The electric field direction is considered linear polarization when the electric field vector (E-vector) associated with the EM wave maintains the same direction as it propagates through space. Dipole antennas and monopole antennas commonly used in UAV–GCS communication are examples of linearly polarized antennas. In the case of linear polarization the transmitting and receiving antennas must be polarized in the same direction for maximum power transfer.

In the case of elliptical polarization, the electric field vector traces out ellipses (Figure 12.9) in a plane perpendicular to the direction of propagation. The polarization in this case may

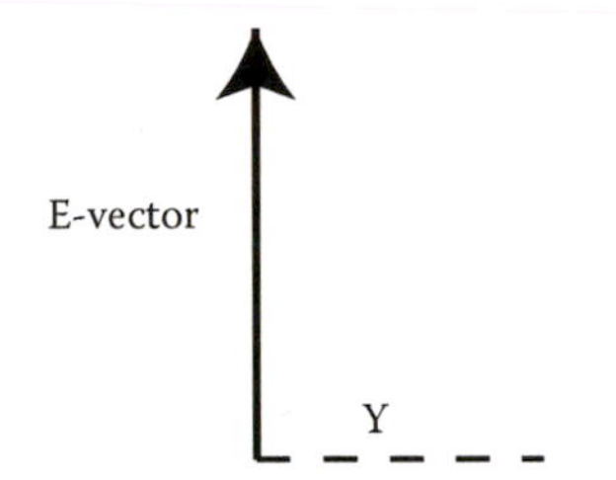

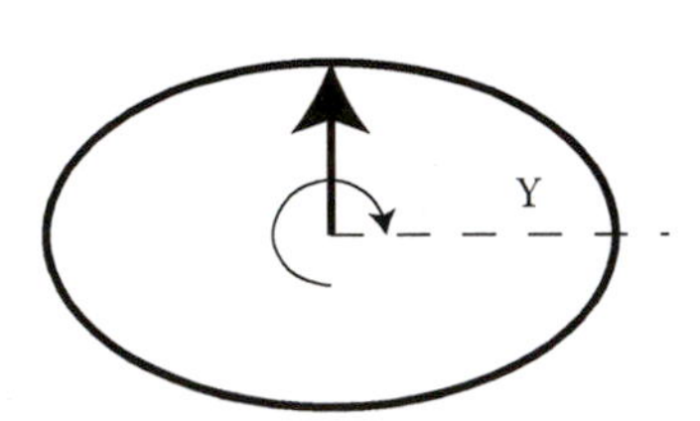

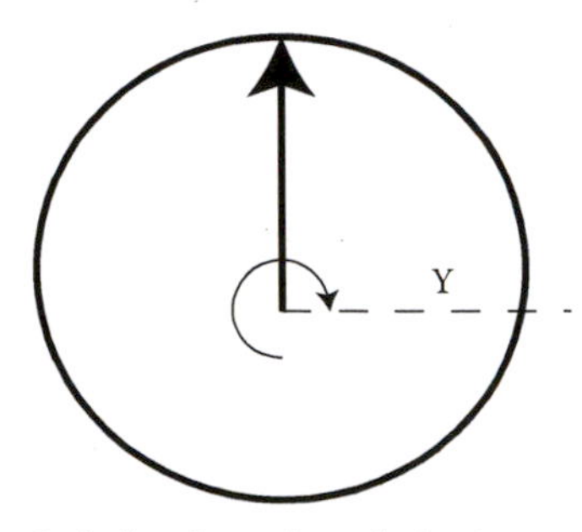

FIGURE 12.9
Linear, elliptical, and circular polarization.

be either right handed (clockwise) or left handed (counterclockwise). Circularly polarized antennas have E-vectors that trace out a circle in a right-handed or left-handed sense. In the case of elliptical or circular polarization it is not enough to have the matching polarization between transmitting and receiving antennas; the sense must match as well. As an example, circularly polarized patch antennas are commonly used in UAVs for receiving GPS signals from satellites.

12.3.4 Receiver

After passing through the channel the EM signal arrives at the receiver where it is converted by the antenna to electrical energy. While passing through the space between antennas the signal becomes very weak and needs to be amplified in a *low-noise amplifier* (LNA) very close to the antenna. For the signal to be picked up and demodulated by the receiver a certain signal-to-noise (S/N) ratio called the receiver sensitivity is required. Both these items are discussed below, along with the dispreading of the frequency hopped signal.

12.3.4.1 Signal-to-Noise Ratio

We begin our discussion about S/N ratios by taking a look at the sources of noise in the wireless communication system. First there are the electronic devices themselves known as *thermal noise* and is due to the thermal agitation of electrons. This noise sometimes known as white noise is independent of frequency. So for a system that permits B Hertz of electronic signals to pass through it (aka bandwidth) at a temperature of T Kelvin, the noise power (N) generated in watts is given by the relation

$$N = kTB \tag{12.4}$$

where k is the Boltzmann constant having a value of 1.38×10^{-23} joules per Kelvin. Note that joule is a unit for energy and is equal to the product of power in watts and time in seconds. Other items impacting noise are the mixing of two different signals causing unwanted energy production at sum and difference frequencies (or their multiples) called *intermodulation noise*.

Crosstalk, another form of noise, can impact a wireless communication system when antennas pick up unwanted signals from space, although highly directional antennas can avoid this for the most part. *Impulse noise* can occur from both natural and man-made sources and is experienced as irregular pulses and spikes. Lightning and pulses from electromagnetic pulse (EMP) weapons are examples of natural and artificial sources of impulse noise (Stallings 2007, pp. 89–91). An advantage of frequency hopping systems is that it has the ability to spread impulse noise across a broader bandwidth upon reception. The S/N ratio is simply the ratio of the signal power to the noise power, and more commonly this is given in a decibel form as follows:

$$\left(\frac{S}{N}\right)_{dB} = 10\log_{10}\left(\frac{S}{N}\right) \tag{12.5}$$

If only thermal noise was considered then,

$$\left(\frac{S}{N}\right)_{dB} = 10\log_{10}\left(\frac{S}{kTB}\right) = 10\log_{10}S - 10\log_{10}k - 10\log_{10}T - 10\log_{10}B \tag{12.6}$$

12.3.4.2 Receiver Sensitivity

The receiver is the minimum power required by the receiver to detect an RF signal and demodulate the data. Typically dBm is used to designate signal power levels as below in communication systems. The signal power (S) in dBm is calculated using the following formula:

$$(S)_{dBm} = 10\log_{10}\left(\frac{S\,\text{in milliwatts}}{1\,\text{milliwatt}}\right) \quad \text{or} \quad 10\log_{10}\left(\frac{S\,\text{mw}}{1\,\text{mw}}\right) \tag{12.7}$$

If the power associated with the noise is N, then,

$$(N)_{dBm} = 10\log_{10}\left(\frac{N\,\text{mw}}{1\,\text{mw}}\right) \tag{12.8}$$

The receiver sensitivity also depends on its clearance from the noise level. As an example in Wi-Fi LAN networks the noise level needs to be cleared by about 20 dB. So if the noise level in a room is –95 dBm, then the signal level should be higher than –75 dBm. If one knows the signal and the noise levels in dBm, then he can also compute the S/N ratio in dB by the following equation:

$$\left(\frac{S}{N}\right)_{dB} = (S)_{dBm} - (N)_{dBm} \tag{12.9}$$

12.3.4.3 Despreading the Signal

The frequency hopping FSK scheme requires the spread signal from the transmitter to be despread at the receiver. This is achieved using the mixer/frequency synthesizer combo that switches frequencies in the same pattern as the transmitter prior to demodulation. An added advantage of using this scheme in the mixer/frequency synthesizer combo is that the noise picked up by the EM wave passing through the channel (air in this case) is spread out across the entire system bandwidth by the dispreading process (Figure 12.10) thereby lowering the noise level of the receiving system.

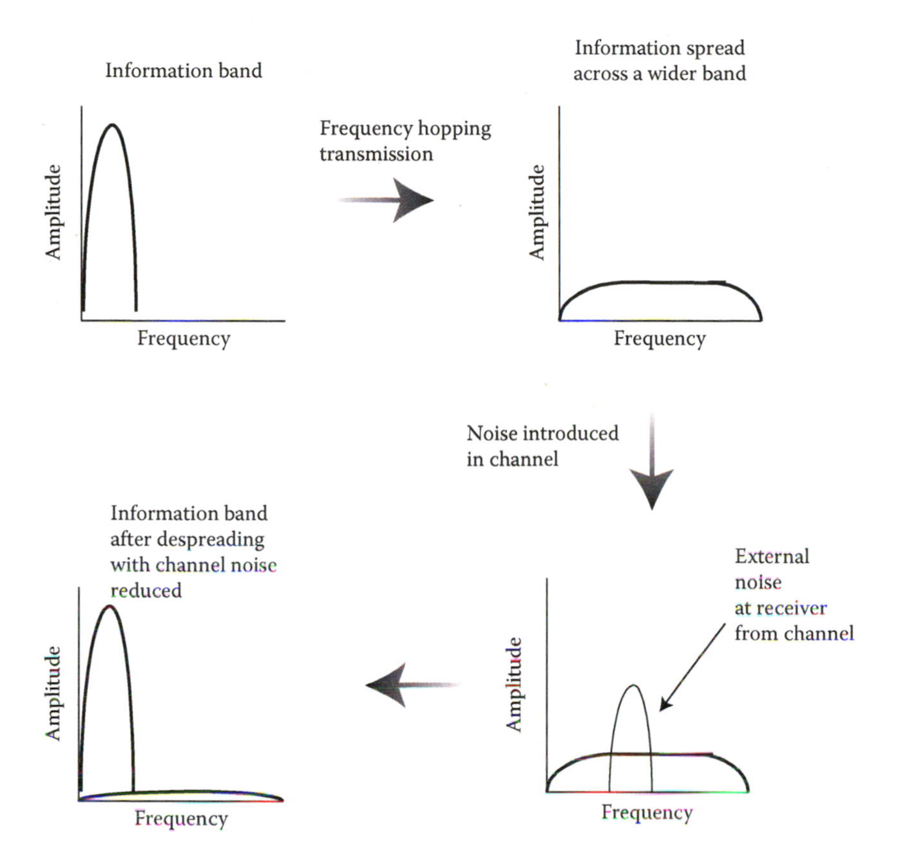

FIGURE 12.10
Impact of despreading on chanel noise.

12.3.5 Demodulation

After the frequency hopping FSK signal has been through the dispreading process, demodulation is done by mixing the signal in two different paths by the two different frequencies and integrating the outputs over the bit interval times. Demodulation results in a binary stream of data. If analog data had been transmitted then the digital output would need to go through another step where a digital to analog conversion takes place.

12.4 System Design

In this section the goal is to develop design principles that allow us to use off-the-shelf parts to design a complete UAV–GCS communication system. This means we will be developing specifications based on a particular application that will guide us in purchasing and assembling the proper receiver and transmitter components. No attempt will be made to design the individual parts since that is out of the scope of this chapter and the focus will be on system integration. We will begin the discussion by first studying the bandwidth requirements of the particular system.

12.4.1 Establishing Bandwidth Requirements

The bandwidth is a range of frequencies of interest. Bandwidth requirements vary within the same receiver or transmitter depending on which component the signal is passing through. The input information constitutes what is known as the *baseband*. When we are talking about analog signals the baseband is established by figuring out the differences between the highest and lowest frequency components. Digital sources usually provide data rates as opposed to bandwidth. Data rates are usually provided in bits per second (bps). The *Shannon–Harley relation* provides a method of calculating bandwidth (B) in Hz from data rate (DR) in bps and vice versa using the following relationship:

$$\mathrm{DR} = 2\mathrm{B}\log_2\mathrm{M} \tag{12.10}$$

where M is the number of levels transmitted.

EXAMPLE 12.1

An analog source of 4000 Hz bandwidth is sampled using 256-level sampling. What is the equivalent data rate? If the transceiver data rate is given as 10,000 bps, will this data rate be sufficient for the source bandwidth?

Using Equation 12.10,

$$DR = 2 \cdot (4000) \cdot \log_2(256) = 64{,}000 \text{ bps} \quad \text{or} \quad 64 \text{ kbps}$$

The data rate far exceeds the limits of the transceiver and as such will not be able to carry this source.

For our next example, we will calculate the bandwidth given the data rate.

EXAMPLE 12.2

An input data rate for a transceiver system is capped at 10,000 bps for a 2-level digital signal. What is the equivalent bandwidth for this baseband signal in Hz?

$$10{,}000 = 2 \cdot \mathrm{B} \cdot \log_2(2)$$

$$\text{or} \quad 10{,}000 = 2.\mathrm{B}.1$$

$$\therefore \quad \mathrm{B} = 5000 \text{ Hz}$$

In a real-world system, achievable data rates through a channel need to be calculated only after taking into consideration the S/N ratio. This is found by calculating the *Shannon limit* using the following relationship:

$$\mathrm{DR} = \mathrm{B}\log_2\left(1 + \frac{\mathrm{S}}{\mathrm{N}}\right) \tag{12.11}$$

In the next example, we will calculate the DR given an S/N ratio for a channel. This example will also help us specifying the bandwidth requirements for antennas on our link.

EXAMPLE 12.3

While selecting an antenna for a system you are required to specify a bandwidth assuming that you are using a modulation scheme that requires the modulated signal to only have the same bandwidth as the baseband signal if the data rate is 100,000 bps. The given $(S/N)_{dB}$ is 20 dB.

At this bandwidth how many levels are needed for the digital data?

$$20 = 10\log_{10}\left(\frac{S}{N}\right)$$

$$\therefore \quad \left(\frac{S}{N}\right) = 10^{(20/10)} = 100$$

Hence the S/N ratio is 100.

$$100000 = B \times \log_2(1+100)$$

$$\text{or} \quad B = \frac{100,000}{\log_2(101)} = \frac{100,000}{(\log_{10}101/\log_{10}2)} = \frac{100,000}{6.658} = 15020\text{ Hz}$$

The minimum bandwidth required is therefore 15.02 kHz. Using Shannon–Hartley relationship, we get

$$100,000 = 2 \times 15,020 \times \log_2 M$$

$$\text{or} \quad \log_2 M = \frac{100,000}{2 \times 15020}$$

$$\text{or} \quad M = 2^{(100,000/2\times 15,020)} = 10.048$$

In order to achieve the 100000 bps data rate using a 15.02 kHz bandwidth, a 11-level digital signaling would be required.

In the case of FSK the bandwidth of each channel is given by $2\Delta f + 2B$ and not simply by B as in example 12.3, where Δf is the frequency difference between the high frequency and the low frequency used in modulating 1s or 0s; B is the baseband bandwidth. For the frequency hopping FSK modulation this is the bandwidth of each channel, and the total channel bandwidth depends on the number of channels available.

EXAMPLE 12.4

A frequency hopping FSK modulation is operating using a system bandwidth demarked by 902–928 MHz. If the difference Δf between the high and low frequencies of modulation is 4 kHz what is the maximum number of channels that can exist when the baseband signal has a bandwidth of 15 kHz?

From example 12.3, B = 15 kHz

$$\text{Channel} = 2\Delta f + 2B = 2 \times 4000 + 2 \times 15,000 = 38,000\text{ Hz}$$

$$\text{Number of channels} = \frac{928,000,000 - 902,000,000}{38,000} = 684\text{ channels}$$

12.4.2 Link Design

Simply put the key question that one hopes to answer from the link design is whether enough transmitted power is getting through to the receiver given that distance the signal has to travel after accounting for all losses along the way. Figure 12.11 presents an illustration of a link and some loss considerations.

12.4.2.1 Reflection Antenna–Cable Junction

If the reader is unfamiliar with the concepts of electrical impedance and characteristic impedance of a cable he may move on to the next section since the effect of reflection at the antenna–cable junction is going to be ignored in our discussion (assuming a well-matched condition exists). However there can be a significant reflection at the antenna–cable junction if the cable and the antenna are not well matched. Moving forward without much detail, let us treat the concept of electrical impedance as a measure of opposition to circuit current when a voltage is applied. Impedance can be a complex quantity. The reflection coefficient (Γ) can be calculated by knowing the antenna impedance (Z_{ANT}) and the cable characteristic impedance (Z_0) by the following relation:

$$\Gamma = \frac{Z_{ANT} - Z_0}{Z_{ANT} + Z_0} \tag{12.12}$$

Now if one carefully matches the antenna impedance to the cable characteristic impedance Γ tends to be very small. If reflection was considered, assuming that the transmitter cable carries the transmitted power (P_T) to the antenna would be partly reflected as given by the relation below:

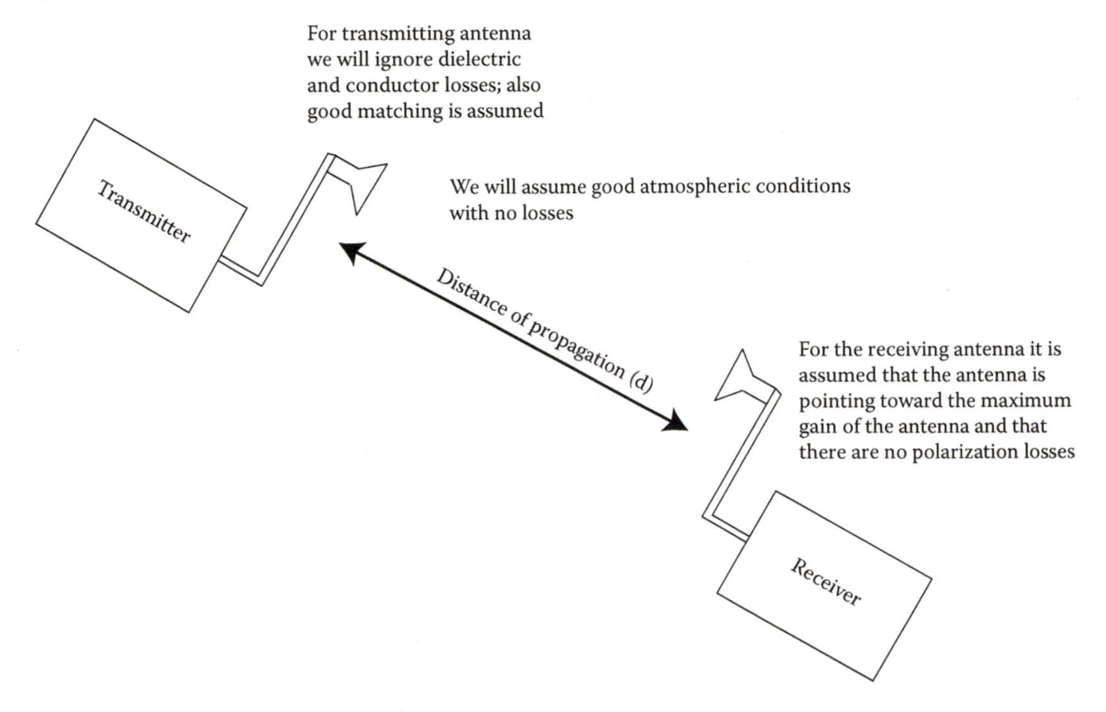

FIGURE 12.11
Considerations for calculating received power at receiving antenna.

Power coupled to the antenna via transmitter cable

$$= P_T(1-|\Gamma|^2)$$

$$\approx P_T \quad \text{when} \quad \Gamma \approx 0$$

In our discussion of link design we will move on assuming we have a good match and all the transmitter power (P_T) is coupled to the transmitting antenna. However, one should remember that there can be a serious coupling problem if the matching is not good.

12.4.2.2 Losses at the Transmitting Antenna

Some small dielectric and conductor losses occur at the transmitting antenna due to a dielectric covering designed to protect the antenna from exposure to the elements (*radome*) and conductor losses. In many cases antennas themselves are partly constructed with dielectrics. These losses are difficult to calculate by analytical techniques but are measurable. In our discussion moving forward, these losses will be ignored. So at this stage, we can say the transmitter power P_T is also the power being transmitted.

12.4.2.3 Losses due to Free Space Propagation

So far, we have managed to ignore matching losses and antenna losses by making requisite assumptions. Propagation losses cannot be ignored and they form the bulk of the losses of the link.

Let us assume if the receiving antenna is aligned in the transmitting antenna's direction of maximum gain (G_T), then the power density (P_D) at the receiving antenna located at a distance d from the transmitting antenna can be calculated from the following relation (Blake 2002, pp. 520–527):

$$P_D = \frac{P_T G_T}{4\pi d^2} \quad \text{watts/m}^2 \tag{12.13}$$

One way to look at Equation 12.13 is to note that product $P_T G_T$ is the power that an isotropic antenna would require to create the same power density P_D at the receiver (as created by the directive antenna). Otherwise stated, we will get the power density of an isotropic source radiating $P_T G_T$ power into free space by simply dividing the radiated power by the surface area of a sphere of radius d. The product $P_T G_T$ is also sometimes known as the *effective isotropic radiated power* (EIRP).

EXAMPLE 12.5

A directional antenna has a gain of 7 dBi (decibel isotropic) at 900 MHz. If the transmitted power is 0.5 W, what is the power density 6 miles from the antenna in its direction of maximum gain?

First, we need to convert the gain from decibels to a dimensionless number (G_T).

$$7 = 10 \times \log_{10}(G_T)$$

$$\therefore G_T = 10^{(7/10)} = 5.012$$

Now using 12.13 we can compute the power density 6 miles away (9656 m).

$$P_D = \frac{0.5 \times 5.012}{4 \times \pi \times (9656)^2} = 2.139 \times 10^{-9}$$

watts per meter^2 or 2.139 nanowatts per meter^2

12.4.2.4 Power Received at the Receiving Antenna

As the EM wave travels through free space it creates a power density P_D given by Equation 12.13 at the receiver. The question is how much of this power is accepted by the receiving antenna. When the EM wave arrives at the receiving antenna, the antenna is considered as an effective receiving aperture (A_{eff}). This effective aperture receives power that is a ratio of the power received by the antenna (P_R) in watts and the power density (P_D) in watts per meter square. If the receiving antenna has its maximum gain (G_R) pointing at the transmitters maximum gain, then the effective area of the receiving antenna is given by

$$A_e = \frac{\lambda^2 G_R}{4\pi} \tag{12.14}$$

$$P_R = A_e P_D \tag{12.15}$$

$$P_R = \frac{A_e P_T G_T}{4\pi d^2} \tag{12.16}$$

EXAMPLE 12.6

If a receiving antenna receives power from the transmitting antenna as in Example 12.5, what is the power received (P_R) by the receiving antenna?

The wavelength (λ) at 900 MHz is

$$\lambda = \frac{c}{f} = \frac{3 \times 10^8}{900 \times 10^6} = 0.333 \text{ m}$$

$$A_e = \frac{\lambda^2 G_R}{4\pi} = \frac{0.333^2 \times 5.012}{4 \times \pi} = 0.044 \text{ m}^2$$

$$P_R = A_e P_D = 0.044 \times 2.139 \times 10^{-9} = 9.412 \times 10^{-11} \text{ W} \quad \text{or} \quad 94 \text{ pW}$$

Equation 12.16 can be further expanded to the following form:

$$P_R = \frac{\lambda^2 G_R P_T G_T}{16\pi^2 d^2} = \frac{\lambda^2 P_T G_T G_R}{16\pi^2 d^2} \tag{12.17}$$

Equation 12.17 can also be rearranged to a form that derives the ratio of the power received by the power transmitted:

$$\frac{P_R}{P_T} = \frac{\lambda^2 G_T G_R}{16\pi^2 d^2} \tag{12.18}$$

Equation 12.18 can be rewritten by substituting where $\lambda = c/f$,

$$\frac{P_R}{P_T} = \frac{c^2 G_T G_R}{f^2 16\pi^2 d^2} \tag{12.19}$$

Now if we are interested in using kilometer and MHz for distance and frequency, we need to make the following substitutions in Equation 12.19:

$$d = 10^3 \times d_{km}$$

$$f = 10^6 \times f_{MHz}$$

$$c = 3 \times 10^8$$

Hence, Equation 12.19 can be written as

$$\begin{aligned} \frac{P_R}{P_T} &= \frac{(3\times 10^8)^2 G_T G_R}{(10^6 \times f_{MHz})^2 16\pi^2 (10^3 \cdot d_{km})^2} \\ \text{or}\quad \frac{P_R}{P_T} &= \frac{(3\times 10^8)^2 G_T G_R}{(10^9)^2 16\pi^2 (f_{MHz})^2 (d_{km})^2} \end{aligned} \tag{12.20}$$

$$\begin{aligned} 10 \times \log_{10}\left(\frac{P_R}{P_T}\right) &= 10 \times \log_{10}\left(\frac{(3\times 10^8)^2}{(10^9)^2 16\pi^2}\right) + 10 \times \log_{10}(G_T) \\ &\quad + 10 \times \log_{10}(G_R) - 20 \times \log_{10}(f_{MHz}) - 20 \times \log_{10}(d_{km}) \\ &= -32.442 + (G_T)_{dBi} + (G_R)_{dBi} - 20 \times \log_{10}(f_{MHz}) - 20 \times \log_{10}(d_{km}) \end{aligned} \tag{12.21}$$

The path (L_{path}) loss in decibels can be computed by finding

$$\begin{aligned} L_{path} = 10 \times \log_{10}\left(\frac{P_T}{P_R}\right) &= 32.442 -)(G_T)_{dBi} - (G_R)_{dBi} + 20 \times \log_{10}(f_{MHz}) \\ &\quad + 20 \times \log_{10}(d_{km}) \end{aligned} \tag{12.22}$$

Once path loss is computed from Equation 12.21 and the power output of the transmitter is known, the received power can be calculated. Note that in such case dielectric and material losses at the transmitting antenna, atmospheric losses, and polarization losses have been neglected. However, polarization losses at the receiver can be minimized by playing with the antenna structure's positioning and alignment. Generally the losses may add 5–10 dB extra to the path loss and can be accounted for by adding on the extra amount to the minimum clearance from the receiver sensitivity.

EXAMPLE 12.6

The receiving and the transmitting antennas have the same gain at 3.5 dBi. The distance between them is 5 km. The power at the transmitter is 0.5 W at 500 MHz. How much power is received at the receiver?

First, we will find the path loss.

$$L_{path} = 32.442 - (G_T)_{dBi} - (G_R)_{dBi} + 20 \times \log_{10}(f_{MHz}) + 20 \times \log_{10}(d_{km})$$

$$L_{path} = 32.442 - 3.5 -$$

$$3.5 + 20 \times \log_{10}(500) + 20 \times \log_{10}(5) = 93.401 \text{ dB}$$

But

$$L_{path} = 10 \times \log_{10}\left(\frac{P_T}{P_R}\right) = 93.401$$

or $$\left(\frac{P_T}{P_R}\right) = 10^{93.401/10}$$

$$\therefore P_R = P_T \cdot 10^{-9.3401} = 0.5 \cdot 10^{-9.3401} = 2.285 \cdot 10^{-10} \text{ W} \quad \text{or} \quad 22.85 \text{ nW}$$

12.4.2.5 Power in Decibel Milliwatt

It is very common to have power levels provided in decibel milliwatt or dBm form.

Simply put,

$$(\text{Power})_{dBm} = 10 \cdot \log_{10}\left(\frac{(\text{Power in milliwatts})}{1 \text{ milliwatt}}\right)$$

EXAMPLE 12.7

What is the input power from Example 12.6 in dBm?

$$(\text{Power})_{dBm} = 10 \cdot \log_{10}\left(\frac{(0.5 \cdot 1000)}{1}\right) = 26.99 \text{ dBm}$$

12.4.2.6 S/N Ratio at the Receiver

The S/N ratio at the receiver can be calculated if one knows the particular noise figure of the receiver. The noise figure (NF) is simply the ratio of the S/N ratio at the input by the S/N ratio at the output.

$$NF = \frac{(S/N)_i}{(S/N)_o} \tag{12.23}$$

In decibel form,

$$NF_{dB} = 10 \times \log_{10}(NF) \tag{12.24}$$

If the input power at the receiver is known and if one considers that the thermal noise (N = KTB) is found, then one can find the S/N ratio at the receiver with the help of the noise. This implies knowing the bandwidth and the temperature.

$$NF = \frac{(S/N)_i}{(S/N)_o} = \frac{P_r/kTB}{(S/N)_R} \quad (12.25)$$

$$\text{rearranging,} \quad \left(\frac{S}{N}\right)_R = \frac{P_r/kTB}{NF} \quad (12.26)$$

$$\left(\frac{S}{N}\right)_{RdB} = 10 \times \log_{10}(P_r) - 10 \times \log_{10}(kTB) - NF_{dB} \quad (12.27)$$

If the received power is given in dBm then

Equation 12.27 is rewritten as follows:

$$\left(\frac{S}{N}\right)_{RdB} = (P_r)_{dBm} - 10 \times \log_{10}(1000\ kTB) - NF_{dB} \quad (12.28)$$

EXAMPLE 12.7

If the power received at the input of the receiver is the same as that calculated from Example 12.6, what is the signal-to-noise ratio at the receiver at 300 K for a bandwidth of 1 MHz (using $NF_{dB} = 10$)? (Boltzmann constant, $k = 1.38 \times 10^{-23}$ Joules/K).

Using Equation 12.28,

$$\left(\frac{S}{N}\right)_{RdB} = 10 \times \log_{10}(P_r) - 10 \times \log_{10}(kTB) - NF_{dB}$$

$$= 10 \cdot \log_{10}(2.285 \cdot 10^{-10}) - 10 \cdot \log_{10}(300 \cdot 1.38 \cdot 10^{-23} \cdot 10^{6}) - 10$$

$$= 37.419\ dB$$

12.4.2.7 Calculation of Signal-to-Noise Margin from Receiver Sensitivity

Receiver sensitivity (RS_{dBm}) is the minimum power level required by the receiver to demodulate signal information. This information is provided by the manufacturer. A decent clearance from this value (>10 dBm) is desired since the noise level in different systems can vary. With RS_{dBm} known we can calculate the margin using the following equation:

$$\text{Margin in dBm} = (P_r)_{dBm} - RS_{dBm} \quad (12.29)$$

EXAMPLE 12.8

If the receiver sensitivity is –85 dBm what is the signal-to-noise margin from the problem in Example 12.7?

$$\text{Margin in dBm} = (P_r)_{dBm} - RS_{dBm}$$

$$= 10 \times \log_{10}(2.285 \times 10^{-10} \times 1000) - (-85)$$

$$= 18.589 \text{ dBm}$$

The obtained value is quite a good margin.

12.5 Summary of Design Principles

The first step would be to decide the data rate required for the application we have in mind. If we are using off-the-shelf components, manufacturers will have the specifications about what data rates are manageable. Section 12.4.1 has general rules for establishing data rate requirements for a given application when FSK modulation is used where there were only two levels (M = 2). In other digital modulation schemes different values of M may need to be used such as M = 2 for binary phase shift keying (BPSK) and M = 4 for quadrature phase shift keying (QPSK). Very briefly, BPSK uses the same frequency and opposite phases to modulate, while QPSK uses the same frequency with four different phases.

Currently frequency hopping FSK is most widely used in UAV commercial applications but future applications may employ BPSK and/or QPSK. Loss calculations are not affected by modulation techniques (FSK, PSK, or QPSK).

The worst-case link lengths need to be estimated and link losses worked for them. Section 12.4.2 covers much of this discussion. A look at Equation 12.22 not only tells us how to compute the path loss but it also gives us the variable with which to play in order to decrease path loss. Since we really do not have as much control over the ISM band frequencies that we use or the worst case distance between the receiver and transmitter, selecting higher gain antenna always reduces the path loss. It has to be remembered that the EIRP must not exceed FCC guidelines.

12.6 Associated Problems from EMI Interference, Jamming, and Multipath

12.6.1 EMI Interference

EMI interference occurs when an unintended signal finds its way to the receiver at the frequency of concern. Although this might come into the receiver through a variety of pathways such as conduction, radiation, crosstalk, ground, and power lines (Gerke and Kimmel 2005, pp. 3–8). The EMI (sometimes also call RF interference) can, in some instances, carry more power than the intended signal and overwhelm the intended signal while passing through the receiver stages. While spread spectrum techniques like frequency hopping are secure from having information read by unauthorized entities, it cannot protect from interference especially when the interfering source has higher or comparable power levels.

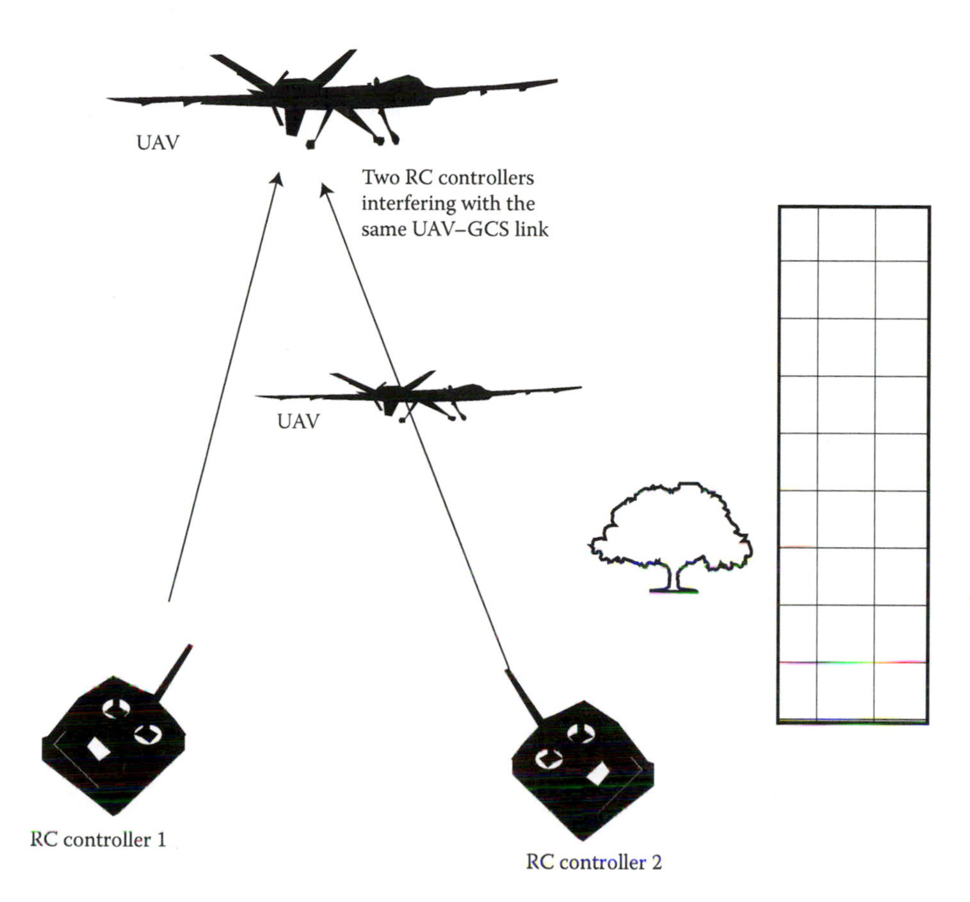

FIGURE 12.12
EMI interference for small UAV.

With the future of the UAVs looking brighter, EMI problems are going to be increasingly more challenging. From the current regulatory trajectories, it seems that commercial applications are likely to require the operator stay in visual contact with the aircraft at all times. In a crowded airspace this can easily lead to an antenna picking more than one LOS signal intended for someone else (Figure 12.12). The growing problem of EMI interference is problem that will need increased attention.

12.6.2 Jamming

Jamming problems are similar in some respects to the EMI problem we just discussed. A jamming signal is disruptive whether this is in contested airspace or not. The jamming signal carries a power level that overwhelms the receiver making it difficult or impossible to navigate or complete an intended application. Whether it is the GNSS signal being jammed or the UAV–GCS link, this poses a threat to UAV operation. In the future, as newer applications demand more and more UAV platforms, the problem of illegal jamming for commercial advantage will also create problem that are normally associated with contested battlefield scenarios. Currently researches have focused on the design of spatial processing antennas that would cut down jamming sources (Heng et al. 2014).

12.6.3 Multipath

We have discussed the concept of multipath and the different pathways a signal can take to a receiver and how some of these signals arrive at the receiver after bouncing around (Figure 12.4). The problems from multipath stem from the fact that two or more copies of the signal arrive at the receiver and sometimes interfere destructively by opposing the phase of the direct signal; sometimes it can increase signal amplitude by being in phase. A GNSS receiver which also encounters this problem is able to address this through signal processing techniques, by which signal arriving at later interval can be ignored by the processor (Kaplan and Hegarty 2006, p. 703). However when the multipath signals arrive closer to the direct signal it still creates errors. The solution to this UAV–GCS problem may lie in a similar solution to the one being proposed by Heng et al. (2014) in which the antenna is designed to cut off the multipath signal through special processing.

REVIEW QUESTIONS

1. Give an example of a data link for UAV–GCS communication specifying frequency and modulation techniques? Use the classification systems presented in this chapter to describe the type of link.
2. Research the Inmarsat Network. What are the potential advantages to using this system?
3. How does antenna transmit energy into space with charges to carry them?
4. What is a simplex connection? What is a duplex connection?
5. Describe FSK modulation? Why are we studying this type of modulation?
6. What is the function of the power amplifier in the transmitter?
7. What is a spread spectrum technique? What are the advantages of using spread spectrum technology?
8. What is a communication channel? Give examples of three different channels?
9. What is a radiation pattern of an antenna?
10. What is meant by radiation intensity?
11. What is the far field?
12. What is antenna directivity?
13. What is antenna gain?
14. What kind of losses exists in antenna and its radome? (Just discuss loss mechanisms in the antenna material.)
15. How many types of polarization did we discuss?
16. What is right-handed circular polarization?
17. What is linear polarization?
18. What is the importance of the S/N ratio? How is channel capacity influenced by its S/N ratio?
19. What is thermal noise? How do you calculate it?
20. What is impulse noise? What are the sources of impulse noise?
21. When do we need an ADC?
22. When do we need a digital-to-analog convertor?

23. How are bandwidth requirements established for a data link?
24. What is data rate?
25. In the Shannon–Hartley relation (Equation 12.10), what is the significance of M?
26. What is the Shannon limit?
27. What assumptions are being made that lead to neglecting reflection loss at the transmitting antenna–cable junction?
28. What is meant by EIRP?
29. What is noise figure (NF)? What does it tell us about an electronic system?
30. What assumptions are being made that lead to neglecting the polarization loss at the receiving antenna?
31. What is EMI interference? How is this a threat to UAV–GCS communication?
32. How is EMI a challenge to the future of UAV–GCS communication?
33. What is meant by jamming?
34. What problems are we running into while trying to solve the problem of multipath?
35. Very briefly describe the steps to designing a data link.

REVIEW PROBLEMS

1. An analog source of 4000 Hz bandwidth is sampled using 16-level sampling. What is the equivalent data rate? If the transceiver data rate is given as 1000 bps, will this data rate be sufficient for the source bandwidth?
2. While selecting an antenna for a system, you are required to specify a bandwidth assuming that you are using a modulation scheme that requires the modulated signal to only have the same bandwidth as the baseband signal if the data rate is 10,000 bps. The given $(S/N)_{dB}$ is 20 dB. At this bandwidth how many levels are needed for the digital data?
3. A frequency hopping FSK modulation operates using a system bandwidth demarked by 902–928 MHz. If the difference Δf between the high and low frequencies of modulation is 4 kHz, what is the maximum number of channels that can exist when the baseband signal has a bandwidth of 60 kHz?
4. A directional antenna has a gain of 5 dBi (decibel isotropic) at 900 MHz. If the transmitted power is 100 watts, what is the power density 6 miles from the antenna in its direction of maximum gain?
5. If a receiving antenna receives power from the transmitting antenna as in review problem 4, what is the power received (P_R) by the receiving antenna?
6. The receiving and the transmitting antennas have the same gain at 5 dBi (decibels-isotropic). The distance between them is 15 km. The power at the transmitter is 100 watt at 500 MHz. How much power is received at the receiver?
7. If the power received at the input of the receiver is from review problem 7, what is the S/N ratio at the receiver at 300 K for a bandwidth of 1 MHz (using NF_{dB} = 10)? (Boltzmann constant, $k = 1.38 \times 10^{-23}$ Joules/K).
8. If the receiver sensitivity is –85 dBm what is the signal-to-noise margin in review problem 7?

References

Balanis, C. 2005. Radiation mechanism. In *Antenna Theory Analysis and Design*. 3rd ed., 7–16. Hoboken, New Jersey: Wiley.

Blake, R. 2002. Free-space propagation. In *Electronic Communication Systems*. 2nd ed., 520–527. Clifton Park, New York: DELMAR.

Fahlstrom, P. and J. Gleason. 2012. Data-link functions and attributes. In *Introduction to UAV Systems*. 4th ed., 191–204.

Gerke, D. and B. Kimmel. 2005. EMI, noise, and interference—A different game. In *EDN Designers Guide to Electromagnetic Compatibility*. 1st ed., 3–8. W. St. Paul, MN: Kimmel Gerke Associates, Ltd.

Heng, L., T. Walter, and G. Gao. 2014. GNSS multipath and jamming mitigation using high-mask-angle antennas. *IEEE Transactions on Intelligent Transportation Systems* 16 (2): 741–750.

IEEE, Standards. 1983. IEEE standard definitions of terms for antennas. (Accessed March 29, 2015).

Kaplan, E. and C. Hegarty. 2006. *Understanding GPS Principles and Applications*. 2nd ed. Norwood, Massachusetts: Artech House, Inc.

Sevgi, L. 2007. The antenna as a transducer: Simple circuit and electromagnetic models. *IEEE Antennas and Propagation Magazine* 49 (6): 212.

Stallings, W. 2007. Transmission impairments. In *Data and Computer Communications*. 8th ed., 89–91. New Jersey: Pearson.

Wagenen, Juliet. Inmarsat parrot bebop drone takes flight for journalists. In Satellite TODAY News feed, ST Briefs, Telecom [database online]. 2015 [cited March 19 2015]. Available from http://www.satellitetoday.com/telecom/2015/03/11/inmarsat-parrot-bebop-drone-takes-flight-for-journalists/ (accessed March 19, 2015).

13

Command and Control

Nathan Maresch

CONTENTS

13.1 Introduction

The command and control* functionality of unmanned aircraft systems allows operators to know what is happening on board the aircraft while flying. Through the command and control link, an operator may also send updates to the aircraft's systems and modify the flight plan (Gundlach 2012). Depending on the mission at hand, some unmanned aircraft are flown completely autonomously without transmitting or receiving communications from the ground station throughout the entire flight. For almost all applications, however, two-way communication with the aircraft is desired. Focus will be given to radio frequency (RF) data links, because they are the most common form of communication for unmanned systems. Other mediums such as tethered cables and beams of light are sometimes used for information transfer (Austin 2010). This is true for both the uplink—the commands from the operator to the aircraft—and the downlink—the status information sent from the aircraft.

Put simply, the RF spectrum comprises the large range of frequencies found between those we can hear and those we can see, beginning with the audio frequencies (3 kHz, a

* The UAS command and control subsystem/functionality is sometimes denoted by the shorthand designation, "C2." This convention will also be observed throughout this chapter.

wavelength of 100 km) and ending near the beginning of the infrared light frequencies (300 GHz, a wavelength of 1 mm). In the United States, the civil regulatory authority for this spectrum is the Federal Communications Commission (FCC). Part of the job of the FCC is to manage the radio spectrum in an efficient way, assigning specific frequencies (channels) to users, and regulating their power output. This management is to prevent interference from other users transmitting on the same frequency in the same locality; everyone gets their own radio channels—not a particularly easy job to manage. If someone simply decides to transmit on a particular frequency without authorization, he may face significant fines or penalties, especially if he interferes with someone who has a license for that frequency. Typically, the FCC allocates parts of the spectrum for certain uses—such as the bands used for AM/FM music radio. To date, there are no bands set aside for use exclusively by unmanned aircraft. Licenses are issued on a case-by-case basis by availability, though a number of specialized bands are being considered for unmanned aircraft use (Transportation 2013). There are also certain bands that do not require a license to use as long as the equipment and operation complies with a number of rules set forth in Part 15 of the FCC rules (the FCC rules may be found in Title 47 of the US Code of Federal Regulations [CFR]). These unlicensed bands are becoming increasingly popular for use with small unmanned aircraft systems (sUAS). Using these bands, the radio equipment must hop between one frequency to another very rapidly over a large range of frequencies (spread spectrum). The transmitter and receiver must be synchronized; the receiver must hop using the same pattern as the transmitter for proper reception. Many times a specific pattern in use may appear completely random to an outside observer, but the communicating devices remain synchronized to each frequency change; the pattern is pseudo-random. The benefit of using this method is that multiple users may operate in the same band without causing interference to others; different users employ different hopping patterns. This communication method is relatively immune to other sources of interference as well, because noise occurring at one frequency results in losing only a small portion of a spread-spectrum transmission. In fact, the unlicensed frequencies are within part of the spectrum that previously was specified only for RF-noisy industrial, scientific, and medical (ISM) uses; microwave ovens operate in one of these bands. More advanced spread-spectrum transceivers are able to detect existing sources of interference (such as a microwave oven running nearby) and avoid those channels completely (Blake 2002). Wi-Fi is an example of spread-spectrum technology in action. A visual representation of frequency hopping is provided in Figure 13.1, which depicts the spectrum measurement of an unlicensed command and control link over about 60 seconds time. A section of the RF spectrum is plotted along the horizontal axis. Elapsed time is plotted along the vertical axis. Each horizontal segment indicates an average amount of RF energy sampled during an interval of a few

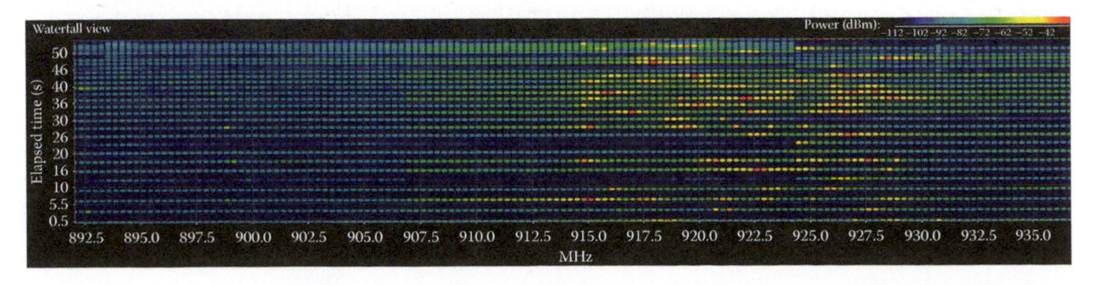

FIGURE 13.1
Spectrum waterfall view indicating frequency hopping over time. (Image courtesy Nathan Maresch.)

seconds. Most of the RF energy is concentrated at differing frequencies during each measurement interval. In contrast, a solid consistent vertical concentration would indicate a traditional fixed frequency transmission.

13.2 sUAS Navigation Systems

Regardless of the underlying navigation technology used for controlling the unmanned aircraft, the primary method for an operator to create navigation commands for the aircraft is to use a graphical map with waypoints to mark the course of the aircraft. Each waypoint has an altitude associated with it. The aircraft flies from waypoint to waypoint, or may perform other maneuvers such as circle (orbit) about a waypoint. Figure 13.2 shows an example of a ground-control station interface in use, where the aircraft has just started flying to the first waypoint. The autopilot methods behind waypoint navigation vary between different types of autopilots. Large amounts of sensor and location data must be combined and processed to navigate and intercept an intended flight track. Many autopilots use Kalman filters (or extended Kalman filters) to process the navigation information.

13.2.1 Line of Sight Communication

Virtually all civil unmanned aircraft use line of sight (LOS) RF communications for their command and control data link. LOS in the context of RF signals simply refers to a direct path between one antenna and one or more others. The distance between LOS antennas

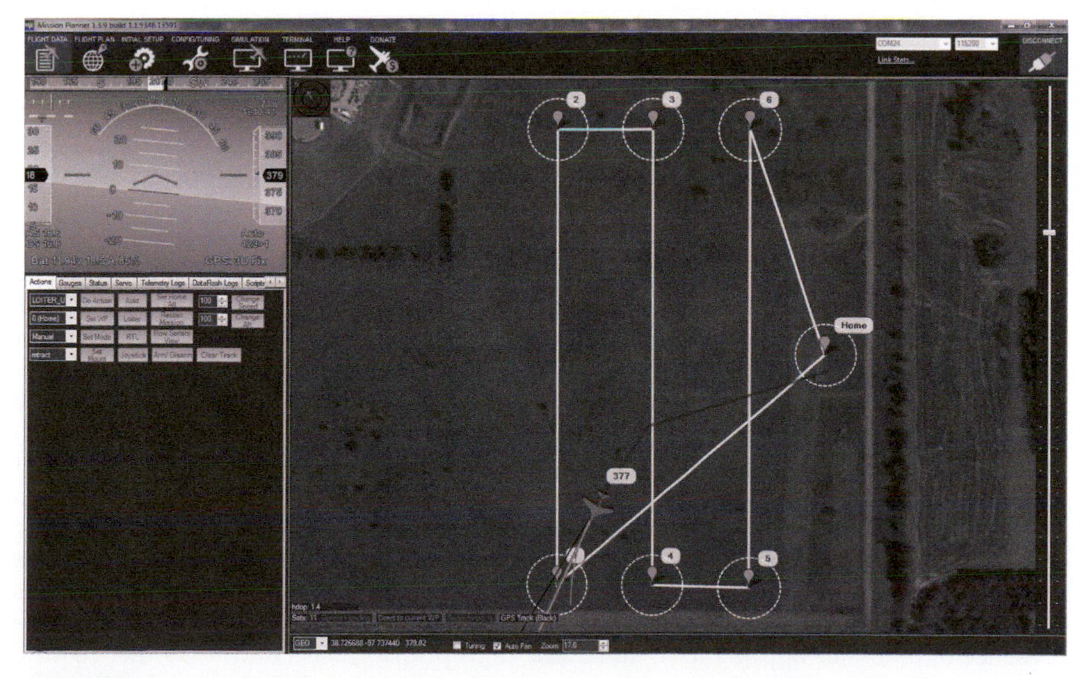

FIGURE 13.2
Ground control system interface with waypoints time. (Image courtesy Nathan Maresch.)

in some cases may be many miles, certainly much further than the human eye can see. Commonly called space waves, these RF frequencies are generally all those above 30 MHz. Those below 30 MHz usually either follow the terrain of the earth, or are reflected off the earth's atmosphere to be received at another geographic location; conversely, because LOS frequencies are by definition direct, buildings, mountains, or even the curvature of the earth itself may become an issue for signal reception. While the specifics are beyond the scope of this chapter, depending on many factors such as transmission power and receiver sensitivity, the higher the antenna above the earth, the further the signal may propagate. A higher "gain" directional antenna at the ground station (or if practically feasible on board the aircraft) may increase the usable operating distance as well. Though there are power limitations for licensed transmissions and usually greater limitations for unlicensed transmissions, there is generally enough transmission power available to accommodate most flights within visual LOS without the need for high-gain antennas. Also, note that there may be regulatory limits for the amount of gain an antenna may have.

As briefly mentioned earlier, unlicensed frequencies are popular for sUAS. The most common unlicensed frequencies are 915 MHz, 2.45 GHz, and 5.8 GHz, all of which are LOS frequencies. Sometimes, all three are used on the same aircraft. For example, 915 MHz may be used for operator command and control of the aircraft, 2.45 GHz for external pilot control, and 5.8 GHz for a video or other payload downlink. Besides the reasons mentioned earlier for the popularity of spread-spectrum technology, these specific frequencies (and similar traditional licensed frequencies) are also popular because of the compact nature of the antennas required to transmit and receive them. These frequencies are high enough that the required antennas may be easily accommodated by most sUAS. Recall that as frequency increases, the wavelength decreases. As the wavelengths become shorter, so also may the dimensions of the corresponding antennas.

13.2.2 sUAS Autopilot Systems

Autopilot systems for sUAS generally consist of a number of core components integrated into the device itself, as well as some external sensors and devices. Based on the instructions and information it has, the autopilot outputs command directly to the aircraft flight controls. Figure 13.3 shows the main elements that the autopilot communicates with. At the heart of the autopilot is a processor or microcontroller. This reads inputs such as sensor values and command inputs, and performs computations such as determining the amount of deflection required for a heading correction. Once computed, it commands the flight controls to deflect an appropriate amount. The controller usually operates an inner loop and an outer loop. The inner loop operates at a much faster rate, compensating for wind and keeping the aircraft in the air. The outer loop focuses on the navigation of the aircraft, and may be updated much more slowly. It may be surprising to learn that regardless of which loop the controller is focusing on, it cannot do more than one basic task at a time. Everything that it performs is done in sequence, with a clock pulse signaling to go on to the next item on the list. This clock is merely a very fast timed pulse, signaling at rates from many millions of times per second, to over a billion times per second for higher-end systems. The purpose of the clock is to keep everything synchronized. This clock signal is derived from the regular vibrations of a quartz crystal vibrating in the presence of a voltage applied to it. To handle high-priority tasks and tasks that require attention at specific intervals, the processor uses "interrupts." This is a mechanism that literally interrupts the current list of things to process to perform others, before resuming

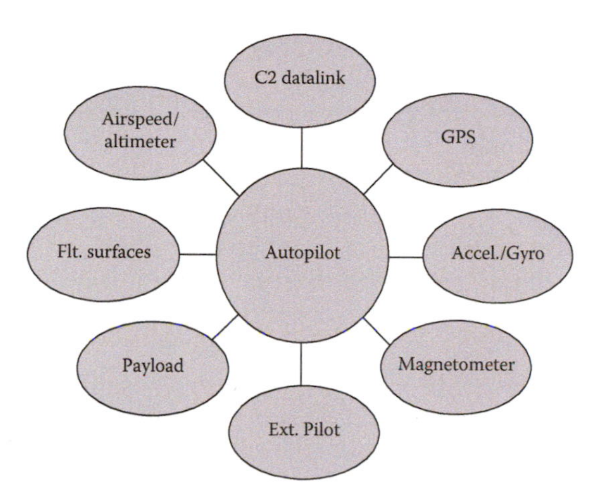

FIGURE 13.3
Main autopilot interfaces. (Image courtesy Nathan Maresch.)

once again. An interrupt may be triggered by an internal timer, performing important tasks at certain intervals, or by an external signal. For example, an interrupt may be used to trigger the reading of the current location of the aircraft from a GPS receiver when new data is available.

Depending on the design of the autopilot system, the various sensors that interface with the main processor may be mounted on the autopilot circuit board itself or exist as separate systems that plug in. Figure 13.4 shows the layout of the major components found on the circuit board of an open-source autopilot system. Some autopilots offload certain processes to a second processor. In some cases, this secondary controller can handle the

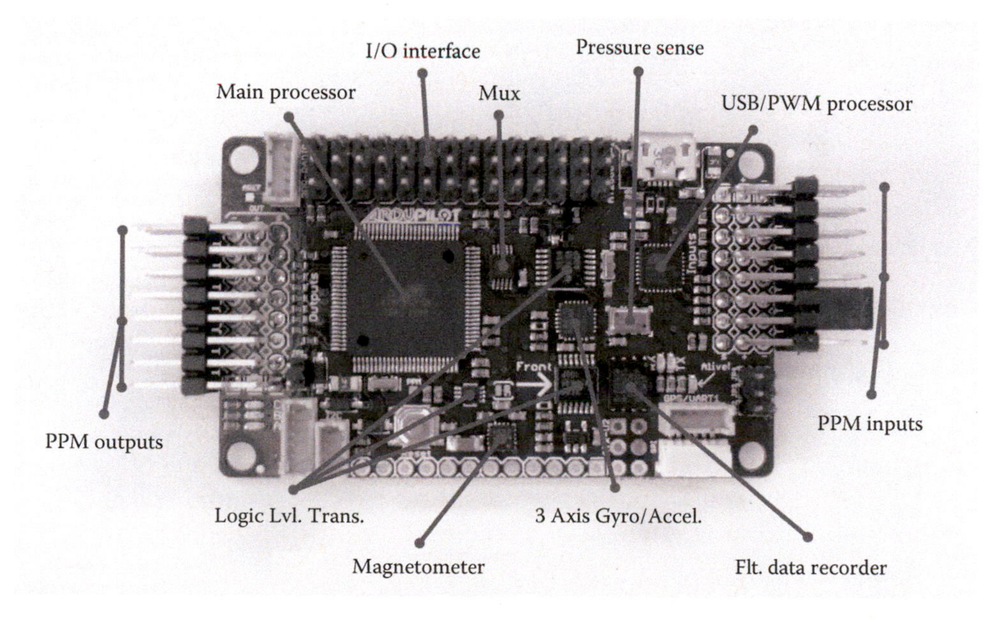

FIGURE 13.4
Autopilot components. (Image courtesy Nathan Maresch.)

reading of the external pilot inputs, among other tasks such as the encoding of the outputs to the control-surface servos. If the main processor fails, the secondary processor can route the inputs from the external pilot directly to the servos, acting as a multiplexer. A multiplexer is basically a switch allowing a choice of selection between multiple sets of inputs—one set is routed to the output. In this way, the external pilot may fly the aircraft manually and safely land. For systems without this functionality integrated into the autopilot, an external multiplexer may be added so that the external pilot can switch control from the autopilot to their controller in the rare event that an autopilot system failure occurs. On the other hand, if external pilot signals are not received, then control is automatically routed to the autopilot.

The external pilot controllers for sUAS are generally the same as those used for hobbyist radio-controlled aircraft. The flight control servos on board the aircraft (and throttle servo or electronic speed controller (ESC) on an electric aircraft) expect a pulse that exists for a specific window of time. For a servo, a pulse of 1.0 ms corresponds to full deflection in one direction, 1.5 ms center deflection, and 2.0 ms full deflection in the other direction (with any pulse/deflection in between). The rate that these pulses are sent to the servos varies between 50 and 400 Hz depending on the system and type of servos. The technology used for transmitting the pulses between the external pilot transmitter and the receiver on the aircraft also varies; Pulse Code Modulation (PCM) and Proportional Pulse Modulation (PPM) are common methods, and modern transmitters also use spread-spectrum technology. Autopilots commonly have the provision for an external pilot to control the aircraft through the command and control data link itself. This usually consists of a controller or joystick plugged into the ground control station or operator station. There is usually more delay with this method due to the additional processing associated with encoding and decoding the data packets, and autopilot translation of the data into servo pulses.

Data link modems attach to the autopilot to transmit telemetry to the ground station, and receive command and control information from the operator. They are typically a transceiver connected to one or more antennas. Figure 13.5 shows three different command and

FIGURE 13.5
Various sUAS command and control transceivers. (Image courtesy Nathan Maresch.)

control transceivers. In rare cases, there may be a separate receiver and transmitter operating on different frequencies (Gundlach 2012). Various protocols are used for data transmission, but specialty protocols designed just for unmanned aircraft are becoming common. Generally, an information packet is transmitted and an acknowledgement is returned by the receiver. If an error occurs, the packet may be retransmitted. A number of techniques exist for detecting errors, ranging from a simple checksum to a cyclic redundancy check (CRC). A checksum is the sum of all the zeros and ones that go into a particular data packet. The sum is divided by a fixed number, and the remainder is transmitted with the packet for the check. If this addition and division process results in the same remainder when the packet is received, then it is valid. There are potential issues with using a checksum. If, for example, a zero gets changed to a one somewhere, and somewhere else a one gets changed to a zero, the sequence will still be detected as valid. A CRC is a bit more advanced, dividing the packet data by a cyclic polynomial at each end, and using the remainder once again. CRCs can detect more types of errors than a checksum can, but no error checking method is completely perfect. Some data link modems take advantage of extra data link capacity by adding extra information called error correcting code (ECC), or forward error correction (FEC). Using error correction, a significant percentage of errors may be fixed at the receiver without having to request the data be re-transmitted. There are also provisions for avoiding collisions—the transmission of data by different systems at the same time (Blake 2002).

Even if there is a relatively small amount of activity on any side of the data link, there are still status packets being transmitted back and forth. If the aircraft does not receive a transmission from the ground control station for a preset amount of time, a lost-link procedure is executed. The operator may customize the lost-link timeout and behavior before the flight. Generally, multiple actions are taken based on the length of blackout time—the first may be to help to reestablish communication, such as canceling the existing flight plan and loitering near the ground control station. If a connection is not reestablished for an additional length of time the system may take further action, such as performing an autolanding. In some cases the lost-link action may be triggered in reverse direction as well. Some systems may be programmed so that if the ground station does not receive status information from the aircraft, it may transmit a command to initiate the lost-link procedure just in case the aircraft is still able to receive commands (the easier, less secure way is to intentionally stop transmitting and let the aircraft trigger the procedure). If at any time the link is reestablished, the operator may cancel the lost-link flight plan and resume as necessary.

Users of wireless data links should be aware of the signal strength for the link they are operating. One procedure that is beneficial to perform before flight is a reduced power check. Reducing the power helps one find out if there are issues or problems with antenna connections and radio equipment. Some radios can be configured to reduce the transmitting power on both ends enough for the check, other times an attenuator is added between the antenna and the transceiver. An attenuator is a matched resistor that dissipates a portion of the RF of power. Some attenuators are fixed, and some may be adjusted to dissipate more or less power; in the case of a transceiver, the attenuator reduces the transmitting power before reaching the antenna, as well as the received power before it reaches the receiver. It is important to note that an attenuator must be installed and removed with the equipment off. If the system is still able to operate with reduced power when separated by a number of meters, then it follows that it should function at greater distances with full power. The signal strength is presented to the user in different ways depending on the system. Sometimes the strength is displayed as a simple 0%–100% range. Other systems provide the raw measurement of received power to the user. This is usually displayed in decibels referenced to 1 milliwatt of power (dBm). The received power is almost always less than 1 mW, so the dBm reading will be negative (in contrast, transmitters almost always output more than 1 mW, so the dBm is positive). Generally –40 dBm is considered a very strong signal, and –100 dBm a very weak one (the more negative, the weaker the signal). Most receivers will not be able to tell the difference between the signal and ambient noise at about –100 dBm. Different models of receivers (and the associated link data-rate settings) will vary in their signal sensitivity, so acceptable lower values will also vary; moreover, the ambient noise changes depending on environment, so some safety margin is required to maintain a positive signal to noise ratio (SNR) regardless of the sensitivity of the receiver. Note the acceptable range of signal strength values associated with your specific equipment before flying. Some data link modems will also give an indicator of the quality of the data received, usually some ratio of erroneous data to error-free data.

It may be surprising that at the time of this writing, few civilian data links for sUAS are encrypted or secured. Generally, a unique ID, serial number, or network address is all that is used to determine if one is controlling the correct aircraft. Sometimes the ID must match between the aircraft and ground station transceivers, other times it is specified or selected from a list of aircraft on the operator screen.

Besides transmitting telemetry through the data link, some autopilots may have a built-in flight data recorder. This consists of an onboard memory—in some cases even an SD card that may be removed for easy download. A copy of the flight data is saved in this memory, as well as transmitted through the data link. If there is a lost-link situation or the aircraft is lost, the flight data may be downloaded later to aid in analysis or troubleshooting. In some cases, this memory is useful for diagnostic functions and precise measurements because the user may select the type of data to be collected, as well as the rate at which it is collected. More information may be saved in this memory than most data-links are able to transmit back. For example, much faster accelerometer data may be saved for vibration analysis purposes. Depending on various factors, the recording of additional data may affect autopilot flight performance due to the extra processing required to save the data.

The autopilot receives magnetic compass information through the use of sensors called magnetometers. Because of their sensitivity to magnetic fields, magnetometers are primarily mounted external to the autopilot and aircraft electronics, though they are occasionally incorporated onto the autopilot circuit board. In most cases, the magnetometer includes a sensor for each aircraft axis. Not only can they be used to detect the compass heading of the aircraft, but sensing in all three axes helps determine the general attitude of the aircraft. Magnetometers must be calibrated to the earth's magnetic field in the region in which flights are to be flown, as there is significant magnetic variance from one region to another. Some can also compensate for certain local magnetic fields generated within the aircraft. The easiest compensation to apply is called "hard iron" compensation. This takes into account nearby sources of fixed magnetic fields, such as a magnet or any other consistent magnetic field. An offset is applied to correct for these fields. "Soft iron" magnetic fields are harder to compensate for. In many cases, soft iron compensation is included, but because of the high level of processing required, it is not always found as a feature. Any ferrous metals contribute to soft iron distortion. These (not otherwise magnetized) metals are influenced by the magnetic field of the earth itself, and the distortion read by the magnetometer will change as the aircraft changes orientation relative to the earth's magnetic field (Cork 2014). It is important to note that compensation for both hard and soft iron errors assumes the sources of distortion are fixed in relation to the magnetometer; if an interfering object is in motion relative to the magnetometer, these compensation techniques will not succeed.

Barometric altimeter and airspeed information are received by pressure transducers. The sensors that measure static pressure may be discrete devices mounted to the autopilot without any air plumbing, or for more precision, include a port to attach a tube to aircraft static ports. Barometric pressure sensors are sensitive enough to measure very small changes in pressure. In fact, some sensors are able to detect a change of pressure over only 6 inches of height change. It is important to calibrate the pressure altimeter prior to flight. Systems from different manufacturers vary on how the calibration is done. Some systems have the same interface as a manned aircraft, with a manual altimeter setting input. While the manual process requires knowledge of what to input, the benefit is that it may usually be updated while flying as with manned altimeters. Other systems may automatically calibrate when power is applied, though recalibration may be required if a significant amount of time passes before flying. Systems like this generally must be on the ground in order to be calibrated, and usually indicate altitude in reference to ground level.

A differential pressure transducer is commonly used for airspeed measurement. This measures the difference in pressure between the static ports and the ram air from the pitot tube to indicate airspeed. Many times, temperature is also measured for true airspeed indication. The pressure produced by a pitot tube flying through the air is extremely small—the measurement range for airspeed sensors is usually only a fraction of 1 pound per square inch (PSI). In most cases, these sensors may be "zeroed" during preflight checks. When testing airspeed sensors, do not make contact with a pitot tube to blow into it or use compressed air. Simply pressing one's finger firmly to a pitot tube is sufficient for most quick air leak checks, as the trapped air is compressed enough by your finger to be detected. Figure 13.6 shows the end of a combined pitot/static tube with the differential pressure sensor.

Autopilots rely on the global positioning system (GPS) for primary navigation; GPS is essential to the navigation of almost all unmanned aircraft systems. Most systems have one or more GPS receivers, which update the aircraft position multiple times per second. It is rare to find a GPS antenna mounted to an autopilot itself—it is usually mounted remotely. Figure 13.7 shows various GPS receivers. A front and backside view of one is shown at

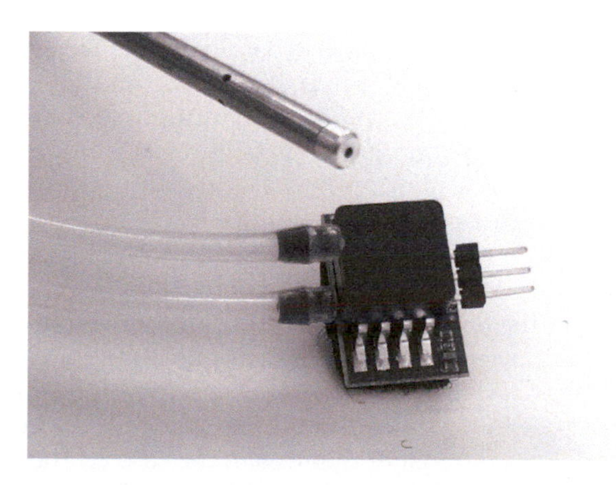

FIGURE 13.6
Pitot/static tube with sensor. (Image courtesy Nathan Maresch.)

FIGURE 13.7
Various GPS receivers. (Image courtesy Nathan Maresch.)

the top of the figure. Note that some have the ceramic GPS antenna integrated into the receiver itself, while others are designed to have a remote antenna attached via a cable. The receiver must have a lock on at least four satellites to obtain a three-dimensional position. The signals from the satellites are extremely weak. In fact, the signals are weaker than the ambient noise (noise floor); they have a negative SNR. A GPS receiver uses algorithms that look for patterns in the noise consistent with what it expects from the GPS signals, and then isolates the signals from the noise. Needless to say, the GPS antenna should have a clear view of the sky at all times. If the receiver is in motion, the velocity and direction of movement can be determined based on previous GPS readings. This is why the GPS aircraft heading is incorrect or not available when the aircraft is sitting on the ground. The GPS receiver itself does this processing.

Because GPS signal distortions are caused by various effects such as passing through different layers of the earth's atmosphere, correction signals are generated by stations at a known location to augment the original information and increase positional accuracy; the distortions may then be subtracted out. There are various acronyms or abbreviations associated with this type of augmentation. You may hear it referred to by DGPS, SBAS, WAAS, and many others. Sometimes the stations themselves transmit the correction signal on a local level, other times satellites retransmit them, as in the case for the wide area augmentation system (WAAS) (Cork 2014). There are subscription-based services that provide proprietary corrections to improve the positional accuracy of GPS as well. Systems like this may improve GPS accuracy by many times. Some unmanned aircraft need even better accuracy for maneuvers such as automated landings, and employ real-time kinematic (RTK) GPS. This uses a station at a known location similar to augmentation systems, and establishes a lock on a specific portion of the carrier wave of the GPS signal. Because the carrier portion of the signal is a much higher frequency than the modulated code inside, greater positioning precision is possible with the faster timing. These systems again require data to be transmitted from the base, to the receiver aboard the aircraft. Without regular transmissions from the base, the remote receiver will quickly lose lock on the carrier and revert to using standard GPS. With RTK corrections, the positional accuracy may be down to one inch or less.

To determine the basic movement and attitude of the aircraft, the autopilot makes use of accelerometers and gyros. These are usually found on board the autopilot itself. Accelerometers detect acceleration along an axis, where gyros detect angular velocity about an axis (Woodman 2007). These sensors require calibration or initialization. It is important to note that because accelerometers by definition measure acceleration, they will indicate the acceleration due to gravity. Accelerometers and gyros found on most autopilots are not used alone for primary long-term navigation. The reason for this is that the small accelerometers and gyros suitable for these systems do not have enough precision to be useful over time; errors are additive without some correction. These sensors are relatively low-cost microelectric mechanical systems (MEMS) devices fabricated using the same methods and materials as those employed by the integrated circuit industry (Cork 2014). Many times on the outside, a MEMS device looks exactly the same as any other solid-state chip mounted on a circuit board, and is just as small—even with all the elements for three accelerometers and three gyroscopes (sometimes even three magnetometers as well) in the same package. It is important to remember that MEMS are mechanical devices, and subject to damage and failure. Though damage is extremely rare, care should be taken to avoid sharp drops and falls. They should be checked regularly for proper operation.

Autopilots generally require tuning to properly control and stabilize a particular airframe. The flight control systems commonly consist of a proportional, integral, differential (PID) closed feedback controller for each set of flight surfaces. Each of these three components has numeric settings, called gains. The gains are established during initial test flights, and usually do not require further tuning once set. The proportional gain applies a correction that is proportional to the magnitude of the error in the system. As the amount of error decreases due to the correction, so does the correction itself. Performance may become an issue with smaller errors, as the corrections, while decreasing take longer to complete. Figure 13.8a shows a simulation of proportional gain response. The horizontal line is the set-point where we are trying to get to, and the curved line is where we are in time relative to the set-point (the controller response starting at the lower left corner). If not concerned with performance, proportional control alone may be acceptable, though small corrections also present another problem besides speed. Because of nonlinear forces opposing the correction, and due to the fact that a nearly-zero error results in a nearly-zero correction force, proportional feedback alone may not be able to fully correct for a disturbance—there may be a dead band near the desired set-point called steady-state error. Oscillations will result if proportional gain is excessive (Kilian 2006).

The integral component is added to decrease or eliminate steady-state error. This adds correction based on accumulating previous error. This can, however, introduce instability in the system. The system can now easily over-correct, and create oscillations because no braking is applied until after overshooting. Figure 13.8b shows the dangerous, unstable response of having too much integral gain. If this response were to happen while tuning in flight, the aircraft would likely shake apart.

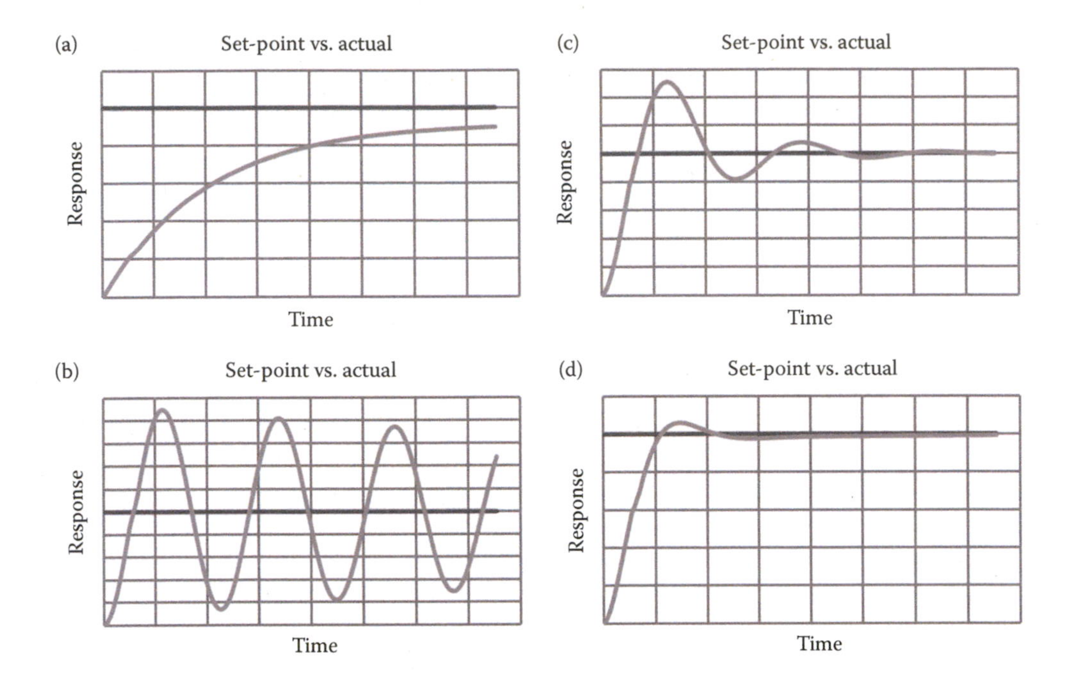

FIGURE 13.8
(a) Slow response plus steady-state error. (b) Dangerous unstable oscillations. (c) Excessive overshoot. (d) Properly tuned. (Image courtesy Nathan Maresch.)

Finally, to reduce overshoot the derivative portion of the PID is added for some extra braking effect just before reaching the intended target. It also gives extra force at the beginning of the movement, trying to predict the future error based on the rate of change of the current error. Figure 13.8c shows derivative gain added, but the integral portion is still too high. There is still too much overshoot. Too much derivative gain can slow down performance due to the damping effect. There are various techniques for tuning PID controllers, but tuning in general may be a challenge because changing one term also affects the other two. The desired response is fast and precise, without too much overshoot or instability/oscillation. Figure 13.8d illustrates a desired response. Some autopilot systems have built-in methods to automatically tune the system via an ongoing process of gain tweaking. Because of the ability to correct for a changing dynamic, this is called adaptive control (Kilian 2006). Adaptive methods vary from fairly simple algorithms to highly complex forms of artificial intelligence. In some cases, the automatic tuning (or even marginal manual tuning settings) may cause system instability when a dynamic such as a wind velocity changes. Because of this, it is important not to save manual gain settings close to a point where the system becomes unstable.

13.2.3 IMU/INS Stabilization Systems

A small UAS inertial measurement unit (IMU) incorporates (usually MEMS) gyros and accelerometers to sense and input rate of change and rotation data to the inertial navigation system (INS), which, in turn, uses this data to compute velocity, attitude, and position of the aircraft without the need for external sources of real-time information. These systems estimate their current position from a known heading and starting point. From this starting point, inertial systems detect acceleration (and therefore speed) of the aircraft in each axis and rate of rotation about each axis (and therefore heading) to perform dead reckoning. For some inertial systems, these three accelerometers and three gyros are all that is required; however, many times a magnetometer is added to detect the magnetic compass heading and help reduce the rotational error of the gyro system. In some cases, GPS is also added to correct errors that accumulate (Cork 2014). The fusion of data from these sensors is processed with a computing device to determine relevant navigation information such as current position, velocity, and heading. Stereo-optical sensor data (optical odometry) may be incorporated with the INS data as well through the use of an extended Kalman filter. This provides accuracy close to that of GPS (Kelly et al. 2007).

Sophisticated inertial navigation devices contain more precise sensors for measurement than MEMS/GPS sUAS autopilots generally do; for example, the drift rating on a fiber optic gyro (FOG) in an INS system may be a fraction of a degree per hour or better. It is typical for the MEMS accelerometers and gyros on board sUAS autopilot systems to have position errors of many meters or drift 30° per hour or more (if uncorrected). Generally, the more precise accelerometers and gyros within dedicated inertial navigation systems are much larger, consume more power, and are more costly than are their autopilot counterparts (i.e., IMU/INS integrated into the autopilot). As a result, dedicated INS systems are rarely found on sUAS; they are typically found aboard manned aircraft and large unmanned aircraft.

13.3 Large UAS Navigation Systems

Larger unmanned systems contain many more subsystems and redundancies than smaller systems do. They typically contain components that are standard on manned aircraft, but not usually found on sUAS. Some of this equipment includes emergency locater transmitters (ELTs) and and both very high frequency (VHF) and ultra high frequency (UHF) audio communication relays (so that operators can talk to manned aircraft as if they were on board). Military systems also carry Identification Friend or Foe (IFF) systems, as well as signals intelligence packages. They may contain different types of navigation systems, such as both an autopilot system and an INS system. Usually, one or more of these systems is equipped with an identical or similar redundant unit in case one system fails. A high-performance gimbaled sensor package with an electro-optic camera, infrared camera, and laser rangefinder/designator is standard. Other types of sensors such as lidar and synthetic aperture radar may be found.

13.3.1 BLOS Communication

Beyond line of sight (BLOS) communication generally refers to the types of radio signals that can travel over the horizon. This includes ground waves below 2 MHz, which follow the terrain of the earth, and sky waves between 2 and 30 MHz, which bounce off the ionosphere to another location. Command and control communication using these frequencies is rare for a number of reasons. For one, the required bandwidth is generally unavailable at such low frequencies.

Aircraft command and control via satellite communication is commonly designated beyond LOS because of the ability to operate beyond the capability of a single LOS station; it is not however beyond LOS from a technical standpoint because LOS microwave frequencies are used. LOS signals are transmitted from a station to a satellite relay in space, and back down to a receiver on the aircraft. Satellite communication is rare in the civilian realm due to the high costs involved. Large unmanned aircraft commonly use this for long endurance, faraway missions. Operators can control an aircraft from halfway around the world, and it is common to do so. The drawback is that significant delay or latency of up to a number of seconds may be encountered. Sometimes, the relatively precarious launch and recovery of the aircraft is handled by a local crew communicating directly with the aircraft where the delay is minimal, and for the mission they transfer control to the crew operating the sat-com link. In other cases, the entire flight may be conducted via satellite.

Large unmanned aircraft usually have multiple satellite transceiver systems. For high data-rate, wide-band communication, a movable directional antenna is used aboard the aircraft to maintain a lock on a satellite in space. This is characterized by a large hump on the front of many larger aircraft; the hump usually being a radome to protect and provide unobstructed room for this large dish to move about and keep a satellite lock regardless of the movement of the aircraft. For lower data-rate satellite command and control communications, a smaller fixed antenna on the top of the aircraft is sufficient for effective communication (Gundlach 2012). Two examples of lower data-rate satellite links are the Iridium and Inmarsat satellite networks. The antennas and hardware for some of the lower data-rate satellite communication systems are small enough to be used on small civilian unmanned aircraft.

Instead of using a satellite, another aircraft may act as a relay for BLOS command and control communication. This aircraft is equipped with extra radio equipment to relay information between the remote aircraft and the ground station. It may simply follow a loiter waypoint, or have a more complex flight plan to maintain proper distance and altitude between these two. The main mission for this relay aircraft is to preserve signal strength between both the remote aircraft and the ground station. To further increase the usable LOS range, as long as there are no physical obstructions, directional tracking antennas are sometimes used at both the ground station and on board large unmanned aircraft. Due to size, it is uncommon to use tracking antennas on small unmanned aircraft.

13.3.2 Alternative Navigation

Before GPS was available and INS systems were larger and more expensive, other methods for unmanned aircraft navigation were common. One method incorporated stations with tracking antennas maintaining a lock on the aircraft. The bearing of the aircraft could be determined by locking on to the telemetry signal coming from the aircraft. Special timing information was also encoded into the signal to determine the distance. Another method involved identifying geographical features by comparing previously acquired terrain imagery with images taken in flight. Because a literal comparison was made between the two, this was called direct reckoning (Austin 2010). A lot of research effort has been put into a similar, more modern method that does not require pre-acquired imagery, called simultaneous localization and mapping (SLAM). This method uses cameras (or similar sensors) to form an internal map of the visual surroundings, and determines its relative position within. This technology is fairly challenging to implement due to the large amount of real-time processing required. While already useful in indoor environments, recent research shows utility for outdoor settings as well (Wang et al. 2013).

13.3.3 Auto-Takeoff and Auto-Landing Systems

For sUAS, auto-takeoff and auto-landing systems are fairly straightforward. Fixed-wing aircraft commonly rely on launchers for auto-takeoff. The autopilot usually brings the throttle to full before launch, or senses the forward movement of a launch and brings the throttle up as a result. For landing, there are various strategies, but setting a constant descent slope is common, with various parameters available for the final flare and landing. RTK GPS is used for adding precision to the final flare and touchdown point for some systems.

Besides having the automatic takeoff and landing systems integrated into the autopilot system itself, large unmanned aircraft sometimes have separate dedicated systems to handle this. These usually interface with the autopilot system, and are called automatic launch and recovery (ALR) systems. Again, some employ RTK GPS for precision landing. These systems may land the aircraft autonomously in the event of a communications system failure. An equipped aircraft sometimes includes a special transponder that a tracking system near the landing area locks onto. This is especially useful for approaches in high-dynamic landing situations such as shipboard operations and during adverse weather conditions. Other ALS systems are primarily optical-based, employing dual camera systems for landing terrain/precise target object tracking.

13.4 Additional C2 Topics

In closing this chapter, two additional C2 topics remain to be covered. These will be discussed in the following subsections.

13.4.1 Open-Source Systems

Open-source software and hardware have become very popular in recent years in the realm of small unmanned aircraft. Open-source communication standards such as the Micro Air Vehicle Communication Protocol (MAVLink) have made it easier to develop interoperable ground station software and autopilot systems. An operator with a compatible autopilot can test out and choose the best ground station software for their application because the two communicate using the same standard. Since the protocols themselves are open source, the user may wish to further customize one for their specific use— first in developing custom telemetry data, and second in displaying this data by modifying the system software user interface.

To define what "open source" is, there are two important types of code in the software world that we need to talk about first. One type is called source code. Humans can read and understand this type of code, and this is what a software developer uses to write programs using a programming language of their choice. Developers can usually add their own comments or notes about what a certain section of code they have written does; these comments can help the developer remember, as well as help others understand what that section does. This source code is kept on hand, and tweaked and improved for future versions. Once satisfied with a particular version, the developer feeds the code into a compiler. This throws away all the comments and translates all of the human-readable code into object code. Generally part of the end product, the object code or machine code is made up of the raw zeros and ones that a machine (microcontroller, computer processor, etc.) can understand. Everything from firmware to operating systems to the apps that people download to their smartphones is made up of bits of object code. If one were to open an object code file as a text document, it would appear as gibberish. If we wanted to make a change to it to add a feature for example, we would not know where to start—it would be nearly impossible with object code.

When the source code is made available, either in the public domain or via a license, then the program is open source. If a user wants to learn about how their software works, they may view the source code to find out. Users may have the ability to change or modify the way the software works to make improvements or introduce new features depending on the particular license. Contrary to popular belief, open source is not always free; the software license spells out any terms that go along with it. As long as one complies with the license, it is usually fine to charge for open-source software—whether it is your own work or not. The person who buys it usually has the freedom to give it away for free however. Sometimes the original developer sells it, and people buy it from them in order to support the developer's work. In many cases, a community of developers volunteer to work together to create more complex programs that would be difficult to create alone. This is usually made available free of charge, while preserving the freedoms that go along with having the source available. Most open-source autopilot ground control software and the firmware code that runs on the aircraft belong to this category. The associated licenses many times state that the license and terms must remain in the event that you make your own modifications. For example, it would be against the terms of most open-source licenses to take some open-source autopilot code and compile it with your own special features, and then sell commercially as part of your own complete autopilot package (mentioning nothing about source code to the customers).

Open-source hardware is similar to open source software. Developers of open-source autopilots and related hardware make the design files and schematics for this hardware available to the public. Users may use the schematics for troubleshooting purposes, or build their own if they wish. Companies can take the design and begin building and selling the hardware, even if the original developer was already doing so. Sometimes, one group creates an improvement that is eventually incorporated into hardware assemblies from others. While this means that there may be various versions of hardware by different manufacturers, there is great potential for improvement between versions due to the diversity of design expertise.

13.4.2 The Human Element in C2

The most important aspect of operating an unmanned aircraft is safety. For current sUAS operations, many systems rely on external pilots (EP). External pilots have the ability to take over control, and ensure split-second reaction time in the event of an impending collision or emergency. Communication between this person and the autopilot operator is critical. While the external pilot has control, the autopilot operator may need to modify the flight plan or other settings to ensure safety. Additionally, an aircraft handoff may need to be made between two different operators/ground control stations. Generally, provision is made for the new operator taking over to establish connection before the original operator disconnects; communication is the key between these parties (Gundlach 2012). A visual observer (VO) scanning the sky for traffic and communicating with the operator and external pilot is also important to maintaining the safety of the flight. In many cases, there is also a separate payload operator (PO). The communication between the air vehicle operator (AVO) and payload operator is important to the mission at hand. The payload operator may need the aircraft to be maneuvered to a certain area to keep objects or individuals of interest in view of the camera or sensor. It is useful for the payload operator to have the current flight plan in view to know how to maneuver the camera, and when, for example, the camera may be obscured by landing gear during a turn.

In the future, collision avoidance systems or similar technology may reduce the requirement for having external pilots. Detection systems using high-resolution cameras may pick out relative airborne movement, show it to the operator, and prompt them for an action. A single individual may operate systems that currently require separate air vehicle operators, external pilots, payload operators, and launch/recovery crew; regardless, the ultimate goal is to operate unmanned systems with at least the same level of safety as manned aircraft.

The ground control stations (GCS) that are used to monitor and control the flight vary in size and complexity, from large control centers to mobile units, to small tablets and mobile phones. Operator interfaces for unmanned aircraft continue to evolve and advance. Though the basic flight data are essentially the same between aircraft, the display of this information will vary from system to system. The only standardized information found on some operator screens is the inclusion of the "six-pack"* instruments found on manned aircraft; some systems have dropped this inclusion due to the screen space premium. Each interface style and design has its own unique set of strengths and weaknesses. Regardless of the amount of information displayed to the operator, it is important that critical information is prominently displayed and easy to understand and comprehend. For an operator interface, there is usually a delicate balance between operator workload/ease of use and functionality/access to information. Some interfaces are very intuitive and easy to use but have relatively few features. Others have a higher learning curve but include a larger feature set. The perfect human interface is both easy to use and full-featured. A display with a lot of "clutter" is not necessarily a bad thing as long as the information is consistently located on the display, and important flight information is easily located. Other features that improve the human interface include the consistent use of color-coding and graphical indicators. For example, one interface may simply display the system voltage as a number on the screen (Figure 13.9a). An improved interface may color-code it for proper operating range (Figure 13.9b); an even further improvement may be a graphical indicator with color-coded operating range along with the actual readout (Figure 13.9c). As space becomes a premium, it may only be practical to display a limited set of information in this way. Other features such as a warning annunciator system (with indicators fixed in location), tabs or drop-downs, as well as the selective use of "drag and drop" commands may also improve functionality while maintaining ease of use. An intuitive display of map data and flight information is essential to any interface; three-dimensional indication

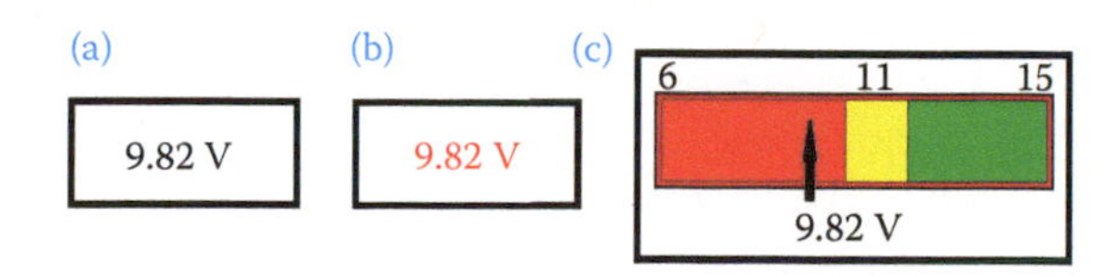

FIGURE 13.9
Voltage status indicator variations: (a) Black text. (b) Red low voltage text. (c) Color-coded red, yellow, green indicator. (Image courtesy Nathan Maresch.)

* Predating the glass panel cockpit, a "six-pack" of round, "steam-gage" primary flight instruments consist of an airspeed indicator, attitude indicator, altimeter, vertical speed indicator (VSI), heading indicator, and a turn coordinator. The reference here is to a similar arrangement of virtual instrument faces displayed on a ground station computer screen.

of topography and flight plan information is also very useful. Color-coding of the flight plan itself may also be used to determine certain information such as a "live" indication (the flight plan is current and active on board the aircraft). Future systems will likely have much more standardization between interfaces. They will be easier to learn and operate, and improve flight efficiency and safety.

13.5 Conclusion

The command and control subsystem/functionality is essential to the operation of any UASs. Through the C2 subsystem, manual control and autonomous flight is achieved and telemetry exchanged between the aircraft and ground station enabling the monitoring of system health and aircraft flight performance. Through the command and control link, an operator may update the aircraft's systems and modify the flight plan (Gundlach 2012). Generally, as UASs become larger, the C2 system becomes more capable and more complex; it also becomes larger, heavier, and more expensive. Despite the capability of current C2 systems, some human oversight and intervention is necessary to maintain the necessary level of safety and, depending on the mission, large and small UAS flight crews may include four people, or more. As C2 systems evolve and mature to become more compact, less expensive, and more capable, it is likely that the requisite level of human intervention will correspondingly decrease, just as occurred in manned flight where transport crews have shrunk from five (pilot, copilot, navigator, radio operator, and flight engineer) to two as the result of improved technologies.

DISCUSSION QUESTIONS

13.1 Define what is meant by the terms, "uplink," "downlink," and "radio frequency." Give the extremes, in terms of frequency and wavelength, of the RF portion of the electromagnetic spectrum. How are frequency and wavelength related? What is spread-spectrum transmission and how does pseudo-random frequency generation enhance this technique? Give examples of applications of spread-spectrum technology. What are ISM frequencies?

13.2 What are waypoints and what role do they play in UAS navigation?

13.2.1 What is meant by LOS transmission? How can the curvature of the earth interfere with LOS communication? What could be done to overcome this limitation? Describe "space wave" RF frequencies? What are the most commonly used frequencies for sUAS? Which of these are LOS frequencies? What is the relationship between wavelength and antenna size?

13.2.2 Describe sUAS autopilot architecture and operation. What is an "interrupt?" Why is the multiplexer important? What is used by an external pilot (EP) in conjunction with the multiplexer to control the sUAS? Describe autopilot modem architecture and operation. Give the various ways that modem information packets are tested to ensure they arrive uncorrupted. What is a "lost link" and what can trigger that event? What happens to the sUAS in the event of a lost link? What is the purpose of an attenuator? What is a magnetometer and what information does it provide to the autopilot? Describe "hard iron" and "soft iron" interference and tell why these may be of concern. Define "pressure transducer" and tell what information these provide to the autopilot. Describe the use and autopilot processing of GPS signals. What is the minimum number of GPS satellites that must be received for a 3D solution? Describe what is meant by DGPS, SBAS, WAAS, and RTK GPS. Describe MEMS sensors and tell what information these provide the autopilot. What is autopilot tuning? Describe operation of a PID controller. What are gains? Proportional gains? What happens if proportional gain is excessive? What is adaptive control?

13.2.3 Describe an IMU and INS and their relationship. What IMU or INS components are MEMS devices and what do they sense? Why are magnetometer and/or GPS inputs sometimes used in inertial navigation systems? What is the purpose and function of a Kalman filter? Compare and contrast the IMU/INS system integrated into sUAS autopilots to the discrete INS found on larger UASs.

13.3 How do the C2 systems of sUAS differ from those of larger UASs?

13.3.1 Describe BLOS communication. Is the use of satellite communications technically BLOS? Explain your answer. What is the disadvantage of using BLOS satellite communications with sUAS and advantages for larger UASs? Why is it common to land BLOS-capable UASs using pilots at the aircraft's home base rather than controllers located in ground stations half a world away? Describe differences between high and low data-rate BLOS systems. Why are tracking antennas generally not used on sUAS?

13.3.2 Describe alternative navigation systems.

13.3.3 What are automatic takeoff and landing and ALR systems? Describe these.

13.4.1 What is MAVLink? What is meant by open-source autopilot/ground station hardware and software and firmware? Describe differences in source code, object code, and machine code? What does a compiler do? Describe open-source licensing and distribution practices.

13.4.2 What are AVO, EP, PO, and PIC? Describe the human element in the C2 sUAS subsystem. Discuss operator interface systems. Discuss those qualities, attributes, and characteristics considered to be desirable and undesirable in a ground station display/human interface.

References

Austin, R. 2010. *Unmanned Aircraft Systems: UAVs Design, Development and Deployment*. West Sussex: John Wiley & Sons Ltd.

Blake, R. 2002. *Electronic Communication Systems*, 2nd Ed. Albany: Delmar.

Cork, L. 2014. Aircraft Dynamic Navigation for Unmanned Aerial Vehicles. PhD diss., Queensland University of Technology. http://eprints.qut.edu.au/71396/1/Lennon_Cork_Thesis.pdf.

Gundlach, J. 2012. *Designing Unmanned Aircraft Systems: A Comprehensive Approach*. Reston: American Institute of Aeronautics and Astronautics, Inc.

Kelly, J., S. Saripalli and G.S. Sukhatme. 2007. Combined Visual and Inertial Navigation for an Unmanned Aerial Vehicle. http://www.robotics.usc.edu/publications/media/uploads/pubs/540.pdf.

Kilian, C. 2006. *Modern Control Technology*, 3rd Ed. Clifton Park: Delmar Learning.

Transportation, United States Department of. 2013. Unmanned Aircraft System (UAS) Service Demand 2015–2035: Literature Review & Projections of Future Usage. Technical Report, Ver. 0.1. http://www.ntl.bts.gov/lib/48000/48200/48226/UAS_Service_Demand.pdf.

Wang, C.-L., T.M. Wang, J.H. Liang, Y.C. Zhang and Y. Zhou. 2013. Bearing-only visual SLAM for small unmanned vehicles in GPS-denied environments. *International Journal of Automation and Computing* 10(5), 387–396.

Woodman, O. 2007. An Introduction to Inertial Navigation. Technical Report, No. 696. University of Cambridge Computer Laboratory. http://www.cl.cam.ac.uk/techreports/UCAM-CL-TR-696.pdf.

14

Unmanned Aircraft Subsystem Integration

William H. Semke

CONTENTS

The integration of subsystems into unmanned aircraft is what allows the aircraft to carry out their intended mission. In most cases, the aircraft is a platform from which one can collect data, make observations, or make deliveries to a desired location or target. Unmanned aircraft provide an ideal platform in situations where rapid deployment may be useful, in environments that may be hazardous, or in missions with extended durations. Examples of the missions where they can be used include search and rescue operations, long duration surveillance, military operations, precision agriculture, sense-and-avoid technologies, and many other applications that utilize the unique abilities of unmanned aircraft systems (UAS).

The discussion of subsystems, and in particular UAS payloads, integration comes from experience and processes developed and utilized at the Unmanned Aircraft Systems Engineering (UASE) Laboratory at the University of North Dakota [1,2]. The intent is to provide the reader an overview of the process along with practical information regarding challenges and requirements associated with UAS-specific integration. Figure 14.1 shows part of the UASE team and three of the aircraft flown; on the left is the Air Robotics AV-7 that is capable of hand launching, in the middle is the UAV Factory Penguin B that is capable of fast long endurance flights, and on the right is the BTE Super Hauler that is a stable heavy lifter platform.

Payload subsystems come in a variety of sizes and configurations, from extremely complex, with heavy computational capabilities, to relatively simple designs. Each system tries to meet the objectives of the operator. Figure 14.2 shows the MQ-1B Predator Multi-Spectral Targeting System, or MTS-A sensor ball. This payload integrates an infrared sensor, optical cameras, a laser designator, and a laser illuminator into a single package that provides multiple assets for military operations. On the other end of the sophistication scale is an example of a UAS payload system, a flying insect capture system using deployable nets

FIGURE 14.1
UASE team with multiple UAS platforms during a payload flight testing campaign.

FIGURE 14.2
Multi-spectral targeting system, or MTS-A sensor ball, integrated onto a MQ-1B Predator UAS for military surveillance operations. Photo Credit: http://duncandavidson.com

attached to its wings, as shown in Figure 14.3. This system was developed to support a study on the distribution and spread of disease by mosquitos. These two examples, among thousands of others, illustrate the wide range of potential applications of UAS payload systems [3–5].

The myriad of applications of UASs with their respective payloads provides great opportunity for researchers in many fields to implement and exploit the capabilities of UASs. The capabilities are unique to the airframes, and many technical issues arise in the integration of the payloads into the aircraft. This chapter provides guidelines for payload developers to ensure safe and effective payload design and integration.

FIGURE 14.3
Flying insect capture payload installed on the wings of an airborne UAS.

14.1 The Design Process

The integration of subsystems onto a UAS ideally uses a systems engineering approach to design. Payloads typically involve many components that must be integrated together to carry out the desired mission. This approach necessitates multiple steps on the way to designing a complete UAS subsystem. The six main steps involved in a payload design include:

- Concept development and trade studies
- Preliminary design review
- Critical design review
- Fabrication
- System testing
- Flight testing

Each of these steps and UAS-specific highlights are reviewed in this chapter. The design process is a complex and iterative process, and multiple texts address valuable design methodologies that can be implemented [6–8]. This chapter provides an overview of the processes that have been proven to be successful in UAS payload development within the UASE Laboratory with emphasis on several issues that are unique to UAS subsystem development.

14.1.1 Concept Development and Trade Studies

Concept development and trade studies are the first steps in the design process and often the most critical to mission success. This is where the objectives of the system are defined and should include end users of the data and/or system. It is crucial that the designers and engineers understand fully the technical data or capabilities that are desired. At this phase, several conceptual designs are developed and assessed with a preliminary engineering analysis. This step often entails "back of the envelope" calculations to make sure there are no significant challenges that are unreasonable or "showstoppers." In parallel with these activities, trade studies are often initiated to explore potential system components. While searching for solutions, innovative technologies are often discovered, which can be incorporated into the system. During the concept development and trade studies, the following areas are addressed:

- Several conceptual designs are developed
- Design sketches are produced
- Parts and vendors are identified
- New parts/ideas are incorporated into the proposed "final design"
- Engineering analysis is initiated

UAS-specific actions at this phase include determining the type of aircraft that is necessary. The aircraft type is based primarily on range, maneuverability, lifting capacity, and costs. With hundreds to thousands of airframes from which to choose, many times there are several viable options, and the best fit typically comes down to a cost versus benefit comparison. Generally fixed-wing aircraft give longer duration flights and can travel greater distances but lack the maneuverability of rotorcraft. Rotorcraft, however, allow for fixed location hovering and station keeping but lack the flight efficiency of fixed-wing aircraft.

In all aviation technologies, the size, weight, and power (SWaP) requirements of the subsystem must be reduced as much as possible. The SWaP requirements dictate the airframe and its subsequent performance during missions. On the large end of the spectrum is the Northrop Grumman Global Hawk with the ability to carry over 3000 pounds of payload, as shown in Figure 14.4. On the other end is a micro-UAS that is only capable of carrying specialized payloads—only a few grams in mass—as shown in Figure 14.5.

FIGURE 14.4
The Northrup Grumman Global Hawk UAV is capable of heavy lifting and long, high-altitude missions. Photo Credit: Wiki "Global Hawk 1" by U.S. Air Force photo by Bobbi Zapka-http://www.af.mil/shared/media/photodb/photos/070301-F-9126Z-229.jpg. Licensed under Public domain via Wikimedia Commons-http://commons.wikimedia.org/wiki/File:Global_Hawk_1.jpg#mediaviewer/File:Global_Hawk_1.jpg

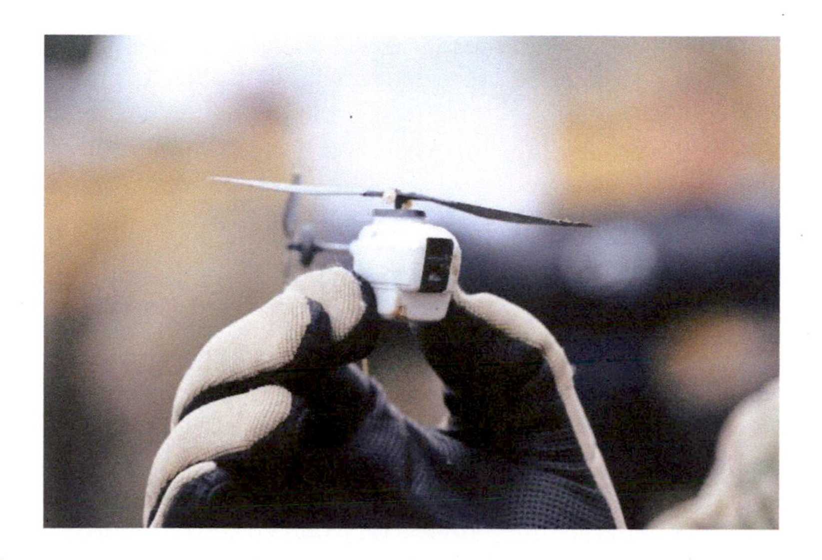

FIGURE 14.5
A Black Hornet nano helicopter illustrating a UAS that is capable of carrying a few grams in an extremely small, lightweight system. Photo credit: "Black Hornet Nano Helicopter UAV" by Richard Watt-Photo http://www.defenceimagery.mod.uk/fotoweb/fwbin/download.dll/45153802.jpgMetadata source: http://www.defenceimagery.mod.uk/fotoweb/fwbin/fotoweb_isapi.dll/ArchiveAgent/5042/Search?FileInfo=1&MetaData=1&Search=45155077.jpg. Via Wikimedia Commons-http://commons.wikimedia.org/wiki/File:Black_Hornet_Nano_Helicopter_UAV.jpg#mediaviewer/File:Black_Hornet_Nano_Helicopter_UAV.jpg

Once several concepts are generated and the performance versus cost benefits explored, this is the best time to have additional discussions with the end users. Often, the desires of the end users may not coincide with the technical abilities and/or budget of the project. Now is the time to discuss what modifications may be appropriate. For example, the end user may desire a long-range radar system, but the only currently available systems capable of achieving the requested range require immense size and weight using today's technology. So either a compromise must be made between requirements and capabilities or the project is at standstill and cannot move forward until technological advances improve the capabilities or an innovative solution is found. It is important that all parties involved must remain acceptable to change and adaptable to continued technological advancements while striving to meet the requirements. Unfortunately, sometimes the technical barriers are insurmountable with airborne capable technologies and alternate systems are deemed more appropriate. In these cases, the process of an airborne system ends at this point. However, with slight modifications and/or additional technical development, the capabilities of airborne systems are immense and can most times meet the true needs of the end user.

14.1.2 Preliminary Design Review

The preliminary design review (PDR) is the next milestone in the design phase. This is a formal review of the design where all parties have another chance to review the progress and offer recommendations. This includes engineers and other technical experts along with the scientists and other end users who are looking for data or a useful tool. The design concepts should have had the appropriate engineering analysis completed with sound engineering practices. Design drawings of the system should also be prepared to illustrate the system layout. This phase must serve as the last chance to make significant design changes to the system. A list of the deliverables of the PDR is provided:

- Engineering analysis to verify design concepts
- Preliminary design drawings
- Preliminary parts list
- Preliminary vendor list
- Cost budget estimate
- Design simplifications and cost savings
- Timeline for completion

Throughout this stage of the design, aircraft Interface Control Documents (ICDs) are extremely useful. Some commercial grade UASs have excellent, readily available documentation, while other aircraft only provide very limited resources. The documentation level may significantly influence the decision on what airframe to implement and/or type of integration to be chosen. Useful information may include wiring schematics along with capacities, mounting locations and types, physical dimensions, weight and balance information, and many other useful design parameters.

UAS-specific actions at this phase include power and weight budgets, data storage, and airframe flight worthiness. Typically, both power and weight resources are scarce in UASs, so accurate accounting of each is critical. Therefore, care should be taken to choose the most efficient and lightweight components available.

Power estimates of each component with the respective efficiencies are tabulated. This budget will help define whether the system can be self-powered or needs to use aircraft resources. In general, a self-powered system with its own battery simplifies integration and aids in the test and evaluation phase. However, using aircraft resources reduces the weight of the payload and can add life to the mission. The drawback to using aircraft resources is that they require detailed airframe interface information. This information is included in detailed ICDs of the aircraft. Many aircraft developers and operators are reluctant to provide this information and would prefer stand-alone systems for safety and proprietary concerns. Using aircraft resources reduces the availability of the resources to flight operations and any unforeseen problems could result in potential aircraft malfunction. When stand-alone systems are used, most malfunctions result in only loss of data, not loss of aircraft.

A weight budget, accounting for all components, is also created accurately. The payload lifting capacity of the aircraft must be sufficient to carry the load, but proper balance must be maintained for safe flight operations. The total weight of the aircraft must be no greater than the maximum weight allowed for the make and model of the aircraft and the center of gravity must be within the allowable operational range of the aircraft [9]. The balance of the aircraft is often the limiting factor, depending on the aircraft type. Some significant weight components that are often overlooked are wiring and mounting hardware, which can account for a relatively large fraction of the total. Sensor packaging and payload layouts are also introduced during this phase. The method used and mounting locations chosen have significant impact on maintaining safe weight and balance for the aircraft.

Another UAS-specific issue is the potentially huge amount of data that may be collected during flights. Decisions regarding onboard data storage versus telemetry to the ground are made at this phase. There is a limited data link between the aircraft and the ground that

FIGURE 14.6
An externally mounted missile system attached to the wing of a UAS. Photo Credit: TSGT Scott Reed, U.S. Air Force-http://www.dodmedia.osd.mil/; VIRIN: DF-SD-06-14785.

restricts the amount of data that can be transferred. Many types of compression schemes can be helpful to reduce the incredible volume that may be created. Though these methods help in reducing the amount of data to a greater extent, the techniques often reduce the quality of the data obtained, especially in high definition imagery. Without a data link, the other alternative is to store the data onboard the aircraft and retrieve it when it lands. This is often the easiest method to implement and also reduces the amount of hardware required to be flown. However, the drawback is that the data are not real-time, therefore not appropriate for applications where instantaneous information is required. In many cases, onboard data storage does meet the needs of the end user and becomes the preferred method of data acquisition due to the simplicity and weight savings that go along with it.

One other consideration during this phase of the design process is the assessment of the payload to ensure the flight worthiness of the aircraft. Besides meeting the weight and balance requirements, any additions and/or modifications to the aircraft and their impact on the flight operations have to be assessed. Examples include wing pods or any other device that changes airflow during flight or may cause a flutter condition to develop, as shown in Figures 14.6 and 14.7. Most often, it is recommended that no structural or aerodynamic alterations be made to the host aircraft. However, if it is deemed necessary to do so, the impact must be analyzed for safe flight. This analysis is often complex and requires thorough modeling and testing [10]. Therefore, if it is not absolutely necessary, it is not recommended.

14.1.3 Critical Design Review

The critical design review (CDR) is the next phase of the design process. This is the final paper design and analysis that is done prior to ordering and acquisition of parts and components for the fabrication phase. The design is finalized and all the stakeholders have a final review prior to fabrication. A complete description of the design specifications and

FIGURE 14.7
An external wing pod designed to carry radar equipment providing a clear field of view without interference from the airframe and engine.

capabilities are presented for final comment. A timeline for completion is also provided that will guide the schedule along with allowing the end users sufficient time to prepare for the product. Many times this is the phase that must be completed before the distribution of funds to support the fabrication of the system. A detailed list of items included in the CDR includes

- Completed engineering analysis to verify design concepts
- Final design drawings
- Final parts list
- Final vendor list
- Final cost budget
- Final timeline for completion
- Preliminary testing to support design assumptions

The UAS-specific actions at this phase include the finalized power requirements and weight and balance calculations. It is also the time when the parts and vendors, which provide the quality and traceability needed for UAS applications, are identified. While the Federal Aviation Administration (FAA) does not currently dictate the exact standards for parts and materials, many of the best practices developed in manned aviation are good guidelines for application on UASs. Some of the more relevant requirements are locking connectors, aircraft or equivalently rated hardware, traceable materials, fastener back-off protection, and proper circuit protection.

Electrical connections are essential for operation and some sort of locking mechanism is needed to ensure the connectors stay connected during flight. There is substantial vibration during flight operations and any connection that becomes loose can lead to catastrophic results. Therefore, several types of LEMO-style push-pull connectors, as well as Powerpole-style connectors have been implemented in our designs over the years. These types of connectors have been proven to be reliable and effective in flight operations. Other types of connectors also work well, but the user should verify that they are not susceptible to loosening or disconnection during flight.

The hardware and materials used in the payload fabrication should be of the highest quality to ensure defect-free construction and problem-free operations. All hardware should be aircraft (AN) grade or equivalent. AN-grade hardware meets high levels of performance and can be traced to the place and date of origin. These fasteners are free from defects, which can occur in lesser grade fasteners, and have extremely high reliability. This makes them an ideal choice in fasteners, especially in flight critical components. They also often come with a back-off prevention method. The use of traceable materials is also recommended to ensure high-quality, defect-free stock from which to manufacture components. This results in dependable parts that meet the design specifications and perform well over the entire lifetime of the payload.

The electronic components to be used should all be evaluated for temperature ratings. It has been observed that elevated temperatures are a common occurrence in small UAS payload systems that do not have any built-in provision for active or passive cooling designed into the system. The small aircraft typically have an enclosed area that houses the payload and over time, elevated temperatures are often encountered. These result in electronic failures, unusual operations, or shutdowns. Conversely, cold temperatures encountered during winter operations and potentially high elevations also produce failures and shutdowns. In both cases, care needs to be taken in the selection of components and proper testing procedures conducted prior to operations. Proper testing protocols are outlined in Section 14.1. "Systems Testing" of this chapter.

Electromagnetic interference (EMI) is another significant factor to be concerned with in UAS payload development. Improper wiring or unanticipated EMI can lead to interference with command and control telecommunication links and also interfere with GPS signals. Proper shielding of sensitive electronics in payload systems can help reduce the risk of interference with aircraft systems, or the possibility of aircraft EMI interfering with the payload operations. In practice, the use of conductive enclosures, EMI gaskets, and copper tape can help control EMI effectively. Coax and other types of shielded cables with grounding also reduce unwanted broadcasting antennae and receiving of stray electromagnetic radiation.

Proper circuit protection is required to reduce the risk of electrical overheating and potential fire. By installing a fused connection to the battery, short circuits and other electrical malfunctions are isolated from the rest of the UAS. In addition, external access allowing the payload power to be turned on and off is very beneficial. Traditional aircraft "remove before flight" switches work well for this feature. The ability to turn on the payload with an external switch is important, since the preflight testing and other aircraft preparations may take significant time during which the power available to the payload keeps decreasing. Therefore, the ability to easily turn on the payload near the time of takeoff increases the operation time of the payload system.

14.1.4 Fabrication

Throughout the fabrication of the payload system, high-quality fabrication and workmanship must be maintained. The fabrication should follow the stated design specifications and parts manufactured according to the part drawings created. All wiring should be secured appropriately; this can be done with tie-downs or tacking down the wires so that they are not free to move. This prevents oscillations that may lead to loose or intermittent connections resulting in failure of the payload to operate effectively. Back-off prevention should also be used on all threaded fasteners to prevent loosening and possible separation. This can be accomplished by several methods, such as by using wire tie-downs, locking washers, deformable threads, nylon thread inserts, or liquid thread locker. The environment in which UASs operate is often very dynamic and these vibrations can cause components to loosen, leading to payload failures and loss of data collection opportunities.

The final product is only as good as the fabrication process; therefore, attention to detail and utilization of proper manufacturing procedures and equipment are required. After producing a quality design using solid engineering principles, the fabrication should result in a high-quality system that will operate as expected. Poor fabrication methods may result in ill-performing systems with many integration issues. The workmanship is also a reflection on the overall perception of the system, and a poor quality appearance may give the impression that it is not a high-quality system. Therefore, it is recommended that experienced technicians perform or oversee the fabrication of the payload system.

14.1.5 System Testing

System testing is the next step in the preparation of UAS payloads for flight. Laboratory testing greatly improves the performance of the payload systems and helps ensure their safe and effective operation in the field. Potential problems are much more easily discovered, and corrected, in a laboratory setting. In this environment, access to the payload and analysis equipment is more readily available and will not impact the limited flight opportunities that exist in the field.

EMI testing is one of the most critical aspects to be aware of and understood. First and foremost, the command and control link to the aircraft cannot be compromised. Many commercial autopilot systems use 900 MHz or 2.4 GHz transmission frequencies for the command and control link. The aircraft must also continually communicate with the GPS to know its position; therefore, the payload cannot interfere with GPS frequencies. In the original GPS design, two frequencies are utilized: the L1 channel at 1575.42 MHz and the L2 channel at 1227.60 MHz. Preliminary lab testing using a spectrum analyzer and receiving antennae and "sniffer" probes evaluate the radiated emissions from the payload when powered and operational, as shown in Figure 14.8. If relatively large amplitude signals are emitted from the payload near any of the command and control points or GPS frequencies, additional shielding and/or suppression is recommended. If there are no significant levels of EM radiation near the frequencies of interest, the EMI testing can move onto the next level. The next level involves installing the payload into the aircraft and supplying power in full operational mode to assess any potential interference from the payload to the airframe, or from the airframe to the payload. Either of the interference situations can be a significant problem that requires attention. The payload must not interfere with the aircraft operations, but if the aircraft interferes with the payload, the mission operations may be sacrificed and not return any useful data.

Vibration testing should also be done on the completed payload system. Typically, vibration testing is done to evaluate the survivability of payload in a harsh vibration environment. The testing is intended to discover potential faults resulting from the dynamic environment experienced in flight. The faults often discovered include connecter dislodgment where electrical connections are not completed or unreliable, fastener loosening where structural components do not stay rigidly attached, and failure or destruction of

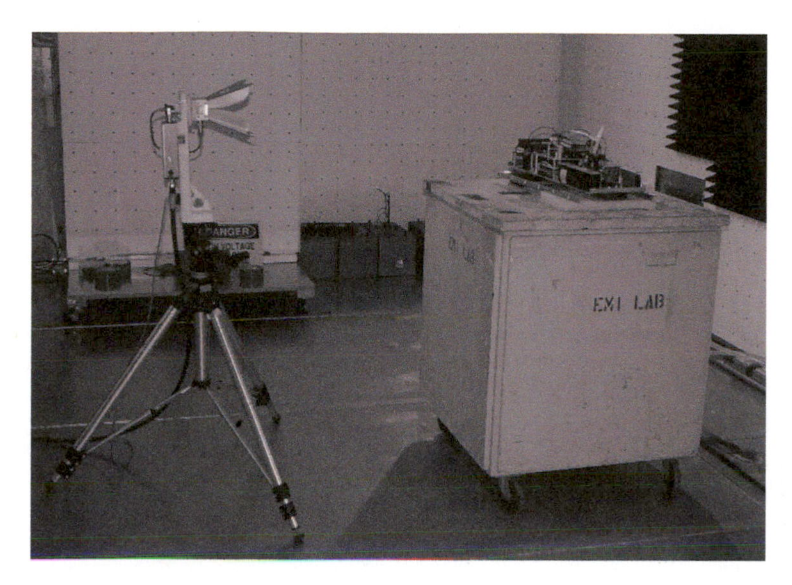

FIGURE 14.8
An electromagnetic field test being conducted to measure the radiated emissions from the payload to check for potential interference with command and control links.

vibration intolerant components in the system. Any one of these failure modes, or any other, can lead to a catastrophic failure of the payload during operation. The vibration levels encountered in different types of aircraft vary considerably due to a variety of reasons, including fixed-wing or rotor-type aircraft, size and weight of the aircraft, electric motor or gas-powered engines, conventional landing gear or belly-land configurations, and many other factors [11]. Therefore, information regarding the vibration environment in the ICD for the aircraft should always be used when available.

A base-line vibration excitation test that may be used is the Protoflight Minimum Acceptance Vibration Test Levels (MAVTL) specified by the NASA Payload Flight Equipment Requirements and Guidelines for Safety-Critical Structures [12]. The MAVTL values are shown in Figure 14.9. This testing level exercises the equipment to a level of excitation that exposes dynamic faults in the system. The random profile is preferred to a swept sine approach because it more evenly excites the structure over a relatively broad frequency range and does not overly excite a single, potentially resonant, frequency of the structure. This broad excitation frequency encompasses most of the excitation frequencies experienced in many types of UASs.

The payload system is placed in the flight configuration on the mechanical shaker and excited for 1 min at the random excitation levels indicated in the MAVTL test procedure. A vibration testing system is shown in Figure 14.10, illustrating the shaker, amplifier, and controller necessary for environmental vibration testing.

After the system has undergone vibration excitation, it is powered up and thoroughly tested for performance. Any issues that arise should be documented and investigated until a reasonable fix is produced. In some cases, it may be appropriate to operate the system during the test to evaluate the systems performance while the excitation is active. Other payloads only require that the system be able to withstand the vibration and continue operations when activated. The appropriate test depends on the needs of the payload and the environment in which it is operating.

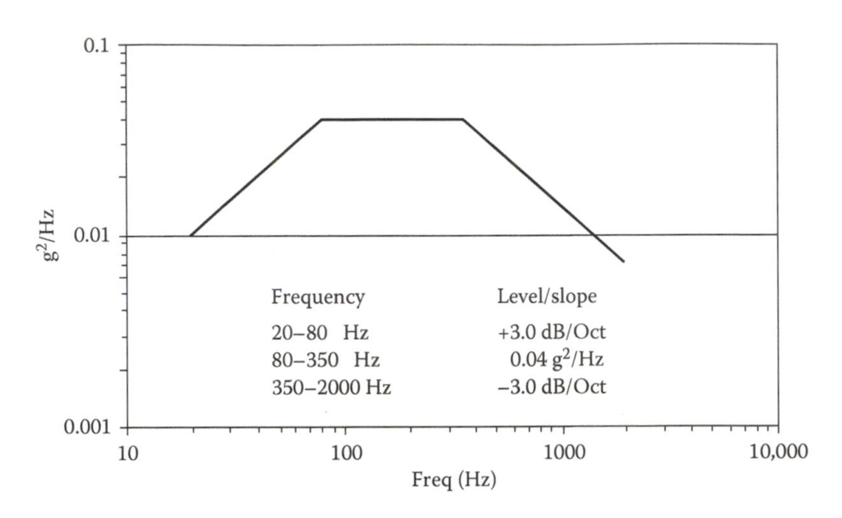

FIGURE 14.9
Minimum Acceptance Vibration Test Levels (MAVTL) specified by the NASA Payload Flight Equipment Requirements and Guidelines for Safety-Critical Structures.

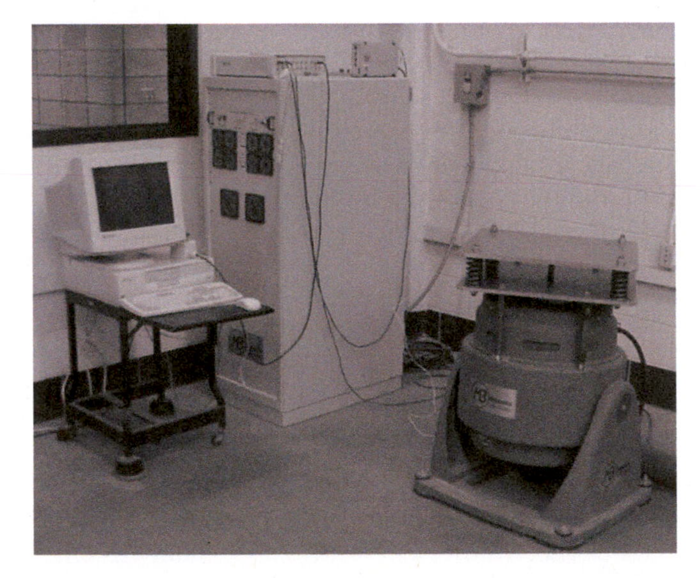

FIGURE 14.10
Vibration testing equipment, including the electrodynamic shaker, amplifier, and computer controller.

Thermal testing in the environmental conditions likely to be encountered during flight operations is also recommended. Many electronic components have safe storage temperatures that are greater than the reliable operational temperatures. For example, both hot and warm environments may interrupt operation of the payload systems during flight. In past experience, problems most often have been due to the elevated temperatures inside the enclosed payload bay of an aircraft on warm sunny days. With the electronics operating, internal temperatures over 40°C may be encountered, resulting in limited mission performance. Cold temperatures can be encountered during operations in winters or in high-altitude missions experienced by some payloads. To assess the ability of the payload to operate in the anticipated environment, a thermal chamber (Figure 14.11) is an ideal way

FIGURE 14.11
A thermal altitude testing chamber used to simulate operating conditions that may be encountered by UAS payloads.

to subject the system to adverse conditions in a controlled manner. The payload is placed into the chamber and commanded to the appropriate temperature while it is running. Continued exercise of the payload should be done throughout the test to assess the performance. Once the desired temperature is achieved and the equipment comes to a thermal equilibrium without issue, the test can be terminated. In addition to thermal testing, many of the chambers also allow for altitude simulation. Most often, small UAS payloads are operated at relatively low altitudes, but some may be utilized in high-altitude missions. In these cases, appropriate altitude testing should also be performed.

The final test procedure that is unique to UAS testing is the mobile truck test. This test exercises the equipment by providing a moving platform with a true GPS data stream along with a vibration environment in self-powered mode. This procedure is an excellent tool to utilize before the expense of flight operations. Typically, flight testing is very limited due to associated costs, logistics, and availability, so truck testing helps to identify any problems in the full operational mode prior to flight. Shown in Figure 14.12 is a test rack built to hold payloads in the back of a pick-up truck. The payload system should be powered in the same fashion as in flight, with battery packs or simulated power from the vehicle. In this case, the rack system allows the payload to operate in a rotated position so that image collection of scenes could be captured. This test often exposes payload issues that would not have been discovered by other laboratory bench testing. The combination of a moving vehicle to exercise the GPS equipment and operating the entire system in a mobile, dynamic environment is close to actual flight testing that can be achieved relatively easily without restrictions that come with flight. It is highly encouraged that the mobile truck test be performed on every payload system prior to flight testing. Flight opportunities are not readily available, and therefore the payload should be thoroughly tested to take advantage of the flights that arise.

14.1.6 Flight Testing

The last and final phase of payload development is the flight testing, which is arguably the most enjoyable as well. While enjoyable, there are several facets of preparation that have

FIGURE 14.12
A payload system being installed for a mobile truck test. This testing platform closely simulates the flight environment in a relatively easy and inexpensive manner.

to be completed for this phase. These will help ensure a successful and productive flight testing experience. The three deliverables that should be addressed in this phase are the following:

- Detailed test plan
- Flight plan
- Equipment list

A detailed test plan is required to properly assess the performance of the payload. With the limited flight opportunities and limited flight time, an effective and efficient test plan is critical. Things that need to be included in the test plan are the settings to be made on the equipment and proper battery and data storage capabilities. In addition, the required altitude and flight lines should be calculated for proper data collection. Any specialized support equipment should also be prepared for flight testing. If any interfaces with the aircraft are required, it is important to describe the interface completely and confirm that it is compatible with the airframe. With a detailed test plan and proper prior testing, the entire process of integration in the field should be easily accomplished. Field integration of a payload into a UAS for flight testing is shown in Figure 14.13.

Proper flight planning with the aircraft operator will mean that the desired data will be obtained. The operator will help establish safe flight operations within the regulations that most closely match the desired values. At times, the payload developer's wants may not be achievable under regulatory restrictions or aircraft performance. For example, a maximum altitude restriction may be applicable in the flight area being used or the desired airspeed is less than the stall speed of the aircraft. Therefore, it is important to work with the aircraft operator to establish acceptable parameters that most closely meet the wants of the payload developer with regard to the rules and capabilities of the

FIGURE 14.13
The field integration of a UAS payload into the airframe for flight testing.

aircraft. This planning allows for more effective flight testing, as the appropriate flight planning can be done prior to operations.

The final preparation for flight testing is to compile a complete equipment list of all the necessary gear and supplies required for the flight operations. This should include all flight hardware, spare parts, test and evaluation tools, and any other supplies necessary for operations and repairs. It may also include personal items that can help make the flight operations more enjoyable. This includes items such as sunglasses, proper clothing, sunscreen, insect repellant, a camp chair, snacks, and water. A detailed list will help guarantee that you have all the necessary supplies in the field and flight operations will meet the mission objectives in a safe manner, as shown in Figure 14.14.

FIGURE 14.14
The start of safe flight of a UAS after utilizing "best practices" in the payload development outlined in this chapter.

14.2 Concluding Remarks

This chapter is meant to provide guidelines that have been proven to be successful for UAS payload development. As regulations and airspace operations continue to develop in this rapidly evolving industry, increasingly specialized regulations and requirements may be imposed. These new rules may become standards in legal operation and development. Therefore, it will be prudent to use the information provided herein in compliance with all the state, local, and federal regulations that may apply.

DISCUSSION QUESTIONS

14.1 How do intended subsystems influence air vehicle design? What are the important factors to consider when pairing subsystems with platforms?

14.2 Summarize the relationship between the concept design, the preliminary design review, and the critical design review.

14.3 Describe how a thorough design review can either help or hinder the operational test and evaluation process.

14.4 What is the market risk in designing aircraft for specific payload subsystems?

CHAPTER QUESTIONS

14.1 Name the six steps in the design process for successful UAS payload development.

14.2 Use the web to find two different fixed-wing UASs with the same wingspan, one using an internal combustion engine and one using an electric motor, and compare the flight times and payload capacity.

14.3 Use the web to find two multirotor aircraft with a different number of rotors and comment on the capabilities of each system.

14.4 Name six types of missions that data collection using a UAS is preferred over a manned platform.

14.5 Name a mission where a real-time data stream is required.

14.6 What type of hardware (screws, nuts, and bolts) should be used in UAS payloads?

14.7 Name two types of fastener back-off prevention methods that may be used in UAS payloads.

14.8 Name two frequency bands that many UASs use for command and control.

14.9 What is the highest vibration excitation frequency used in NASA MAVTL testing?

14.10 Name the three deliverables for UAS payload flight testing.

References

1. Lendway, M., Berseth, B., Trandem, S., Schultz, R., and Semke, W., Integration and flight of a university-designed UAV payload in an industry-designed airframe, *Proceedings of the Association Unmanned Vehicle Systems International (AUVSI)*, Washington D.C., February 7–9, 2007.

2. Semke, W., Schultz, R., Dvorak, D., Trandem, S., Berseth, B., and Lendway, M., Utilizing UAV payload design by undergraduate researchers for educational and research development, *Proceedings of the 2007 ASME International Mechanical Engineering Congress and Exposition*, IMECE2007-43620, Seattle, WA, November 11–15, 2007.
3. Wehner, P., Schwartz, J., Hashemi, D., Stock, T., Howell, C., Verstynen, H., Buttrill, C., Askelson, M., and Semke, W., *Evaluating Prototype Sense and Avoid Alternatives in Simulation and Flight*, AUVSI's Unmanned Systems North America 2013, Washington, DC, August 12–15, 2013.
4. Lemler, K. and Semke, W., *Delivery and Communication Payloads for UAS Search and Rescue Operations*, AUVSI's Unmanned Systems North America 2012, Las Vegas, NV, August 5–9, 2012.
5. Dvorak, D., Alme, J., Hajicek, D., and Semke, W., *Technical and Commercial Considerations of Using Small Unmanned Aircraft Systems in Agriculture Remote Sensing Applications*, AUVSI's Unmanned Systems North America 2012, Las Vegas, NV, August 5–9, 2012.
6. Larson, W. and Wertz, J., *Space Mission Analysis and Design*, Third Ed., Microcosm Press, El Segundo, CA, 1999.
7. Ullman, D., *The Mechanical Design Process*, Fourth Ed., McGraw-Hill, New York, NY, 2010.
8. Dieter, G., *Engineering Design*, Third Ed., McGraw-Hill, New York, NY, 2000.
9. Aircraft Weight and Balance Handbook FAA-H-8083-1A, *U.S. Department of Transportation, Federal Aviation Administration*, Flight Standards Service, Oklahoma City, OK, 2007.
10. Semke, W., Lemler, K., and Thapa, M., An experimental modal channel reduction procedure using a Pareto chart, *Proceedings of the International Modal Analysis Conference (IMAC) XXXII: A Conference and Exposition on Structural Dynamics*, Orlando, FL, February 3–6, 2014.
11. Semke, W., Stuckel, K., Anderson, K., Spitsberg, R., Kubat, B., Mkrtchyan, A., and Schultz, R., Dynamic flight characteristic data capture for small unmanned aircraft, *Proceedings of the International Modal Analysis Conference (IMAC) XXVII: A Conference and Exposition on Structural Dynamics*, Orlando, FL, February 9–12, 2009.
12. Payload Flight Equipment Requirements and Guidelines for Safety-Critical Structures (March 29, 2002). NASA Doc. SSP 52005, sect. 7, p. 7.

15

Detect and Avoid

Dallas Brooks and Stephen P. Cook

CONTENTS

The MITRE Corporation*

"When weather conditions permit, regardless of whether an operation is conducted under instrument flight rules or visual flight rules, vigilance shall be maintained by each person operating an aircraft so as to see and avoid other aircraft."

14 CFR § 91.113, Right of Way Rules (except water operations) Subsection (b).

15.1 Introduction

One of the foundational tenets of flight is the responsibility of the pilot to see and safely avoid other aircraft. From the dawn of manned flight, pilots have been expected to maintain vigilance so as to remain "well clear" from other aircraft.

By the middle of the 1900s, advances in aircraft detection technology (such as air traffic control radar and aircraft transponders) had allowed the air traffic control system to provide more efficient and safe separation between aircraft through purposed technologies—most of which provide more information and more accurate information than a human pilot. Despite these technology advances, however, the ultimate responsibility for separation has remained with the pilot—he or she must maintain vigilance for other aircraft. This concept of the pilot's ultimate responsibility has been a core principle of our aviation regulatory system for decades.

Beyond regulation, this responsibility has been installed and upheld in our judicial system as well. Dozens of court cases have been adjudicated to assess whether pilots have performed the essential function of "see and avoid" in order to remain "well clear" of other aircraft. But for all the emphasis placed upon the linked responsibilities of "see and avoid" and "well clear," neither concept has been defined precisely. Instead, both concepts are left as subjective judgments, are not tied to a specific metric or level of safety, and are left open for interpretation by regulators and the courts. This tenet has remained stable for over a century, despite continual advancements in aviation safety technologies that provide order-of-magnitude increases in aircraft detection, conflict identification, and collision avoidance capabilities.

15.1.1 UAS as a Transformational Technology

The advent of unmanned aircraft systems (UAS) has changed how we approach technology's role in supporting or replacing functions once solely performed by human pilots. Taking a pilot out of the cockpit during extremely dangerous operations (e.g., combat, wildfire support, and radiological leak/spill assessment) has clear advantages—a properly

* Approved for Public Release; Distribution Unlimited. Case Number 15-1403. The author's affiliation with The MITRE Corporation is provided for identification purposes only, and is not intended to convey or imply MITRE's concurrence with, or support for, the positions, opinions, or viewpoints expressed by the author.

equipped and configured unmanned aircraft can perform the same functions, for longer duration, without putting an onboard pilot at risk. But the impacts of the fundamental differences in UASs are not confined to local flight operations—they are felt across the entire airspace system. Removing the pilot from the cockpit means that he or she is no longer able to provide eyes-on, subjective judgment to "see and avoid" other aircraft so as to remain "well clear." This factor, more than any other, has delayed the integration of UASs with manned aircraft in all classes of airspace.

This same factor, however, has a significant benefit as a *forcing function*—for the tremendous capabilities of UASs to save lives and otherwise benefit society cannot be fully exploited until UASs are safely and fully integrated into the fabric of our airspace system. Today, technologies exist that can detect other aircraft and determine potential flight path conflicts long before a human pilot could solely with the naked eye. These technologies have the potential to markedly increase the safety of aircraft operations due to the greater range, accuracy and reliability of the information they can provide to pilots, and their immunity to the stress and fatigue that can negatively affect the pilot's performance. This is particularly true in high-traffic, complex airspace environments such as around busy airports.

15.1.2 Standards as a Driver for UAS Integration

The systems that allow a means of compliance with "see and avoid" and remaining "well clear" are called Detect and Avoid (DAA) systems. The International Civil Aviation Organization defines DAA as "the capability to see, sense or detect conflicting traffic or other hazards and take the appropriate action" [ICAO Annex 2 —Rules of the Air].[1] For DAA systems to be employed safely and effectively, they must be associated with a *performance standard*—a measure of how well, and how reliably, they must work in order to augment or replace that set of human eyes. Standards must also be developed that address how these systems will work in concert with other appropriately equipped aircraft. Ensuring that two such systems cannot, through uncoordinated action, place their respective aircraft in an even greater collision risk situation is paramount. Without such standards, effective DAA systems cannot be safely employed.

Note: "Detect and avoid" and "sense and avoid" (SAA) are used interchangeably throughout this text as they fundamentally refer to the same concepts and capabilities.

The excerpt from 14 CFR 91.113 at the top of this chapter provides the basis for the key regulatory accommodations needed to enable safe, reliable, and affordable technologies that may be substituted for the subjective judgment of a pilot. The most important of these is a quantitative compliance definition of "well clear" that forms the basis of a performance standard for DAA systems. This definition—a quantified specification of the time, distance, or both that must separate aircraft—provides UASs a means to not only meet, but likely exceed the ability of aircraft with onboard pilots to avoid a potential collision. Just as importantly, this same definition supports *the introduction of similar systems into manned aircraft*, allowing pilots the benefit of increased situation awareness as well as reliable alerting for aircraft that may present a collision risk. Progress toward quantified compliance criteria for UASs is detailed later in this chapter.

The second change is semantic—a modification of the term "see and avoid" that changes or removes the word "see," which serves to restrict the process to the human eyeball. Using the more accurate term "detect and avoid" for UASs, or adding a proviso to 14 CFR 91.113 that allows for electronic methods in lieu of seeing would satisfy the regulatory need to open the door to an electronic means of compliance for UASs.

15.2 Regulatory Basis

As referenced at the beginning of this chapter, 14 CFR §91.113, subsection (b) states: "When weather conditions permit, [...] vigilance shall be maintained by each person operating an aircraft so as to see and avoid other aircraft."

It is important to note that this section levies the requirement to "see and avoid" upon all pilots, regardless of what flight rules the aircraft is being operated under.

Subsection (b) further states "When a rule of this section gives another aircraft the right-of-way, the pilot shall give way to that aircraft and may not pass over, under, or ahead of it ***unless well clear.***" [emphasis added]

In addition, 14 CFR 91.111 states that "(a) No person may operate an aircraft so close to another aircraft as to create a collision hazard." This regulation is focused upon the willful unsafe behavior of a pilot purposely flying too close to other aircraft.

For UAS, the challenge is demonstrating compliance with the rules of the air—ensuring that the aircraft, even when not under direct pilot control, is capable of maneuvering sufficiently so as to avoid a potential collision hazard. This challenge may be broken into two main tasks: (1) avoid a collision and (2) remain "well clear." From a systems engineering perspective, these tasks can be viewed as functional requirements that can be decomposed into independent, sequential sub-functions.

15.3 Functions of DAA System

Using the task description for DAA, we conclude that it consists of two major functions: self-separation (SS), the ability of the system to remain "well clear" of other aircraft, and collision avoidance (CA), the ability of the system to prevent a collision with another aircraft.

15.3.1 Self-Separation

Self-separation is the capability of an aircraft to maintain acceptable separation (i.e., remain "well clear") from other aircraft without the need for guidance from an external agent such as air traffic control. Self-separation maneuvers occur at greater times and distances from other aircraft than collision avoidance maneuvers and are intended to be normal, non-obtrusive maneuvers that will not conflict with accepted air traffic separation standards. The intent of self-separation is to ensure the UAS remains "well clear" and preclude, through non-disruptive maneuvering, the need to execute a collision avoidance maneuver. The time and distance minima required for a UAS to remain "well clear" are discussed in Section 15.8.

15.3.2 Collision Avoidance

Collision avoidance, depicted in Figure 15.1, is a DAA system function where a maneuver is initiated to prevent another aircraft from penetrating the *collision volume* (defined as another aircraft being within 500 ft. in the horizontal dimension or within 100 ft. in the vertical).[2] Action is expected to be initiated within a relatively short time or distance between

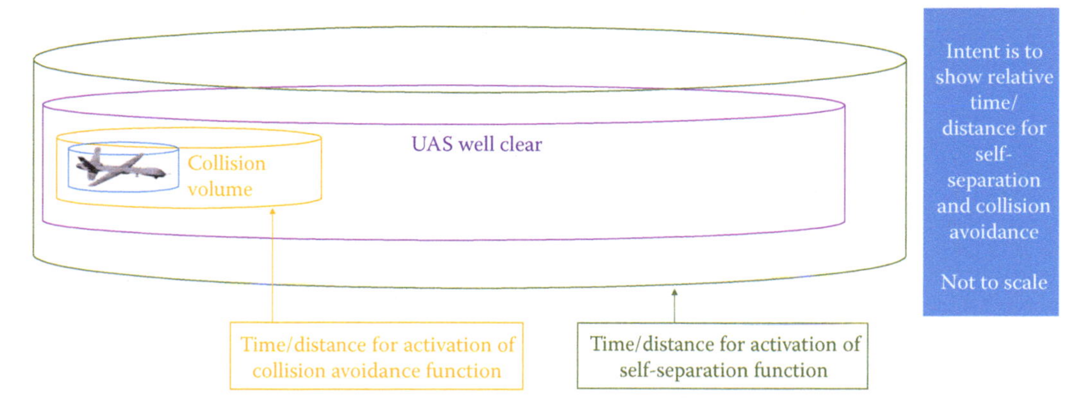

FIGURE 15.1
Relative airspace volumes for the "well clear" and collision avoidance sub-functions.

the two aircraft. The collision avoidance function is a "last-ditch" maneuver intended to engage when all other separation assurance actions fail. Collision avoidance maneuvers may be severe in nature and potentially disruptive to other air traffic control services. The Traffic Alert and Collision Avoidance System (TCAS), in use since the 1990s, uses vertical maneuvers between appropriately equipped aircraft to avoid collisions when other means of separation assurance have not been successful.

Note: "Collision Avoidance System" has a specific regulatory meaning under 14 CFR 125.224 and refers to TCAS. In this chapter, collision avoidance refers broadly to any action that the DAA system takes to avoid a collision after self-separation has been ineffective.

15.3.3 Detect and Avoid: Sub-Functions

To support the core functions of self-separation and collision avoidance, it is necessary to identify and sequence the component sub-functions that support a safe and reliable SS or CA maneuver. In its 2009 report, the Federal Aviation Administration (FAA) SAA Workshop identified eight sub-functions for SAA. In 2011, the U.S. Department of Defense (DoD) SAA Science and Research Panel (SARP) provided its own decomposition which included the addition of three new sub-functions. A breakout of these sub-functions is detailed in Table 15.1.

Note: In both of the approaches listed in Table 15.1, all of the sub-functions apply to both SS and CA. In the specific case of the "Return-to-Course" sub-function, an aircraft which executes a CA maneuver must wait until "well clear" is regained before returning to course.

The primary reason for difference between the two function decompositions was to align the SARP sub-functions with the requirements for DoD-developed SAA systems that were already under development. While the decomposition and nomenclature differ between the two approaches, the underlying processes are largely aligned.

Both the FAA SAA Workshops and the SARP approaches can be grouped using the Boyd cycle tasks of *Observe, Orient, Decide, and Act*. Figure 15.2 depicts the sub-function groupings within the Boyd cycle, with the FAA SAA Workshop function names in italics.

TABLE 15.1
Sub-Functions of the Sense and Avoid Process

FAA SAA Workshops	DoD SAA SARP
Detect	Target detection
Track	Target track
(Not addressed)	Target fusion
(Not addressed)	Object identification
Evaluate	Threat assessment
Prioritize	Threat assessment
Declare	Alert
Determine (action)	Maneuver selection
Command	Maneuver notification
Execute	Maneuver execution
(Not addressed)	Return to course

Source: Adapted from Federal Aviation Administration, *Sense and Avoid (SAA) for Unmanned Aircraft Systems (UAS)*, October 9, 2009 and January 18, 2013; and Office of the Secretary of Defense, *Sense and Avoid Blueprint*, October 26, 2010.

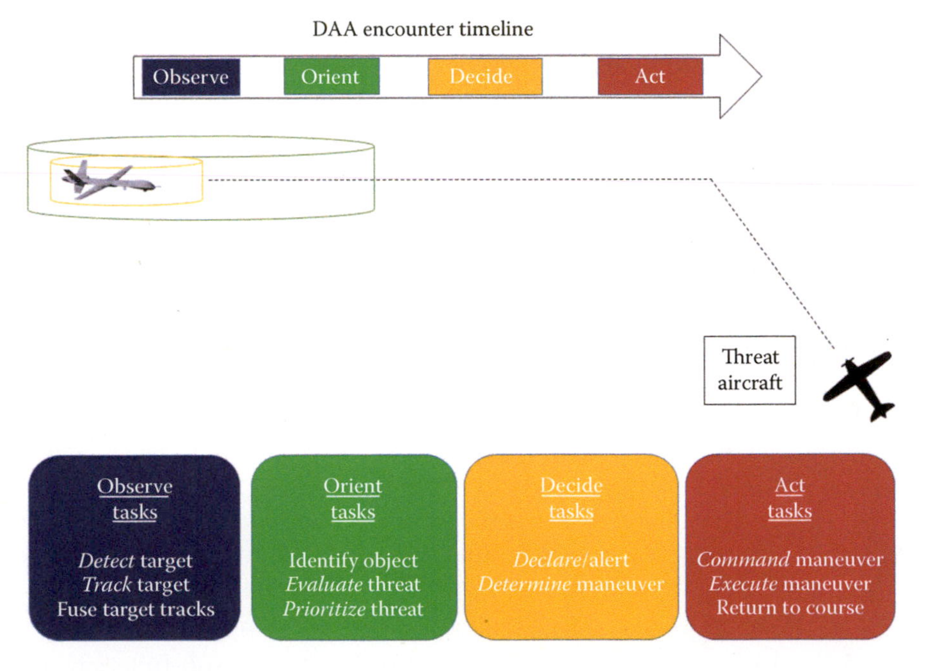

FIGURE 15.2
DAA encounter timeline and task breakdown.

15.4 Process and Functions of a DAA System

This section details the 11 sub-functions of the DAA process depicted in Figure 15.2, categorized by their respective Boyd cycle tasks. Entrance and exit criteria are provided that describe when each sub-function begins and ends along the DAA encounter timeline.

15.4.1 "Observe" Tasks

15.4.1.1 Detect Target

The first step in either remaining "well clear" or avoiding a collision in an encounter with another airborne object is to *detect* the target that poses a conflict. For a UAS, the detect function is accomplished through the use of sensors that detect the presence and location of other airborne objects. These sensors may be on board the aircraft, off-board, or a combination of both. Section 15.7.1 contains a description of the types of sensors that may be used to perform the detect function.

Entrance criterion: UAS on course encounters an airborne object.

Exit criterion: One or more sensors detect the airborne object.

15.4.1.2 Track Target

Once the object is detected, it is necessary for a DAA system sensor to *continue detecting it*. This is referred to as "building a track" which provides position, speed, heading, and/or altitude of the airborne object based on multiple detections from that sensor. There are several algorithms that sensor systems use to attempt to distinguish between targets of interest (such as other aircraft) and extraneous targets (such as birds).

Entrance criterion: Sensor detection of an airborne object.

Exit criterion: Sensor(s) detect the object enough times to build a track with speed, heading, and/or altitude information.

15.4.1.3 Fuse Target Tracks

Once a track has been established, it is vital that this information be fused with other available sensor information to produce a single track for each target. Failure to properly fuse these data results in multiple tracks for the same target, which can confuse a pilot or collision avoidance algorithm. Alternately, if a track is built by one sensor but not correlated (or confirmed) by another, the fusion algorithm must determine the likelihood that the target is actually there. It is the fused information that the pilot or system will use to evaluate the DAA risk.

Entrance criterion: One or more tracks established by processing sensor information.

Exit criterion: A single track provided to the pilot or DAA system.

15.4.2 "Orient" Tasks

15.4.2.1 Identify Object

Presentation of a fused track represents confidence that an airborne object is present and moving in the direction of the track. To validate that the object presents a potential threat, the characteristics of that object are evaluated to assist in classification. The object's size, speed, presence of a transponder, and other factors affect its classification as a threat aircraft, and therefore the strategy in ensuring separation.

Objects that present characteristics not associated with an aircraft (e.g., birds, weather) will invoke other avoidance procedures which are outside the scope of 91.113.

Entrance criterion: A fused track for a given airborne object.

Exit criterion: An identification of the aircraft's characteristics.

15.4.2.2 Evaluate Threat

Once a target is identified as an aircraft, the pilot or DAA algorithm must evaluate the risk of that aircraft either conflicting or potentially colliding with the UAS. This risk is usually measured by determining if the aircraft will violate predetermined proximity thresholds. These thresholds may be measured either in *distance* (e.g., a violation will be closer than a specified number of feet) or *time* (e.g., a violation will occur within a certain number of seconds).

Entrance criterion: An identified aircraft.

Exit criterion: Evaluation of the risk of violating "well clear" or collision volumes.

15.4.2.3 Prioritize Threat

In cases where multiple aircraft are being tracked, the UAS pilot or DAA algorithm must prioritize which aircraft poses the greatest threat and choose the best course of action (maneuver) to reduce that threat.

Entrance criterion: Assessment of the potential risk for one or more threat aircraft.

Exit criterion: Prioritizing which aircraft poses the greatest risk for a "well clear" violation or a collision.

15.4.3 Decide Tasks

15.4.3.1 Declare/Alert

Once a predetermined proximity threshold (either time or distance) is reached, the UAS algorithm must declare to the UAS pilot or flight control system that action is needed to remain "well clear" or avoid a collision. Notifying a human pilot normally involves the presentation of decision aids (such as flashing icons or suggested course tracks and/or altitudes) on a display. For pictorial examples of decision aids on a notional UAS display, see Section 15.7.3.

Entrance criterion: Highest-priority aircraft threat determined.

Exit criterion: Assessment that an avoidance action is needed.

15.4.3.2 Determine Maneuver

Once the pilot or DAA system has been alerted that action is needed, the next step is to determine what the action should be. Changes to aircraft heading, altitude, and speed are options available to the pilot or the system to ensure the aircraft remains "well clear" or avoids a collision. For some DAA systems, the determination of the appropriate action may be left to the pilot. For others, an algorithm may recommend, or in some cases execute, a specific maneuver or range of maneuvers designed to minimize risk. In some cases, coordination with Air Traffic Control (ATC) may be necessary prior to maneuvering the UAS. See Section 15.6 for further discussion on the role of ATC.

Entrance criterion: Declaration that avoidance action is needed.

Exit criterion: Determination of the specific avoidance action needed.

15.4.4 "Act" Tasks

15.4.4.1 Command Maneuver

Once the appropriate course of action has been determined, that action must be communicated to the UAS for execution. This may occur through the remote pilot making a control input to the UAS, or it may be directly commanded by the UAS flight control system.

Entrance criterion: Avoidance maneuver selected.

Exit criterion: Command of avoidance action delivered to UAS.

15.4.4.2 Execute Maneuver

Once the maneuver action has been commanded, the UAS control system must execute it. The timing of that execution is critical—an otherwise correct maneuver, executed too early or too late, might not effectively resolve the conflict. The decision time of the pilot, the communication latency (delay) of the UAS control link, and the aircraft's flight performance dynamics all play a role in determining the acceptable maneuver execution window.

It is important to note that the maneuvering situation remains dynamic. In some cases, the other aircraft may also change direction or otherwise maneuver in an unanticipated manner, decreasing or negating the effectiveness of the separation maneuver. When this occurs, it is possible that the selected maneuver must transition into a more abrupt collision avoidance maneuver.

Entrance criterion: Avoidance maneuver command sent.

Exit criterion: Avoidance maneuver executed.

15.4.4.3 Return to Course

Once the UAS has executed the self-separation or collision avoidance maneuver to resolve the projected conflict, the UAS must now return to its intended course. That course may be

- As filed in the aircraft's flight plan
- As previously directed by air traffic control
- As required to comply with appropriate airspace restrictions

The Return to Course action may be accomplished by either the pilot or the DAA system. The criterion for determining completion of this sub-function is similar to the "clear of conflict" guidance used in the FAA-developed Traffic Collision Avoidance System (TCAS). In some cases, the pilot may need to contact ATC to obtain a new approved route or receive permission to return to the existing clearance.

Entrance criterion: Avoidance maneuver completed.

Exit criterion: UAS is established on its original or amended course.

15.5 The Role of the Pilot

Today's unmanned aircraft systems are actually not "unmanned" at all. For the vast majority of UASs operating today, human pilots either manually control or actively monitor the flight of the aircraft with the full capability to intervene when conditions require. Detect and avoid systems for piloted systems must therefore account for the role and responsibilities of the UAS pilot when addressing potential traffic conflicts. The degree of direct control or intervention that the pilot has in the DAA process is determined by the architecture of the DAA system. The three general types of DAA systems are

1. Pilot in-the-loop
2. Pilot on-the-loop
3. Pilot-independent (automatic)

Each of these architectures allocates the DAA sub-functions to the pilot differently. Table 15.2 provides a summary of how the various sub-functions are allocated under each category.

15.5.1 Pilot in-the-Loop

Per the FAA SAA Workshop report, "a UA pilot, who (actively) controls the aircraft flight path, is said to be 'in-the-loop.' A pilot in-the-loop has the ability to either control the UA directly or has the ability to immediately affect the trajectory of a UA controlled by an onboard computer."

During pilot-in-the-loop operations, the primary responsibility for the Evaluate, Prioritize, Declare, Determine, Command, Execute, and Return-to-Course sub-functions are all allocated to the UAS pilot. This approach has the advantage of leveraging a pilot's

TABLE 15.2
Allocation of DAA Sub-Functions

Allocation of Detect and Avoid Sub-Functions (P = Pilot, U = UAS)			
Sub-Function	**Pilot in-the-loop**	**Pilot on-the-loop**	**Automatic**
Detect target	U[a]	U[a]	U
Track target	U	U	U
Fuse target tracks	U	U	U
Identify object	U	U	U
Evaluate threat	P	U	U
Prioritize threat	P	U	U
Declare alert	P	U	U
Determine maneuver	P[a]	U/P[a]	U
Command maneuver	P	U/P	U
Execute maneuver	P	U/P	U
Return to course	P[a]	U/P[a]	U

[a] ATC may have involvement depending on class of airspace, applicable flight rules, decision time, and other factors.

ability to apply human judgment and right-of-way rules to the maneuver. However, this system has the significant disadvantage of being reliant upon a continuous command and control link between the ground station and the aircraft at all times. When that link significantly degrades or is lost, the aircraft loses its ability to self-separate or execute a collision avoidance maneuver.

15.5.2 Pilot on-the-Loop

A UA pilot, who provides flight path guidance to an onboard computer which controls the aircraft flight path, is said to be "on-the-loop." A pilot on-the-loop has the ability to immediately affect the trajectory of a UA if necessary.

During pilot-on-the-loop operations, primary responsibility for the Evaluate, Prioritize, and Declare a function is allocated to the electronic system. Depending upon the pilot's actions, the Determine, Command, Execute, and Return-to-Course sub-functions may be handled in two ways:

1. If the pilot *intervenes*, the DAA system shifts to a secondary role—notifying the pilot via alerts and recommending action. The DAA system will continue to monitor the situation through the Track, Evaluate, Prioritize, Declare, and Evaluation sub-functions.
2. If the pilot *does not intervene* (e.g., fails to maneuver), the system is capable of executing a self-separation or collision avoidance maneuver. This system provides a useful "last resort" after an attempt to notify the pilot that "action is needed" has not received a response.

Pilot on-the-loop DAA systems have the advantage of keeping human judgment in the equation but taking some of the burden off of the pilot. Another advantage is that they are capable of executing an avoidance maneuver even when the command and control link to the UAS is degraded or lost.

15.5.3 Pilot-Independent

Pilot-independent SAA systems have the capability to manage the UAS flight path without any pilot notification or intervention. Primary responsibility for the Evaluate, Prioritize, Declare, Determine, Command, Execute, and Return-to-Course sub-functions are all allocated to the electronic DAA system. The systems perform the entire DAA cycle without the need for pilot notification or intervention.

As with pilot-on-the-loop operations, the system is capable of executing an avoidance maneuver even when the command and control link to the UAS is degraded or lost. A drawback of the pilot-independent system is that it may not be able to respond correctly to rapidly changing and/or unanticipated scenarios where human judgment is invaluable.

15.6 The Role of Air Traffic Control

For aircraft operating on an instrument flight rules (IFR) flight plan, the air traffic control system normally ensures the appropriate level of separation between other aircraft operating under IFR. However, ATC is not responsible to separate IFR aircraft from visual flight rules (VFR) aircraft, though they may provide traffic advisories and suggest appropriate maneuvering on a time-permitting basis.

As discussed in Section 15.2, the pilot of an aircraft is ultimately responsible to "see and avoid" other aircraft. However, per 14 CFR § 91.123:

> When an ATC clearance has been obtained, no pilot in command may deviate from that clearance unless an amended clearance is obtained, an emergency exists, or the deviation is in response to a traffic alert and collision avoidance system resolution advisory.

The section further states:

> Each pilot in command who, in an emergency, or in response to a traffic alert and collision avoidance system resolution advisory, deviates from an ATC clearance or instruction shall notify ATC of that deviation as soon as possible.

While no regulatory language yet exists that covers DAA maneuvers, the precedent cited for TCAS is clear: deviation from an ATC clearance (or instruction) is authorized in order to prevent a collision. The pilot of a UAS is similarly responsible to notify ATC as soon as possible in the event a DAA maneuver is executed either by the pilot or by the system.

For self-separation maneuvers (collision is not imminent, but maneuver is necessary in order to maintain "well clear"), the most relevant language is contained in 14 CFR § 91.181. This section requires aircraft under IFR to maintain their assigned course, but provides an exception ("does not prohibit") maneuvering the aircraft to pass "well clear" of other air traffic control.

As DAA systems become adopted and certified, clarification in these and potentially other aviation regulations will be needed to ensure a clear understanding of the expectations and procedures for compliance.

15.7 DAA System Components

15.7.1 Sensors

For DAA systems, sensors replace the human eye as the primary method for detecting other airborne objects. Effectively purposed sensors can detect the presence of an object and its relative location to the UAS. When sensors are combined with the appropriate computational systems, this information can be used to calculate the speed and direction of the object (track). This track information provides the basis for the Evaluate Threat, Prioritize Threat, Declare/Alert, and Determine Maneuver functions.

Understanding the capabilities and limits of various sensors is critical to performing safe and reliable DAA. The DAA system must account for potential error in the sensor system(s) in order to compute an appropriate safety margin for DAA decisions and maneuvering. In some cases, using multiple sensors or sensor types can provide additional accuracy and/or reliability.

In DAA systems, sensors are often characterized in terms of energy (active or passive) or modality (cooperative or non-cooperative).

Active sensors direct energy (such as radio, sound, or light waves) toward the target in order to determine its presence and location. Examples of active sensors are radar and LIDAR. Since the amount of directed energy is carefully controlled, active sensors can attain significant precision in determining the relative location of another airborne object. In order to generate this energy, active sensors often require more power, and may be larger and heavier than passive systems.

Passive sensors detect an attribute of the target, such as sound, heat, or reflected light. Examples of passive sensors include acoustic, thermal, and electro-optical. Passive sensors do not need to generate energy to accomplish detection, and thus may be more compact and require less power than active sensors. Passive sensors may be less accurate than active sensors due to variability in the source attributes of the target, such as available light or sound propagation.

Cooperative sensors rely on each participating aircraft to broadcast its location. This is normally accomplished through an interactive process of transmission (interrogation) and response known as *transponding*. This can also be accomplished by aircraft continually broadcasting their location without the need for interrogation. Examples of cooperative sensor systems include Mode C and S transponders and Automatic Dependent Surveillance-Broadcast (ADS-B).

Non-cooperative sensors can detect aircraft which are not equipped with transponders or similar systems. This may be accomplished through the use of active detection systems (such as radar) or through passive systems (such as an optical camera). Table 15.3 provides a summary of common sensor attributes.

Note: Sensor(s) may be located on board the aircraft, or as part of an external system (such as air traffic control radar).

15.7.2 Avoidance Algorithms

For automated DAA systems, computer logic augments or replaces the cognitive processing of a human pilot to determine the best action to avoid a potential loss of separation or collision. The system calculates the relative position and rate of closure for the potential threat aircraft, and mathematically determines a maneuver to resolve the conflict safely. This information may be provided to a human pilot as a suggested course of action, executed automatically by the DAA system, or both.

TABLE 15.3

Attributes of Selected Sensors

Sensor	Modality
Mode C/S Transponder	Cooperative
ADS-B	Cooperative
Optical	Noncooperative, passive
Thermal	Noncooperative, passive
Laser/LIDAR	Noncooperative, active
Radar	Noncooperative, active
Acoustic	Noncooperative, active or passive

Source: Adapted from Lacher, Andrew, Maroney, David and Zeitlin, Andrew (The MITRE Corporation). *Unmanned Aircraft Collision Avoidance—Technology Assessment and Evaluation Methods*, 2007.

Ensuring the safety and reliability of avoidance algorithms requires extensive testing with potentially millions of simulated encounter scenarios before they can be recommended as safe for use. Avoidance algorithms should, whenever possible, recommend a maneuver that is compliant with the appropriate aviation regulations regarding course changes and aircraft right-of-way rules. Per existing regulations, when a collision is imminent, a pilot may maneuver as necessary.

15.7.3 Displays

In general, DAA displays fall into three categories: informative, suggestive, or directive.

Informative displays (see Figure 15.3 for a notional example) are the simplest type of DAA display. Informative displays only provide current information to the pilot on the position (and in some cases, altitude) of surrounding air traffic. They do not provide information on potential avoidance maneuvers, nor predictions of where the displayed aircraft may be going. Informative displays are used to aid an "in-the-loop" human pilot.

Suggestive displays provide the same position information as informative displays. Suggestive displays also provide maneuver options or recommendations to assist the pilot in maintaining "well clear" or avoiding a collision.

Suggestive displays may use a variety of graphical decision aids to assist the pilot's comprehension and decision process. Examples of decision aids include the depiction of "safe zones" or "keep away zones" for the UAS in order to maintain an acceptable distance away from an intruder. A notional suggestive display is shown in Figure 15.4, with the green

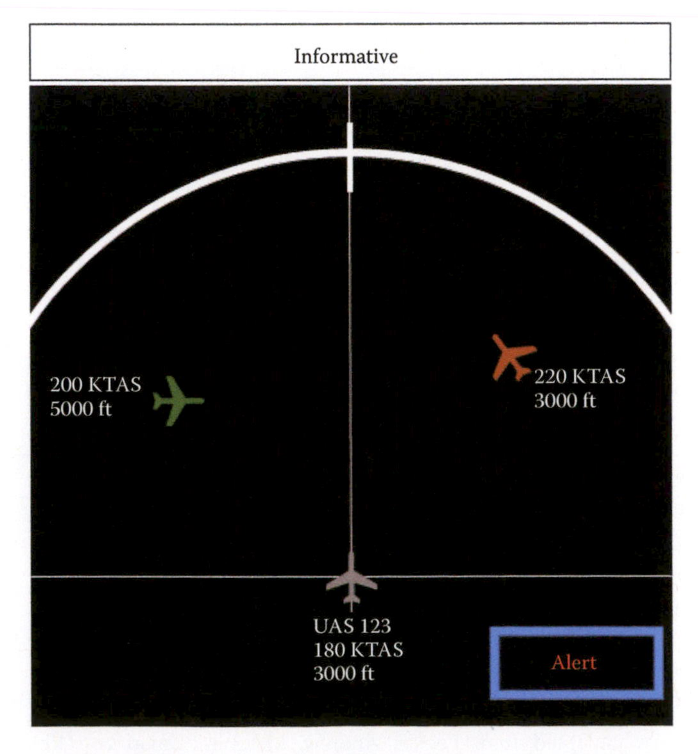

FIGURE 15.3
Notional example of an informative display for UAS.

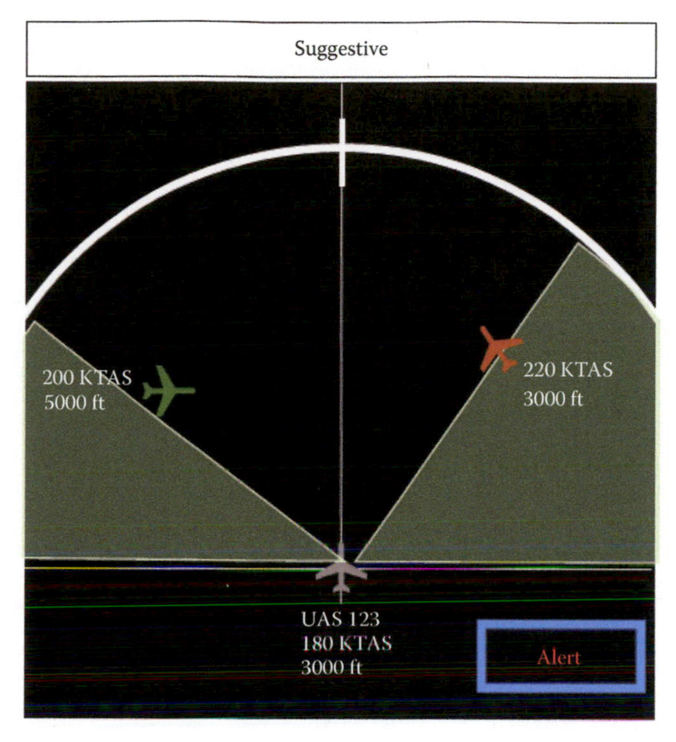

FIGURE 15.4
Notional example of a suggestive display for UAS.

shaded regions showing an appropriate range of horizontal maneuvers for the pilot to remain well clear from the depicted aircraft.

Directive displays provide clear and specific maneuver guidance to the pilot or UAS control system. In the notional example in Figure 15.5, a specific horizontal flight path is drawn on the display, with accompanying text instructions that indicate the recommended maneuver(s) to the pilot (i.e., Turn Right).

This display type requires the use of a tested and validated avoidance algorithm to calculate the optimal path for the UAS to maintain a safe distance from other aircraft.

15.8 Defining Compliance with "Well Clear" for UAS Operations

One of the defining moments in UAS integration was the 2009 creation of the UAS Executive Committee (EXCOM). Originated under Congressional direction, the UAS EXCOM includes senior executives from the FAA, the DoD, NASA, and the Department of Homeland Security (DHS) and "is responsible for identifying solutions to the range of technical, procedural, and policy concerns arising from UAS integration."

In June of 2013, the (formerly DoD) SARP was realigned under the UAS EXCOM, and they turned their attention to the cross-agency issue of defining a means of compliance with "well clear" for UASs. The EXCOM directed the SARP to research the problem and present a recommended compliance definition for "well clear" for UASs within 1 year.

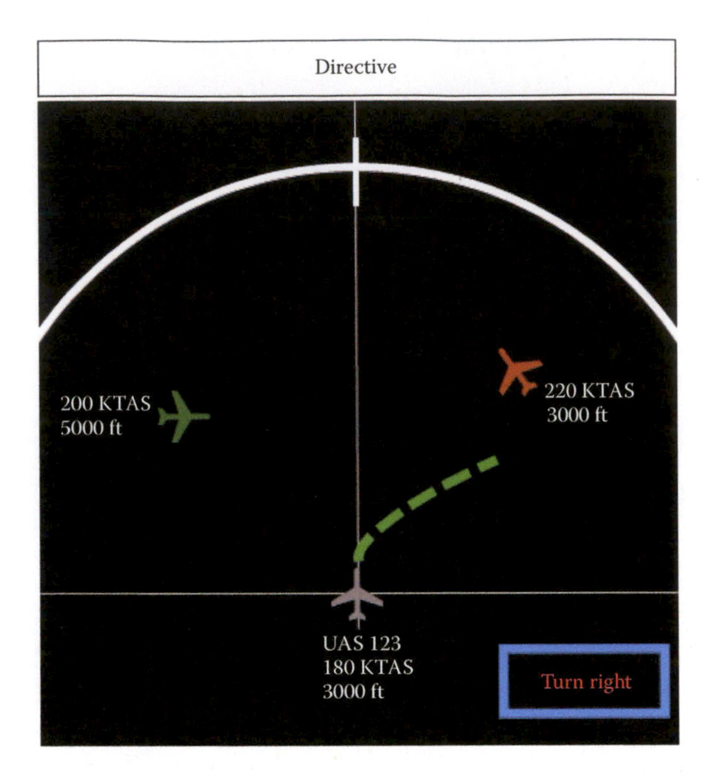

FIGURE 15.5
Notional example of a directive display for UAS.

As a foundation for their approach, the SARP arrived at five guiding principles for the UAS to comply with "well clear":

1. "Well clear" is a separation standard for UASs
2. DAA systems need a quantitative standard, while humans may judge "well clear" subjectively
3. "Well clear" is defined based on minimizing collision risk, but informed by operational considerations and compatibility with existing manned aircraft collision avoidance systems
4. "Well clear" is defined in the horizontal dimension based upon time and distance
5. "Well clear" is defined in the vertical dimension based on distance

With these principles established, the SARP evaluated multiple models developed by the DoD, NASA, and academia in multiple modeling and simulation environments to determine which one provided the optimal balance of collision risk with operational parameters. The result of this evaluation was a model that accounted for both time and distance from other aircraft, and provided a sufficient vertical (altitude) separation to support appropriate avoidance maneuvering.

The SARP's recommendation[3] includes a minimum time to closest point of approach of 35 s, a minimum horizontal distance of 4000 ft, and a minimum vertical distance of 700 ft from other aircraft. A post-recommendation review by an RTCA Special Committee

(SC-228) resulted in re-categorizing the 700-foot vertical distance as a "pilot alerting threshold," and reducing the "well clear" vertical distance threshold to 450 ft. This change was made in order to provide greater operational acceptability and harmonization with existing airspace regulations. The SARP recommendation serves as the basis for the ongoing evaluation and testing needed to support validation, and ultimately adoption, by the FAA.

15.9 Conclusion

For decades, unmanned aircraft systems have held the promise to improve, and in many cases save, human lives by providing air-based capability quickly and cheaply, without putting an onboard pilot at risk. But in order to fully realize this promise, a UAS must be able to operate safely and seamlessly with other air traffic.

Detect and avoid technologies represent the next great leap forward for the integration of the UAS. The use of robust DAA systems enables operation of UASs in virtually all classes of airspace, with a comparable, or even enhanced, level of safety when compared to many types of manned aircraft.

Defining the components of DAA has been proven to be the critical step in focusing the research, development, and test activities that will lead to accepted standards for UAS design, manufacture, and operation. The resultant new generations of UASs will continue to realize the tremendous benefits of unmanned technology—benefits that help ensure a safer, more capable future for everyone.

Acknowledgments

The authors would like to thank:

Stephen George of the Federal Aviation Administration's UAS Integration Office. George created and led the FAA's UAS SAA Workshops, and without his leadership a great deal of the current progress in "detect and avoid" would not have been made.

The participant organizations and individuals of the FAA UAS SAA Workshops from the FAA, DoD, academia, Federally Funded Research & Development Centers and industry.

The past and present members of the UAS SARP, originated under the DoD and now co-sponsored by the DoD, the FAA, NASA, and DHS.

DISCUSSION QUESTIONS

15.1 What are the two functions of a DAA system?

15.2 How can these functions be decomposed into sub-functions?

15.3 What are the advantages and disadvantages of pilot-in-the-loop DAA systems? What are the three components of the DAA system?

15.4 Why do DAA systems need a quantitative means to comply with "well clear?"

References

1. Annex 2 to the Convention on International Civil Aviation: Rules of the Air. International Civil Aviation Organization, 10th Edition, July 2005 (Amendment 44: November 13, 2014), pp. 1–5.
2. DO-185A with Attachment A—Minimum Operational Performance Standards for Traffic Alert and Collision Avoidance System II (TCAS II) Airborne Equipment. RTCA Special Committee 147, December 1997.
3. Cook, S., Brooks, D., Cole, R., Hackenburg, D., and Raska, V. Defining well clear for unmanned aircraft systems. Paper presented to the American Institute of Aeronautics and Astronautics, November 14, 2014.

16

Policy and Public Perception

Ben Miller

CONTENTS

16.1 Introduction

One afternoon in 2008, Ben Miller pitched the crazy idea of flying a little toy-sized unmanned aircraft to Mesa County Sheriff Stan Hilkey and his top leadership at the Mesa County Sheriff's Office (MCSO). To his surprise, that meeting ended with a green light to research the next step and the rest, as they say, is history. Over the next few years, the MCSO would have the distinction of creating and managing one of the very first small unmanned aircraft systems (sUAS) programs in public safety in the United States. Mesa County would achieve many firsts in the use of unmanned aircraft in public safety and in the process would learn volumes, much by trial and error, about the applications of sUAS for the public safety operator.

In 2008, sUAS (defined by the Federal Aviation Administration in numerous guidance documents and policy statements as having a maximum takeoff weight of no more than 55 lbs.) were key elements in the transition of the technology from military to civilian applications. Small computers, the size of a cell phone that could perform all the necessary steps for stable flight and navigation, had only recently become available to anyone outside the U.S. Department of Defense. The military had birthed the concept of small, man-portable and easy-to-fly surveillance drones, and equivalent sUAS were rapidly becoming available to the civilian world in an unprecedented technical revolution driven by hundreds of entrepreneurs, developers, and manufacturers. The MCSO UAS program would begin the process of defining the application of these systems for public safety. The common assumption that the application would be synonymous with that of the military would prove to be false, and in the years to follow, they would develop much of what would become the general applications for the use of sUAS in public safety operations.

Now, 7 years later, the concept of operations is stabilizing and the program is growing more robust by the day. The intentions they began with are vaguely similar to the realities they operate with today. The statistical information collected over the last 6 years of operations suggests that in the long run the sUAS will not create a new capability, but will create an affordable space to carry out many missions accomplished before by other means, but at significantly less expense. The data supports the "democratization" concept of aviation in the use of sUAS in general to perform tasks that were previously not economically feasible. This allows the operator to conduct proven aviation-related operations at a fraction of the cost of manned aviation and in airspace environments not considered navigable in the past. This democratization of aviation has allowed MCSO to collect useful data related to public safety events, thus enabling faster and more accurate tactical and strategic decision making. By 2012, the direct cost of operations for MCSO was determined to be just U.S.$25 per hour. And while the very mention of data and the use of drones by law enforcement has been a sensitive subject among national and international media outlets, the reality is that the access to this data was not enabled by the use of sUAS in public safety, but was just obtained much more efficiently. To that end, it is important to address the issue of privacy as it relates to the use of sUA by local law enforcement.

16.2 Privacy

So what is in a name? You may have noticed that the word "drone" was not used in the introduction. While the term "drone" more quickly identifies the subject, it does little to educate the public about the actual system. Public safety organizations bear the responsibility of being more informative than the use of the word "drone." The word "drone" has been used interchangeably with systems like the MQ-9 Predator-B deployed by the military to carry out lethal strikes against targets in foreign lands. That broad use of the term not only does not explain the specifics, but also causes fear and mistrust among local public safety agencies. Trust is the currency of law enforcement in the United States.

The ability of law enforcement to observe the public from an aerial vantage point has been in practice for many years, and validated by the courts in numerous cases throughout the United States. For example, in *Florida v. Riley*, 488 U.S. 445 (1989), the observation of the cultivation of illegal drugs was observed from the air, and that information was successfully used to obtain a search warrant and prosecute the offender. It is certainly foreseeable that one day the use of sUAS to collect evidence will be challenged in the court system and on that day the court will give little weight to the fact that information was collected by an aircraft without a pilot onboard. An "eye in the sky" is an eye in the sky and the lack, or presence, of a pilot should matter little. The underlying issue is whether the surveillance or observation that led to the arrest was constitutionally permissible under the Fourth Amendment to the U.S. Constitution.

Further, the presence of potentially lethal weaponry on a public safety sUAS is not likely. In the opinion of this writer, the mission of the civilian law enforcement agency in the United States will not allow for this. However, to say that less than lethal capability (rubber projectiles, pepper spray, etc.) will never make it aboard the sUAS may be shortsighted. In fact, the MCSO program considered this early on. The conversation about the legal implication of the use of sUAS has been constant since the advent of the MCSO program, and it will continue into the future. However, the inference that public safety agencies are using military equipment and carrying out military-style drone missions inside the United States is demonstrably false, notwithstanding contrary suggestions by some elements of the mass media and the blogosphere.

The real privacy debate lies in the data, or the information that law enforcement has the ability to collect and retain. The question that should be asked is who collects or receives the data? How long can they store it? Who can they give it to? How can they use it? This is a more holistic approach to the question of privacy. The use of sUAS did not create this conversation but has been used to reinvigorate it. Other forms of data collection such as license plate readers, security cameras, traffic monitors, etc., are equally controversial because of unresolved issues of data retention and use. Legal challenges to the use of all forms of surveillance technology, including camera-carrying UASs, will eventually determine what the limits will be regarding the collection and disposition of data and metadata in the public sector.

16.3 Regulation

In the early days of MCSO's program, the Federal Aviation Administration (FAA) had a very thin process to allow them legal access to the National Airspace System (NAS). The FAA's basic premise was to approach UAS public aircraft operations for a potential violation of the rules found in the Federal Aviation Regulations (FARs) and they therefore required MCSO to obtain a waiver of the relevant rules. Consequently, MCSO sought waivers of FAR Part 91.113(b), whereby all pilots are required to "see and avoid other aircraft."* Originally, the waiver was based on the idea that MCSO could not comply with FAR 91.113(b), as there was no pilot onboard looking out of the front window. To get a waiver, or more officially a Certificate of Authorization or Waiver (CoA), the FAA required their program to demonstrate an alternative means of compliance. MCSO requested permission to fly within visual line of sight of the sUAS, facilitating compliance with the requirement to see and avoid other aircraft under FAR 91.113(b). Of interest to all concerned, MCSO found that by maintaining visual line of sight of the sUAS the operator was not only able to maintain visual separation, but could do it better than sitting in the cockpit of a manned aircraft. The operator of the sUAS has a 4D sense of the airspace, in that he/she can see in three dimensions around their own aircraft as well as hear other aircraft approaching the critical airspace. As MCSO became more comfortable with their understanding of the FARs, they later questioned if the waiver was really necessary.

MCSO originally applied for a CoA on a small "quad" style sUAS called the Draganflyer X6, built by a Canadian company by the name of Draganfly Innovations. It took approximately 8 months to get the approval from the FAA to fly. And, even then, they were only allowed to fly in a small spot at the Mesa County landfill. This permission was called a "training CoA," and operators were authorized to fly daytime flights up to 400 ft. above ground level (AGL) and not farther than the visual line of sight of the operator. With the Draganflyer X6 that was about 400 m.

The FAA required that MCSO provide a letter stating that they had assessed the airworthiness of the Draganflyer X6 and were accepting all liability associated with the

* http://www.ecfr.gov/cgi-bin/text-idx?c=ecfr&sid=3efaad1b0a259d4e48f1150a34d1aa77&rgn=div5&view=text&node=14:2.0.1.3.10&idno=14#se14.2.91_1113

airworthiness of the airframe. Airworthiness is a process that assures that the equipment is ready to fly and is safe to do so. Additionally, a pilot's license is a process by which operators demonstrate that they are ready to fly the equipment in a given environment. Neither of these two processes, derived from manned aviation, applied to the use of sUAS in the NAS, at least not according to the FAA. In the CoA application, MCSO assured that all these issues, and more, were addressed. They explained how their operators would be trained and how they would ensure the aircraft were ready for flight each time they were used. They would also ensure they had contingencies in place if their equipment did not operate as designed or expected. By the time the application for the original CoA was filled out, they had submitted more than 30 pages of documentation and 8 months later they received one of the very first CoAs for the use of UASs by any police agency in the United States. Even after all of this, they were only granted permission to fly in a small area at the local landfill. This process obviously was not going to work. What they had begun to do was train operators and learn just how realistically they could use sUAS in their day-to-day tasks. The training CoA was going to fall short of the goal of full operational capability.

After some pointed conversations with the FAA, the agency addressed these shortcomings with the concept of an Emergency CoA, in that MCSO could call them anytime they needed to fly a mission. The process required MCSO to call the FAA and make the request via a two-page document explaining the situation at hand. The FAA put three stipulations on the issuance of an Emergency CoA: The mission requested could not be performed by a manned aircraft (MCSO did not have any manned aircraft), the requester must have an active CoA in place, and there must be an imminent risk of injury or loss of life. It was felt that by imposing these requirements on the MCSO, the FAA was assuming a law enforcement role in retaining the prerogative to determine imminent risk in not only the use of aviation, but also the law enforcement response itself. The early process to gain permission to fly an sUAS was perceived to be less than ideal in meeting the needs and mission of public safety agencies.

Then Mesa County Sheriff Stan Hilkey directed the team to maintain the high ground, "walk in the front door" and play by the rules. Once they had done so, they could vigorously point out the shortcomings of the process. They did as requested and may have changed the course of history. MCSO would later create strong ties with the U.S. Department of Justice via their alternative aviation program. Through the facilitation of a gentleman named Mike O'Shea, MCSO helped create a new, revised CoA process for public safety agencies that would come after them in using sUAS. The FAA and Department of Justice (DOJ) called this new process the "Common Strategy." The Common Strategy would never be widely accepted, as word had already gotten out to the law enforcement community about the stringent requirements to fly sUAS issued by the FAA. Mesa County's program waited 8 months before getting approval to fly what most people viewed as a toy, weighing in at just a few pounds and small enough to be stored in the trunk of a police car. Furthermore, new requirements after the agreement of the Commons Strategy would further muddy the perspective of the intentions of the FAA toward local law enforcement. For example, after the Common Strategy was accepted, the FAA created a new requirement of having every law enforcement agency requesting a CoA submit a letter from their state's Attorney General declaring that the agency was indeed a subdivision of government. This did not make sense to many involved, although the FAA maintained that the requirement was mandated under federal law, leaving the agency with no choice but to compel what appeared to many to be a frivolous process. (The FAA's interpretation of 49 U.S.C. § 40102(a)(41)(C) and (D) compelling such a letter is found in Advisory Circular 00-1.1A (10)(b), released February 12, 2014.)

While the requirement to secure a waiver for Mesa County's intended operations contained numerous requirements that seemed to defy logic for what they were intending to do, this all was overshadowed by the fact that their flights were conducted in a manner that qualified them as public aircraft operations. Pursuant to published FAA policy, public aircraft operations qualify for exemptions from many of the requirements in the CoA process, including training, airframe certification, pilot qualifications, etc., the exception being the requirement to comply with the operating rules set forth in Part 91 of the FARs.

16.4 Public Aircraft

Qualifications for Public Aircraft Status are set forth in 49 U.S.C. § 40125. The FAA's interpretation of that statute is contained in Advisory Circular 00-1.1A, cited earlier. Public aircraft are not required to possess airworthiness certificates, nor are the operators of public aircraft required to meet the same qualifications of civil and/or commercial aircraft. Essentially, the FAA has little or no jurisdiction over the public operator until the aircraft takes off. Once airborne, all aircraft must comply with FAR Part 91, or what is commonly referred to as the "rules of the road." It is best described as compared to the automobile. If UASs were automobiles on state highways, the state cannot require the public operator to possess a medical certificate, nor set a standard by which they maintain their car, that is, oil changes, tire changes, etc. But, once that car enters the highway the public operator must obey the speed limit, not pass in a double yellow line zone, and so on.

Now, do law enforcement aviation units adopt the same maintenance requirements and pilot certification that the FAA requires of the civil/commercial operator? They most certainly do. Why reinvent the wheel? And how would self-generated requirements by the law enforcement aviation community really differ for manned aviation operations? In the very near future, the FAA will have rules in the Federal Aviation Regulations that direct the safe operation of unmanned aircraft in the NAS. It is foreseeable that the public safety users will adopt these rules in their own operations. However, early adoption will likely be a hybrid, as original sUAS rules from the FAA (in the Notice of Proposed Rule Making process as of this writing) are expected to be too restrictive for law enforcement purposes, that is, not allowing for night flights.

16.5 Public Perception and Education

There were times in the early days of MCSO's program where it appeared that more time was being spent educating the public as to what was being done with sUAS than in actual day-to-day UAS operations. As discussed previously, the "drone" conjures a wide range of public reactions. Much of it is based on a misperception that the vague term allows. The public often identifies drones as tools of war lurking miles above the earth, watching and waiting to attack. In many ways this perception is accurate if the topic is military UAS operations. Military systems were designed to improve upon the shortcomings of the human as he integrates into the mission on the battlefield. For example, the inability of a human operator to stay focused for extended periods of time presents a major challenge to the warfighter. According to Northrop Grumman,* its Global Hawk high altitude long endurance (HALE) UAS can fly more than 10,000 nautical miles and longer than 24 h in one flight. It is also as large as a Boeing 737. While it does not carry weapons, it does carry top-secret high-powered sensors that give it the ability to view the earth below in great detail. Other systems, the MQ-9 Predator B for example, carry weapons. According to the manufacturer of the Predator B, General Atomics,† the aircraft can carry 3000 lbs. of external weaponry and fly for 27 h. These systems not only improve upon the limitations of man, but remove the pilot from danger as well. In many ways, these systems provide more data to the decision maker and allow the military to more precisely apply lethal munitions to the battlefield. It is commonly believed that these characteristics improve the ability to wage war and provide for a reduction in unintended casualties.

However, the lack of an onboard pilot is about the only similarity to these types of UASs and the systems the MCSO program employs. Explaining this to the public is critical in maintaining their trust while incorporating the use of the UAS into day-to-day public safety operations. The public's consternation with the idea of the use of these weapons of war by state and local police is not lost on local law enforcement. MCSO's systems are small and carry small cameras that do not provide a comparable level of detail when compared to the Global Hawk; however, they are used because they are highly cost-effective. The MCSO program has developed a long-term cost of operation of just U.S.$25 per hour. While the cost of operation of a large military UAS is difficult to determine, MCSO's U.S.$25 per hour flight cost is just two percent of the average cost of operation of manned aviation in law enforcement, according to a study by the Bureau of Justice Statistics‡ in 2007.

Maintaining strict transparency standards and explaining the sound reasons for the employment of this technology to the public is critical in the early adoption stages of a UAS program in law enforcement. MCSO dedicated a considerable amount of time briefing the media in order to get this message to the public. The Sheriff's Office hosted meetings with concerned citizens and created a website§ explaining the use and interest in sUAS. In many ways MCSO was very successful in educating the public to the realities of the UAS operations, and the leadership strongly believes that the Sheriff's Office maintains the same level of trust from the public as they did prior to the program.

16.6 Historical Applications and Case Studies (Narrative by Ben Miller)

With equipment in hand, permissions granted, and the trust of the public, we were ready to fly and the stories began. Here are a few key deployments in the program and the interesting lessons learned.

Our very first mission just happened to occur inside our approved training area during the time we were still training on and evaluating the Draganflyer X6, and thus we did not require further approval from the FAA. In October 2009, a child sexual assault case was being investigated by the Mesa County Sheriff's Office and the investigator realized that one potential location for the assault was in the very same area where we were conducting

* http://www.northropgrumman.com/Capabilities/GlobalHawk/Pages/default.aspx?utm_source=PrintAd&utm_medium=Redirect&utm_campaign=GlobalHawk+Redirect
† http://www.ga-asi.com/products/aircraft/predator_b.php
‡ http://www.bjs.gov/content/pub/pdf/aullea07.pdf
§ www.mesauas.com

FIGURE 16.1
Photo taken by MCSO Deputy Casey Dodson with a Draganflyer X6 on October 2, 2009.

training flights inside our original training CoA. The investigator asked if we could fly over the area, collect some photos and determine if there was a likely area the assault may have occurred. We flew a couple of times with the Draganflyer X6. This system carried a high-quality handheld point-and-click camera produced by Panasonic. We flew a few flights over the desert area and took photos (Figure 16.1). We looked at those photos back on the ground and found a recliner chair dumped in the desert (Figure 16.2). Further investigation of the recliner would eliminate it as being involved, but it was useful to use the sUAS to find it. That day, the Draganflyer X6 had some trouble landing, which resulted

FIGURE 16.2
Photo taken by MCSO Deputy Casey Dodson with a Draganflyer X6 on October 2, 2009.

in a trip back to the manufacturer for repairs. We learned a lot about the system, what it was capable of doing and what it was not. The day might have been a little windy.

In the spring of the next year, April 2010, we would deploy on our first search and rescue mission. Our search and rescue coordinator called with a report of a suicidal male who had not been seen since the day prior, and his family reported his .357 magnum handgun was missing.

We were assigned an area of dense vegetation, which may have been a potential location of the missing person. The area was approximately one mile long. We were able to fly from multiple takeoff locations and cover the area by flying to 400 m to both sides of the operator. While flying, the Draganflyer X6 broadcast the video it was collecting back to our control system. From the real-time information, and later review of the recorded video, we were confidently able to inform the search coordinator the subject was not in our assigned area. To get permission to fly we had to request an Emergency CoA from the FAA, which they granted approximately 5 h after the request. As it was, the weather would not allow for flights that day, but the FAA authorized an Emergency CoA for the following 3 days and in that time the weather would eventually allow for the flights. The Draganflyer X6 flew each flight for approximately 10 min and took off, flew, and landed as it was designed. We did communicate back to Draganfly Innovations that we would like to see the X6 fly longer and have better range when transmitting video.

In May of 2011, we would find ourselves on our first structure fire deployment supporting the local fire department in the work to knock down a fire in an old church and assisting them with the following arson investigation (Figure 16.3). The fire department had attempted to stop the fire, but after an injury to a fire fighter it was determined the fire had advanced too far and they allowed it to burn. I responded to the scene, as the sun rose, where mop up had begun and found the incident commander. I introduced myself and told him that I had a small helicopter with an infrared camera and could help him find the hot spots in the building as well as take some aerial photos of the scene. I remember he looked perplexed as he looked around and then asked me where I landed? I laughed and explained that it was an sUAS and I could have it airborne in the next few minutes with a

FIGURE 16.3
Photo taken by Mesa County Sheriff Deputy, May 2011.

FIGURE 16.4
Photo taken by Mesa County Sheriff Deputy, May 2011.

FLIR camera. He laughed and said, "We know it's hot in there." I asked him if he would like to know where it was hot and he was quite excited that I could provide that kind of detail. After the first flight over the burning building with our infrared camera transmitting live video to the ground, we identified a hot spot not visible by the presence of fire. (Figure 16.4) The fire department responded by addressing that area with more water, which increased their confidence that the fire was contained and would not reignite. After we flew the infrared camera, we attached our still camera to the Draganflyer X6 and took aerial photos to aid in the reconstruction the arson investigators now had to do. We used software back at the office to consolidate all the individual photos into one mosaic image of the entire hall (Figure 16.5). I can remember my fascination with arson investigators as they determined direction of travel and other key pieces of evidence from the imagery we had provided.

Another unforgettable mission began late one winter's night when the phone rang. A deputy on our UAS team was calling, relaying a request to look for a 12-year-old boy who had gone missing. The boy's mother had called 911 explaining they had only lived in their current home but for 3 days, that her 12-year-old son was developmentally disabled, and he had not been seen since earlier that day. The outside temperature was now below freezing and the Mesa County Sheriff's Office was providing all available resources to the search, including our forward-looking infrared camera that could see heat, day, or night. I arrived on scene, joining my Deputy and we flew a few flights using our FLIR Tau 640, watching the live video downlink and looking for the boy. As we were working, other deputies continued to talk with the boy's mother, as something just was not adding up. It was soon determined that the boy's dad also lived in town and mom and dad were no longer speaking. A responding Deputy to the dad's house found the boy safe and sound. We learned a few things that night: Most importantly that our mission came before our relationship with the FAA, and that our sUAS could provide a significant tool to a search in the night. While the boy we were looking for was never in our search area, we could easily see numerous search volunteers and MCSO staff. Given that an infrared camera can see heat and display the relative contrast on the screen, the human body is quite apparent in a cold night, especially considering the ambient temperature was below freezing.

FIGURE 16.5
Photo taken by Mesa County Sheriff Deputy, May 2011.

Another reality we noticed is that everything was quiet; there were no other aircraft flying at that time of night, and our sUAS was quite visible in the night's sky with its bright LED navigation lighting.

It occurred to us that if the FAA would have had a better understanding of sUAS at that time, they might have required us to begin our flight testing at night, given the fewer aircraft in the air, the significant contrast of our lighting on the night sky, and people on the ground were, for the most part, indoors. Flying sUAS at night would present a significantly lower risk, at least in our experience.

Another example of how a UAS can be used occurred in December of 2013. Early that day, a parole officer had contacted a parolee who was likely going to have his parole revoked. Upon a knock at the door, the parolee, armed with a high-powered rifle, ran out the back of the residence into the center of a field, finding cover at the base of a tree. This location did not allow for safe cover while deputies approached, so a perimeter was set up and a containment plan put in place. MCSO Special Weapons and Tactics (SWAT) and UAS teams were called to the scene. The UAS team used a fixed-wing UAV from Falcon Unmanned and orbited around the tree at 300 feet above the ground. UAS team members launched the system at 600 m from the subject, attempting to maintain line of sight. During the orbit, the daylight and FLIR cameras were trained on the tree. While it was not immediately visible in the standard definition downlinked video from the UAV, SWAT officers inside their armored rescue vehicle confirmed the location of the subject. We used the Falcon UAV to maintain a watch. Should the subject flee a second time on foot to nearby foliage, the UAV team could then direct SWAT officers to a new containment site. Upon exhaustion of the batteries we landed the Falcon UAV, via its parachute. While floating back to earth the wind blew it inside the perimeter, and it was determined we would not recover it until the situation was safe, as it was too close to the subject with the gun. Looking back, the UAS team should have deployed from farther away. The system was capable of beyond-line-of-sight operations (BLOS) up to 7 km. This would have removed the entire team from the perimeter of the armed subject.

A final example of UAS use may not come to mind so easily, does not contain the Hollywood flare, but may have provided the most useful data to date for our operations. In May of 2014, a large landslide occurred in Mesa County, creating a 3-mile long debris field and taking the lives of three local ranchers. The MCSO UAS team deployed fixed-wing and rotorcraft UASs to the scene. Using Falcon Unmanned's fixed-wing UAS, the lower third of the debris field was photo-surveyed. Collecting more than 700 photos from a 40-minute flight, the photos were then divided into groups and observed by a group of volunteer search and rescue (SAR) personnel, our crowd-sourced support. These SAR personnel would look through their assigned high-resolution photos, looking for signs of the lost ranchers, and then store those photos in a specific location for later review. At the end of the cycle, the selected photos would be viewed by the entire group with UAS team members and further processed. This process did not determine any sign of life or location of the lost ranchers, as it was later determined there were none. That same dataset was then processed into a large geo-rectified orthomosaic photograph that can be laid into GIS software like Google earth. The photos were also used to create a three-dimensional point cloud and digital surface model that allowed responders to determine the volume of the entire debris field. Relating that to historic survey data, the estimate for the volume of the debris field was 40 million cubic yards. Later, the Mesa County Public Works department would continue UAV survey flights with their Trimble UX5 fixed-wing UAS to collect subsequent data sets, looking for comparative movement in the debris field. The MCSO UAS team assisted the Public Works team in attaining their CoA and starting a program of their own.

Similar to the survey process, the MCSO UAS team is now capable of surveying crime scenes with a UAS where photos are taken in either Nadir perspective (straight down) or oblique (at an angle to the target). With this technique, the MCSO UAS team has captured numerous homicide scenes and produced highly accurate orthomosaic and three-dimensional models. In these models, animated recreations of events have been created and submitted as evidence. This process may become the primary application of the UAS in public safety in the future.

16.7 Tactics and Procedures

Since its inception, the MCSO UAS team has created, deleted, and recreated numerous tactics for the use of the UAS in public safety. There have been many lessons learned. Here are a few of the operational procedures that the team has adopted.

The standard personnel profile is one operator in charge (OIC) and one observer. However, this is not always acceptable or necessary. The MCSO UAS team selects and trains staff to become UAS operators and not pilots. This term seems most appropriate after consideration of the language found in the proposed FAR Part 107 (NPRM mentioned earlier). It is also in keeping with the culture created by the team since the program's inception. UASs are used like any other tool designated for specific tasks in public safety. These tools are taught to the officer and they operate them appropriately. MCSO UAS operators do not become licensed pilots of manned aircraft, nor is there an expectation that they achieve such a level of mastery and expertise. UASs in public safety will rarely become a full time job in the future.

In SAR missions, two operational profiles have presented themselves as most efficient. These profiles are system-agnostic. It is usually quite challenging in a SAR mission to know where to start. Commonly, a starting point occurs at the last known location of the person or persons. However, direction of travel from that point is not always apparent and this variable will determine the profile in which one would direct their UASs.

When direction of travel is known, a UAS team can deploy the UAS flying a grid pattern in the direction of travel of the target. That grid pattern can be created taking into consideration the field of view of the camera and altitude. Twenty percent overlap of that field of view is suggested. This allows for certainty of coverage of the search area while maintaining search efficiency. During flight the imagery should be viewed real-time for actionable information; however, it has been the experience of the MCSO UAS team that this can inundate the observer, and therefore it is recommended that a form of image processing be applied to the data stream to supplement the observer.

When direction of travel is not known, it is more efficient to begin to fly consecutively larger concentric circles around the last known location. The radius of the subsequent flight path is determined by field of view of the camera and desired overlap. Observation of the data in this profile does not change from that of the previous profile. When the observer feels the orbit is complete, the radius is advanced to the next concentric circle and so forth until the end of the flight, or evidence of direction of travel has been discovered. Then the mission profile changes to a grid search.

In crime scene mapping, there are also two general mission profiles that our program has found useful in most, if not all, crime scene missions. The first being Nadir image collection, where the camera is placed at 90° down orientation, and the sensor in the camera is parallel to the earth's surface. The UAS flies in a grid pattern over the target area taking photographs in a predetermined interval that allows for desired overlap from image to image.

The second profile is oblique camera orientation where the camera is pointed at the target and the UAS is offset. The UAS flies in a circle around the target area taking photographs in a predetermined interval that allows for desired overlap from image to image. It is suggested that the UAS fly multiple orbits around the target in at least two different altitudes above ground.

Active gimbal stabilization assists in maintaining Nadir, as well as oblique camera orientation by counteracting the movements of the aircraft. It is highly likely the data collected without stabilization will not be useable by post-processing software. Camera setting should provide for faster than normal shutter speeds, and gimbals should also provide for vibration isolation and dampening. There is currently a large offering by industry in photo processing options, but the user should select software that can provide, at the least, both orthomosaic imagery and dense points clouds.

16.8 Limitations of UAS in Public Safety Applications

While UASs offer access to airspace rarely used (ground level to 500 ft. AGL), there are still numerous things they cannot provide and likely never will. For example, a UAS will likely never rescue stranded individuals like an officer working on the end of hoist. There are public safety missions that will always require the presence of a beating heart and the heroic extra effort our public safety operators bring to the equation. When the mission requires large aircraft, large enough to carry people, fully autonomous control systems can easily be replaced or augmented with human pilots. In fact it is questionable if, in that size of an aircraft, autonomy provides greater improvement than the decrease in risk.

UASs in public safety will also never be valued more than the cost of manned aviation. Given the mission, should any specific application of a UAS overrun the cost of their manned counterparts, state and local governments will likely opt for the more cost-effective manned version. Why would a state or local government spend twice the cost of manned aviation on the same capability afforded by manned assets? Public safety organizations are limited by the budgets their service populations provide.

UASs will also never be the fix-all that new technologies seem to be expected to achieve. Public safety in the United States will always require the decision making capability of a human mind, in a service-minded, heroic individual. Our concept of operations list is significantly different today than it was when we began, when all we had to write was our imaginations.

16.9 Conclusion

Today UASs are proving to be a valuable asset in public safety. They have been very slowly adopted, but as the FAA creates regulations that are both clear and applicable to the UAS, the pace of adoption will only increase and "flight authorization" will fall out of the conversation. As adoption increases, so will use and applications, and unfortunately mishaps. Public safety operates on the public's trust. It is imperative, therefore, that trust is considered in all things involving UASs and public safety. In the application, standards, and use of the UAS, public safety organizations should approach UASs with solid research, professionalism, and pragmatism. Humans have an amazing capability to create tools to extend beyond our physical limitations. The UAS can provide significant expansion beyond those physical limitations for all mankind.

DISCUSSION QUESTIONS

16.1 In your opinion, what is the greatest public policy challenge faced by the integration of this technology into our National Airspace System (NAS)?

16.2 Outline a solution pathway related to question number 1.

16.3 Should a society trade a measure of privacy loss with the aim of greater security? Qualify your answer. What risks are associated with your answer?

16.4 Who should be responsible for educating the public regarding UAS integration into the NAS? What are the expected challenges in doing so?

17

The Future of Unmanned Aircraft Systems

Richard Kurt Barnhart

CONTENTS

17.1 Introduction

Heraclitus of Ephesus (c. 535–475 BC), an ancient Greek philosopher, was noted for his observance of the constancy of change in the universe. Nowhere is that constancy more evident than in the high-tech world of unmanned aircraft systems (UAS). The very term itself has been the subject of debate with some sectors including many in the media adopting the term "drone" versus UAS. Other sectors have migrated to the term "remotely piloted aircraft." To proponents of this last term, the usage of "unmanned" in denoting UAS is a liability when attempting to explain publically just what these aircraft do. As such, writing about the future of the UAS is a bit slippery. Therefore, the majority of this chapter will focus on broader and more enduring industry trends as opposed to attempting to make specific predictions in this highly dynamic area.

17.2 Anticipated Market Growth

No discussion of an industry is complete without an examination of both historical precedents, for context, and future trends. It has only been within the last 15–20 years that the UAS has become a significant industry segment that makes it young by any measure. Driven by robust technological advancement, the market for UAS manufacturing and support services that began as a miniscule, barely noticeable segment of the industry, has become, in a relatively short span of time, a major component of the aerospace enterprise. Recent growth has been steady, and predictions point to continued expansion. As is often the case, new segments of industry experience rapid growth, almost immune from the economic cycles that affect more mature industries. This has certainly been, and continues to be, the characteristic of the UAS market.

According to marketresearchmedia.com, UAS market expenditures can be broken down into nine basic segments:

- Research, Development, Testing, and Evaluation (RDT&E)
- Platforms or air vehicles
- Ground control systems
- Payloads and sensors
- Service and support
- Sensor data processing and dissemination
- Training and education
- Data management
- Revenues by UAV groups

By all accounts, the worldwide UAS market forecast continues to indicate strong growth for the foreseeable future. Major UAS market research firms are in agreement that the industry will experience strong growth over the next 5–10 years. The Teal Group Corporation, a team of integrated market research analysts, estimates in their 2014 forecast that over the next 10 years worldwide UAS expenditures will experience a compound annual growth rate ranging from 5% for the military sector to nearly 20% for the civil sector topping U.S.$90 billion. The United States will account for 65% of the research and development portion of those funds and 41% of the total procurement in dollars. While the overall global market forecast is stronger than the previous forecast in 2010, the overall U.S. percentages of those estimates are down by 10%–15% in the 2014 forecast. The downturn reflected in projections for the domestic market indicates an increase in UAS proliferation outside the United States. It is worth noting that, globally, military UAS expenditures continue to dominate economic projections, accounting for over 85% of the total combined

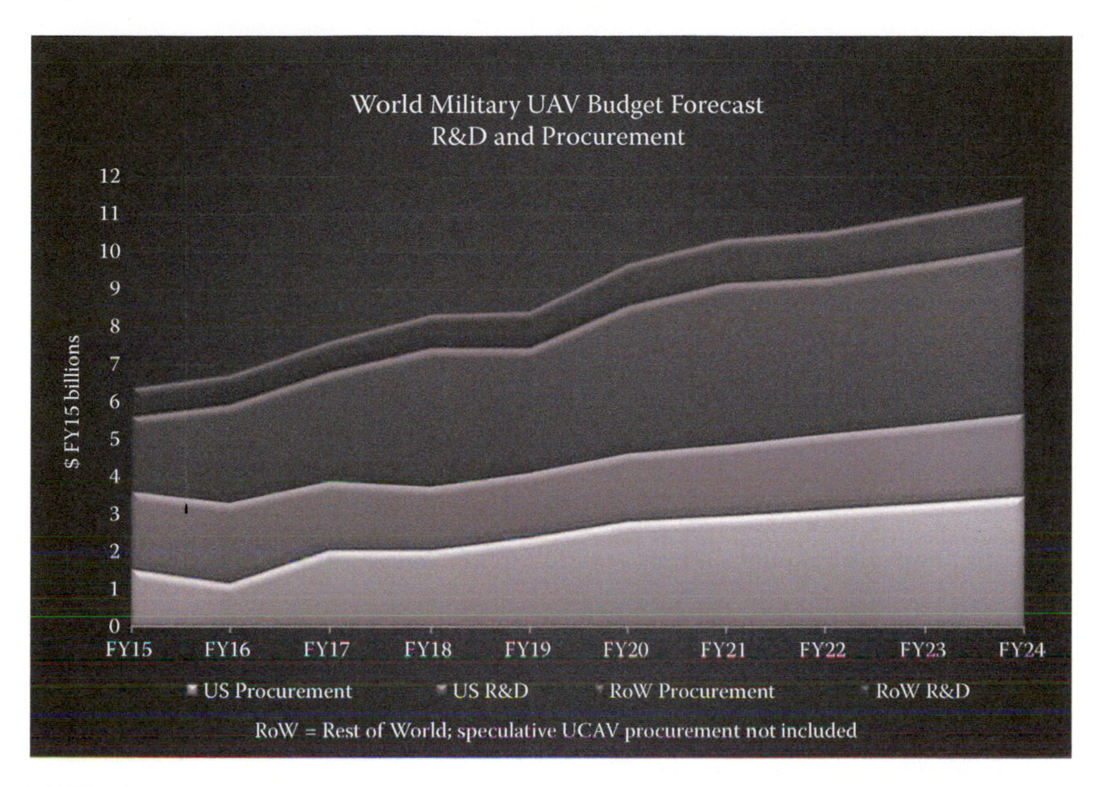

FIGURE 17.1
Teal Group's forecast. (From The Teal Group 2014. UAV Integrated Market Analysis. Web. 17 July 2014. http://www.tealgroup.com/index/php/about-teal-group-corporation/press-releases/118-2014-uav-press-release.)

market. As civil airspace continues to open here and elsewhere, the overall defense portion of this market will undoubtedly decrease as a percentage (see Figure 17.1).

Frost and Sullivan (2014) is another group that continues to predict very strong growth in the UAS market. Their numbers predict that within 5 years the global combined military UAV market will reach a total of nearly U.S.$87 billion by the year 2018 which bests their 2010 6-year forecast by nearly U.S.$25 billion, demonstrating substantial anticipated growth. They also predict a 12% combined annual growth rate in this same market (see Figure 17.2).

17.3 The Future of UAS Market Segments

The FAA has long divided flight operations into two broad categories depending upon whether the responsible entity is public or private. As defined under 14 CFR 1.1, public (or public use), aircraft are those "aircraft owned [or leased for at least 90 days] and operated by the government of a State, the District of Columbia, or a territory or possession of the United States or a political subdivision of one of these governments," or by a branch of the U.S. military. Under 14 CFR 1.1, private aircraft are those that are "not public aircraft," that is, those that are privately owned and commercially operated. These definitions have been extended by the FAA to include UASs. Given that the FAA (2013) will initially integrate

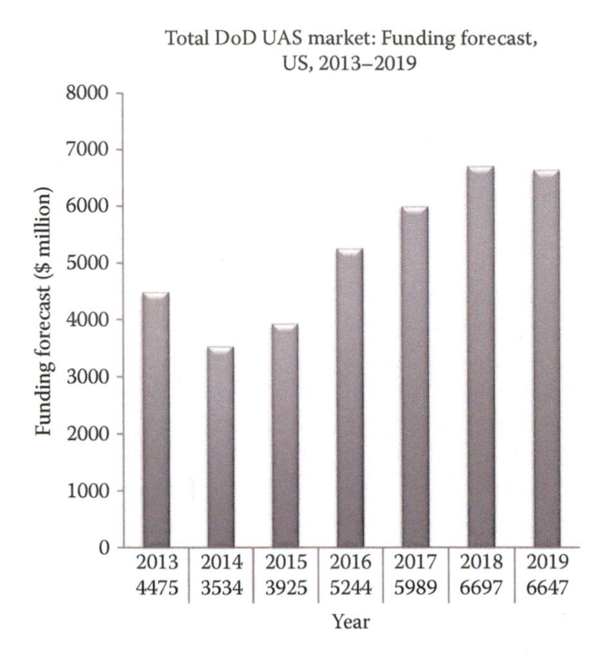

FIGURE 17.2
Frost & Sullivan forecast. (From Frost and Sullivan. Global Military Unmanned Aerial Vehicles Market Assessment Market Research Media. *Market Research Media Premium Market Analysis*. Web. 9 January 2014. http://www.marketresearchmedia.com/?p = 509.)

small UAS (sUAS, i.e., those weighing less than 55 lbs.) into the National Airspace System (NAS), these are the aircraft that are most likely to be privately owned and the first to operate commercially.

17.3.1 Private/Commercial UAS Market Segment

In the first edition of this book, published in 2010, the Private/Commercial UAS market was described as "the dam which is ready to break," and at no time has that statement been more accurate than it is now—albeit somewhat later than many expected at that time. Currently, significant restrictions placed by the FAA on private companies (and others) seeking to operate UASs in the NAS continue. The FAA, however, as directed by Congress in the 2012 FAA Reauthorization Act, has recently established an achievable path forward for private companies to access the nation's airspace, providing the mechanism to operate UASs commercially through an exemption process in the short term and a set of proposed federal rules as the long-term solution. Although long awaited, many in the industry now widely expect that the commercial UAS industry will have an enabling regulation in place sometime in the year 2017, if not before. This rule will apply exclusively to sUAS and is currently proposed to restrict operations to daylight hours and only when the vehicle operator is within visual line of sight of the vehicle (among other restrictions). The exact wording of this "sUAS rule," however, is yet to be determined specifically since the FAA has received public comments requesting that these restrictions be modified or lifted under certain conditions. Up to this point, the primary regulatory mechanisms for most UASs to operate in the NAS is to apply to the FAA for either an alternate type of airworthiness certificate or a Certificate of Authorization

(COA) which establishes a wide spectrum of controls and limitations for specified UAS operations. COAs have typically only been available to public agencies, thereby excluding commercial operations. Until the need is eliminated by regulation, the COA process is expected to continue to play a role in airspace access for those operating UASs defined under 14 CFR as public aircraft. If a proposed operation falls outside the allowances of any regulation, then the implementation of the COA process will again likely be necessary. Thus, the COA process is expected to play a role in UAS operations in the foreseeable future.

The potential for commercial UAS applications seems to have no limit and is growing by the day. As more people and industries become aware of the game-changing potential of this technology new potential applications emerge, all driven by technological enablement as well as by a light at the end of the regulatory tunnel. As Kyle Snyder of North Carolina State University has said, "Show me an industry and I'll show you [one or more applications] for UAS." Universities across the nation are seeing rather dramatic rises in external relationships developing around the commercial potential of this technology. These industries include energy, real estate, public safety, transportation, public resource management, animal health, journalism, TV/Film, and the list goes on.

17.3.2 Public UAS Market Segment

This segment of the market will continue to be dominated by the military, first responders, and higher education-related research/training activity. Universities, both individually and in partnership with other industries, will continue to lead with technological advances and innovation that drive and create economic opportunities. The military will continue to lead by pushing the limits of technology to meet the needs of situational awareness/response and will be responsible for the vast majority of UAS sector spending for quite some time. First responders will continue to help lead the public sector by driving both the need for and the capability to use UASs to save lives, thereby enhancing their public acceptance. This is in addition to all of the potential applications by state and federal agencies/offices to use UAS to both improve efficiencies and increase the quality of services.

17.3.3 Predicates to Future Market Access

The aforementioned expansion of UAS market segments is dependent upon the ability to gain access to the NAS, as well as upon establishment of training and certification standards similar to those that have evolved for the regulation of manned flight.

17.3.3.1 Routine Airspace Access

The term "airspace" applied herein refers to the usable space from ground level out to where the atmosphere thins and space begins. The United States operates the busiest, most complex NAS in the world, yet it is among the world's safest systems. Most people take this system for granted and never think twice about it since it largely consists of an invisible network of aerial highways, preferred routes, urban transition zones, and no-fly zones, all dictated by stringent regulation and policy. The reason this busy, highly complex system is so safe is because access to the NAS is tightly controlled and reinforced through the FAA's certification process, consisting of policies and procedures that apply to operators at all levels (pilots, air traffic controllers, etc.), aircraft, landing/navigation facilities, etc. Without this system of certification, the skies above our heads would be nothing short of chaotic and a very dangerous place. As can be seen, the integration of an entirely new technology into this system can be a complex and often slow task, but one that is necessary to minimize disruption and to ensure that the historic levels of safety to which we all have become accustomed will continue once UASs have been integrated into this system.

As was pointed out previously, the pathway for routine airspace access beginning with sUAS has been proposed, and a mechanism to allow commercial sUAS activity is currently available, albeit on a case-by-case basis for now. In the future, we can look forward to the issuance of a final sUAS regulation, hopefully within 2 years, to be followed by a regulation allowing routine operations for larger UASs. The large UAS regulation will be somewhat more complex and is several years from fruition because of the substantial increase in complexity inherent in operating large UASs in the NAS. These operations will require larger amounts of airspace as the vehicles will need to fly higher and well beyond the visual line of sight (BVLOS) currently found in the proposed sUAS rule. Further, any restrictions on night operations will likely need to be removed to allow these vehicles to reach their full potential. As has been the pattern of the FAA, restrictions will likely be gradually loosened over time as operational track records are established and existing prohibitions either prove to be too limiting or demonstrate themselves to be necessary.

17.3.3.2 Training and Certification

As was mentioned earlier, for many years federal standards have governed all facets of aviation from operator training and certification requirements to aircraft materials certification, to aviation-related manufacturing and aircraft maintenance standards. As UAS standards are developed, established and proven manned aircraft standards are often incorporated or modified where applicable and practical. The size and performance of the vehicle in conjunction with the complexity of the intended mission will drive UAS-related certification. For instance, consider that for the smallest vehicles intended to be operated within visual line of sight, current expectations are that no airworthiness certification processes will be required, and only minimal operator training will be mandated. For larger, more complex vehicles, the implementation of airworthiness standards as well as more stringent operator requirements, including some level of manned aircraft pilot certification, will likely be prescribed. It is also likely that the airworthiness standards, at least for some vehicles, will consist of industry consensus standards following a model similar to the light-sport aircraft certification process the FAA implemented approximately 10 years ago. Currently, work at Kansas State University is underway in conjunction with the FAA to validate consensus standards by applying airworthiness standards developed for sUAS by a group of industry representatives as a part of the American Society of Testing and Materials (ASTM) F-38 committee structure. In this analysis, a sUAS, representative of those aircraft available for commercial operations, was selected for evaluation. To assess the relevancy of F-38 standards, K-State researchers conducted a series of tests replicating the gamut of potential operational conditions and determining at each stage of the evaluation whether ASTM guidelines were appropriate to ensure safe operation and the integrity of the aircraft and associated systems. Other groups are currently working through similar validation processes to develop and evaluate airworthiness standards in areas that include UAS electronics and operator certification.

17.4 The Potential for Career Opportunities

UAS career opportunities in the future will abound as the vehicles and applications become more numerous and as the airspace opens to routine operations. There will be opportunities for UAS operators, sensor operators, and technicians (aircraft maintenance, electronic, and information technology). Larger vehicles with a larger logistical footprint will require more job specialization, whereas smaller vehicle operation is more apt to require the operator to perform multiple functions such as launch/recovery, flying, maintenance, and data collection/dissemination operations. Many anticipate that for operations which can be accomplished by sUAS, the main occupational skill will be in a primary technical area other than those which are aircraft-related (i.e., civil engineering, infrastructure maintenance, etc.) with the UAS operational skill as an add-on. Others believe that the sUAS operator will need to be able to operate across a spectrum of industries sending the imagery for analysis to those with the appropriate skill set. The future reality will likely be a hybrid of these two extremes. Those desiring a career in UAS would do well to select training which exposes them to a wide range of platforms and automation control software in addition to an education which exposes them to the larger challenges faced by the industry to include political and economic challenges.

Many universities are now creating UAS-focused degree programs centering on vehicle operations, system integration, and data collection. Examples include Kansas State University, the University of North Dakota, Embry-Riddle Aeronautical University, Middle Tennessee State University, Purdue University, as well as multiple 2-year colleges such as Sinclair Community College in Ohio, Cochise College in Arizona, among others.

17.5 Emerging Trends in Technology

There are numerous technological trends related to UAS to look for in the coming years. A brief discussion of a few of the most notable of those trends is found in the subsections that follow.

17.5.1 Miniaturization

Driven by advances in materials and processing technology, the size of many platforms will continue to become smaller. Each evolution of electronics technology allows designers to incorporate greater capability into increasingly smaller platforms. The limiting factor in miniaturization is often heat dissipation of the energy; more energy is released as more is being accomplished in these small spaces. In the future, as this problem is solved, it is likely that all the components necessary for vehicle operation in the NAS (navigation, communication, position reporting, etc.) will be located on one small printed circuit board which could be easily removed and placed in another vehicle. As miniaturization technology enables, micro air vehicles (MAVs-wingspan smaller than 6 inches) and nano air vehicles (NAVs- wingspan smaller than 3 inches) will become more prevalent in the future (see Figures 17.3 and 17.4).

FIGURE 17.3
Microchip. Permission is granted to copy, distribute, and/or modify this document under the terms of the *GNU Free Documentation License*, Version 1.2 or any later version published by the Free Software Foundation; with no *Invariant Sections*, no Front-Cover Texts, and no Back-Cover Texts. A copy of the license is included in the section entitled *GNU Free Documentation License*. (From Wikimedia search for Microchip.)

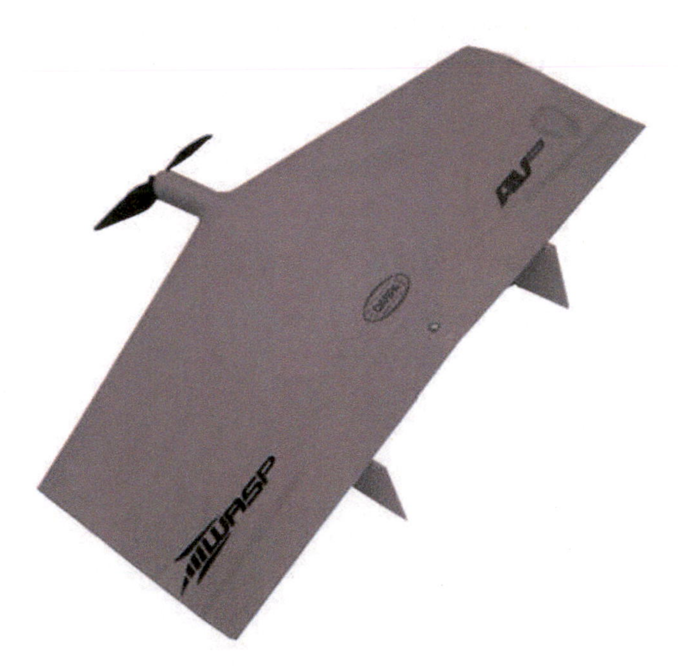

FIGURE 17.4
MAV. This image or file is a work of a Defense Advanced Research Projects Agency (DARPA), an agency of the United States Department of Defense, employee, taken or made as part of that person's official duties. As a work of the U.S. federal government, the image is in the public domain. DARPA website: Information presented on http://www.darpa.mil/ is considered public information and may be distributed or copied unless otherwise specified. Use of appropriate byline/photo/image credits is requested. Additional information may be found in: Wikimedia Commons Search for Micro Air Vehicle.

17.5.2 Power Solutions

In the future, energy to power UASs will continue to be the subject of much research. The requirements to become more ecologically friendly, less expensive, and more capable will stretch the limits of current sources of power into future solutions.

17.5.2.1 Alternative Energy

UASs will be no exception to the move away from fossil-based fuels, and much work has already been accomplished in this area. Hydrogen fuel continues to be one of the most desired sources of fuel for UASs due to its minimal impact on the environment in the form of emissions. Historically, the challenge of using this fuel source on aircraft has been the heavy weight of the equipment necessary to carry hydrogen in compressed liquid form; however, one company is looking to solve this by using hydrogen in pellet form combined with other chemicals which turn to hydrogen gas when heated slightly, thereby eliminating the necessity of the heavy and bulky tank.

Biofuel technology typically involves using the conversion of biomass into useable liquid fuel. Several biofuels have been tested on UASs, but it remains to be seen what role this technology will play in supplying our future energy needs and which fuels those will be. Debate continues on the exact carbon footprint of biofuel technology. However, future advancements will likely continue to reduce that impact making this type of energy more popular in the future.

Developments in reciprocating engine technology include the latest in digital engine control which permits the engine to vary ignition timing and compression ratios to allow these engines to burn a variety of bio- and fossil-based fuels making them highly flexible and adaptable to a range of operating environments and conditions.

Solar technology continues to hold great promise as a renewable fuel source for UASs. Current limitations revolve around limited payloads and the number of solar arrays required to develop sufficient power along with battery and other weight penalties. Efficiencies here will allow researchers to translate more solar energy using less surface area and storing that energy in lighter, more efficient ways. The hope for solar power continues to lie in advances in material technology which allow more energy to be extracted/stored using less weight as well as by advances in electronics which require less power to operate, thereby increasing the utility of solar-powered vehicles. Google's Titan project is one such example. The Titan aircraft has a 15.5 m wingspan and can carry a 32 kg (70 lb.) payload. It is powered by 3000 solar cells and can provide up to 7 kW of power stored in lithium-ion and lithium-sulfur batteries.

17.5.2.2 Electric Options

Currently, electric motor powered UASs are vehicles capable of carrying small payloads and are limited to an endurance of 1–2 h, at most, with the weight of the battery being the largest limiting factor. Advances in lithium-polymer battery technology currently hold much promise for extended battery life, lightest weight, and shape-ability that allows the battery to conform to aircraft design. Future advances in electric UAS will involve replenishment ability from power lines, an electric fuel "tanker" concept which can refuel vehicles in flight, or, as antennae and laser technology develops, the transmission of electricity through the air to recharge onboard batteries (see Figure 17.5).

FIGURE 17.5
LiPo battery. Source: I, the copyright holder of this work, release this work into the *public domain*. This applies worldwide. (Wikimedia Search LiPo Battery.) (From Wikimedia search for LiPo Battery or Lithium Polymer Battery.)

17.5.3 Materials Improvements

It is inherent in the field of aircraft design that the less weight required for the structure of the aircraft, the more payload it can carry. Advances in structural materials will focus in large part on composite technology and will no doubt become lighter and more durable, as well as easier to manufacture, maintain, and repair. Costs for these advanced materials are sure to escalate accordingly; however, the prices for current composite materials may correspondingly go down. Some current limitations of composite aircraft structures include assurance of long-term structural integrity, especially when exposed to abnormal conditions such as in a contaminated/caustic environment, etc. However, advances in nondestructive testing (NDT) technology are offsetting this limitation. One recent advance in composites technology, which will continue to enhance the fabricators' ability to use composite material in more intricate and cost-effective ways, is resin infusion molding (RIM) technology, a technology that allows for parts to be fabricated without the use of an autoclave—something that is often prohibitively expensive for smaller fabrication shops. Also, thin-ply tape technology now allows for ease in the fabrication of complex parts with intricate shapes and curves (see Figure 17.6).

Other material advances now and in the future will allow for the shape of the air vehicle structure itself to become dynamic depending on the needs of the existing flight condition. Principles of aerodynamics dictate that the structural configuration of an aircraft is different when optimized for high-speed flight than for low-speed flight. Typically, altering the shape of a wing (the structural member most typically modified) has required the use of complex and heavy variable-sweep mechanisms to accomplish this task. For example, a wing configured for high-speed flight will be very smooth (laminar) with low camber and, often, aft sweep. Wings for low-speed flight are typically the opposite with high camber, low sweep, and often with technology that disrupts laminar flow in favor of lift-enhancing technology such as vortex generators, flaps, and leading-edge slats. Military aircraft of the past have used complex variable wing sweep technology as well to assist in the transition from high-speed to low-speed flight. Materials in the future will allow for the shape of the structure to be modified as the mission requires without the use of heavy, complex infrastructure even to the point of being able to flex so as to eliminate the requirement to have traditional control surfaces such as flaps and ailerons.

FIGURE 17.6
Composite material – carbon fiber. Source: I, the copyright holder of this work, hereby publish it under the following licenses: This file is licensed under the Creative Commons Attribution-Share Alike 3.0 Unported license. (From *Wikimedia search for* Carbon Fiber.)

The concept of recyclable thermoset plastics is now becoming a reality that will allow certain structural plastics, traditionally discarded once the shape was "set" by heating past a certain point, to be recycled. This will make these materials more desirable for use in certain aircraft applications and will help reduce permanent waste.

Other advances in the materials sciences will allow UAS structures and components to become self-repairing when damaged or the aircraft to become suddenly "invisible" to observers using material that assumes the color and texture of the background behind it. Although not a material per se, advances in adaptive flight control technology will interface with these new materials to allow an aircraft to instantly reconfigure to an optimum "fly-away" state should it become damaged in flight.

17.5.4 Revolutionary Manufacturing

The revolution in 3-D printing technology is allowing for more complex components to be produced "on-demand" using a variety of materials including polymers and metals. This is enabling the concept of "distributed manufacturing" whereby products and parts can be produced and assembled (including by the customer) at multiple simultaneous locations rather than being produced and assembled centrally then distributed through a traditional supply chain. The impact on traditional manufacturing is expected to be significant and revolutionary across multiple industries. As the cost of these 3-D printers continues to decline, it will likely become standard practice to produce and assemble the major components of a UAS at any location (i.e., home). This is closely related to another trend in manufacturing called "additive manufacturing" whereby rather than beginning with a larger piece of material and removing the material not needed for a part, the part or component is produced from the raw material with very little, if any, waste.

17.5.5 Computing and Artificial Intelligence

Continual advances in the capabilities associated with powerful micro-processing are enabling a future airspace system where advanced vehicles are capable not only of autonomous flight but also of operation in an environment where vehicles communicate and coordinate with each other for collision avoidance and sequencing into and out of busy environments with little direct human involvement other than monitoring. Both wide-area communication systems and real-time operating systems are now being developed to enable this future system. Multi-input sense and avoid-enabled aircraft will likely, through the use of artificial intelligence and neuromorphic technology, learn from their environment and from other "machines" via interconnectivity and become capable of intelligent and autonomous decisions. The ethics of such technology are, and will continue to be, hotly debated as the future reality of machines that can outthink human beings in certain dimensions and can even self-replicate comes into clearer focus.

17.6 Future Applications

The technological advances discussed in the preceding sections both enable and drive the evolution of unmanned aircraft and the mission capabilities of the UAS, that is, what they can accomplish and how they will be utilized. Several future applications, novel and evolved, will be briefly discussed in the subsections that follow.

17.6.1 Atmospheric Satellites

One potential application of UAS includes supplementing the array of space-based satellites by using ultra-long endurance high-altitude solar-powered UASs which carry an array of sensors including cameras and communication relays providing similar capability to that of space-based satellites at a reduced cost, albeit serving a more limited geographic area due to the lower altitude. One example of this technology, currently under development, is the Google Titan project that aims to provide wireless Internet connectivity to under-served areas of the planet. The Titan has a 15.5 m wingspan and can carry a payload of up to 32 kg. It is powered by 32 solar cells delivering up to 7 kW of power for payload and propulsion storing excess in lithium-polymer and lithium-ion batteries. The aircraft cruises at an altitude of over 20 km, higher during daylight hours and drifting slightly lower during darkness until the sun's energy allows the aircraft to return to its optimum altitude.

17.6.2 Air Transportation

It is no secret that Amazon plans in the future to make same-day small package deliveries in certain markets when the regulatory and technological frameworks allow. While that will be some time yet, some air carriers have openly investigated the potential for air freight to be delivered on long over-water routes by UASs. When the air traffic control structure is in place to accommodate this technology, these routes would be perfectly suited to prove the viability of large-scale commercial UAS applications given that they would occur primarily over open water with no one on board. Similarly, although many people currently would never board an aircraft for a flight without a human pilot, it is definitely possible, if not likely, that some future generation will not be apprehensive at all in that situation.

For that to occur, the safety case will need to have been demonstrated beyond question for many years having been proven on similar aircraft by freight operations. While this concept is many years in the future yet, we will likely arrive there through a series of smaller steps such as from going from two pilots up front to one while operating in a much more modernized air traffic control system. For one pilot to be sufficient, technology will need to have rendered that single pilot to that of automation manager/monitor only intervening physically in cases of emergency. In the 1940s, the largest passenger aircraft operated with a flight crew of five (not counting cabin crew): captain, first officer, flight engineer, radio operator, and the navigator. Advances in technology have allowed this number to slowly be reduced to two through the years. The captain and the first officer are now the only two pilots and safety is better now than ever. Consider the fact that twin turbojet aircraft now (and have for many years) routinely carry passengers for hire currently in the NAS on corporate aircraft that are approved for single-pilot operations. In this case, certain automation technology must be present and available to the pilot so as to reduce their workload if needed. For larger scale operations carrying more people, it would be important to be able to monitor the health status of that single pilot so that a backup pilot from the ground could be made available in the unlikely event of failure of the onboard automation.

17.6.3 Unmanned Combat Air Vehicle

The concept behind the unmanned combat air vehicle (UCAV) was to design an offensive unmanned aerial weapons delivery platform as opposed to mounting weapons on a platform that was designed for another purpose. Currently, carrier operational trials are underway between Northrop and the Navy and are on track and successful. Many understand that this aircraft will be the standard for the future, and the human will likely be relegated to observing UCAV performance and monitoring systems from the ground. Removing the operator from on board is somewhat controversial with proponents advocating that human limits on acceleration combined with the weight penalty for onboard life-support systems produce a vehicle with less than optimum performance. Detractors continue to argue that computer logic will never adequately replace the human decision-making process, especially the ability to make split-second high-consequence decisions. However, this debate appears to be moot in that investment in this direction is already occurring; not that there will not be automation failures in the future, there most likely will be; however, the rate of those failures/errors will be much less than the rate would have been with a human operator, something that will need to have been proven by quantifiable data (see Figure 17.7).

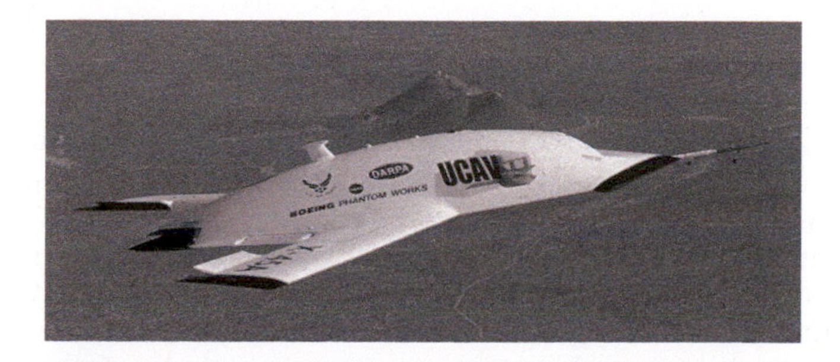

FIGURE 17.7
UCAV X-45A. This work is in the public domain in the United States because it is a work of the United States Federal Government under the terms of Title 17, Chapter 1, and Section 105 of the U.S. Code. See Copyright. (From Wikimedia search for UCAV or Unmanned Combat Air Vehicle.)

17.6.4 Commonality/Scalability

Given the proliferation of UAS technology produced by multiple manufacturers, it is the publically expressed desire of the military to move toward technical commonality which will allow for efficiencies in acquisition, support equipment, training, servicing, and support. As an analogy, some manned aircraft operators choose to operate one common model of aircraft with different variants in order to increase the familiarity of their aircraft operators, servicers, and maintainers with the equipment; much of the training and many of the features become similar allowing those who work with the aircraft to become much more efficient in performing their jobs. Likewise, the military is seeking similar system commonality for many of the same reasons. It is inefficient to attempt to field a multitude of unrelated vehicles. The concept of scalability, closely related to the concept of commonality, allows the features of one vehicle to be "up-" or "down-sized" based on the mission requirements. The technological advances discussed in the preceding sections will enable greater commonality among UASs and allow unmanned aircraft to be more facilely and efficiently scaled, in less time, at lower costs.

17.6.5 Swarming UAS

The concept of "swarming" is largely a military-oriented concept (borrowed from nature) that is making inroads into the potential civil market. In military parlance, the concept of swarming is when an objective is accomplished (i.e., a target is attacked or observed) using multiple simultaneous aircraft, through varied means. It is a technique used to overwhelm a target and subdue it quickly and gain a tactical or strategic advantage. This concept, which is already being tested in military research, would involve the close coordination of multiple independent systems in a relatively small amount of airspace. In other words, these systems would need to display a high degree of interoperability most likely coupled with a higher degree of autonomy in the future. The command and control infrastructure has yet to progress to a point where swarming could be supported, but a move toward this concept will no doubt drive those necessary technical improvements (Flaherty 2014).

17.7 Five Years and Beyond

Since many future concepts are based on technology yet-to-be-invented, this is where the discussion gets a bit more difficult and devoid of specifics. What is known is that scientific and technological advancements are truly mind-boggling, and intelligent minds continue to constantly push to almost any limit when challenges arise and the need and/or opportunity is great. As has been mentioned, machines, which can learn to accomplish complex tasks on their own and also learn from each other, are being designed and fielded. Complex tasks in the future will involve interacting with humans, learning to speak, and generating ideas. Combined with advances in mechanics, structures/materials, and power delivery, the future is anyone's guess. As was mentioned, we may indeed see a future where artificially intelligent machines can repair or replicate themselves, seek their own fuel source, and make decisions that could run counter to their originally intended design.

Another concept sure to continue into the future is the field of unmanned spaceflight. Certainly, many unmanned space missions over the last 40 plus years have demonstrated the advantages of being able to explore space and other planets without having to consider the limitations of human physiology. The successes of the Mars rover projects have demonstrated on a grand scale that unmanned operations are often the best way to accomplish a dangerous mission.

There is much more that could be discussed on this topic and certainly much more depth that could be explored on each topic, but due to the scope of this text we shall leave that for further exploration by the reader. One thing is for sure, as has been said, the future is unlimited and unmanned!

DISCUSSION QUESTIONS

17.1 List the advantages and disadvantages of the term "Unmanned Aerial System." What are some alternate terms?

17.2 Refer to the eight basic segments of the UAS market in Section II of this chapter. Use the internet to list one current development (within the last 90 days) in each area.

17.3 List three challenges of converting ground-based infrastructure (i.e., airports) to joint manned/unmanned use.

17.4 Many UAS systems are designed to be used for surveillance; what challenges may arise with widespread UAS use for surveillance?

17.5 What should be the limits of artificial intelligence as it relates to autonomous decision-making by UASs?

References

Federal Aviation Administration (FAA). *Integration of Civil Unmanned Aircraft Systems (UAS) in the National Airspace System (NAS)*. November 2013. Washington, DC: FAA Communications. www.faa.gov.

Flaherty, N. Aiming for the high life. Unmanned Systems Technology. November 2014. High Power Media.

Frost and Sullivan. Global military unmanned aerial vehicles market assessment market research media. *Market Research Media Premium Market Analysis*. Web. 9 January 2014. <http://www.marketresearchmedia.com/?p = 509>.

The Teal Group 2014. UAV Integrated Market Analysis. Web. 17 July 2014. <http://www.tealgroup.com/index/php/about-teal-group-corporation/press-releases/118-2014-uav-press-release>.

For Further Reading

Singer, P. W. *Wired for War: The Robotics Revolution and Conflict in the 21st Century*. New York: Penguin Press HC, The, 2009. Print.

Epilogue: A Final Word

Richard Kurt Barnhart, Douglas M. Marshall, Michael T. Most, and Eric J. Shappee

Unmanned aircraft systems (UAS) technologies and the uses to which they have been applied have evolved dramatically since the publication of the first edition of this book. Arguably, the most disruptive element of this evolution has been the introduction into the world market of relatively affordable multirotor vertical takeoff and landing aircraft equipped with gimbaled high-definition or multispectral cameras capable of delivering quality images and video from a stable platform. The ensuing "Cambrian explosion" of manufacturers offering a wide variety of systems and components has generated an untold number of entrepreneurs and new categories of UAS operators, to a point that has overwhelmed the regulators. As of this writing, the U.S. Federal Aviation Administration (FAA) has granted over 2500 exemptions to commercial operators, and the number grows by a factor of 10 or more per day, with no end in sight. Much of this exponential growth has occurred even while this second edition was being written, and while the preceding 17 chapters have been intended to provide the reader with descriptions and summaries of the state of the art of the technology, by necessity informed by past history, it is likely that some of the information the authors have shared will be overtaken by new advances and developments even as this edition goes to press.

It is useful, however, to summarize the high points of each chapter in an effort to bring some order to the chaos of this dynamic and exhilarating industry. What follows reflects the editors' takeaways from each chapter.

Chapter 1

This chapter traces the history of unmanned aviation from ancient Chinese kites, through early unmanned aircraft, to the Wright brothers and beyond, accelerated by military usages in WWI and WWII.

It was in the crucible of "the war to end all wars" that aviation came of age and along with this wave of technological advancement came the critical but little recognized necessity of achieving effective flight control.

Unmanned aircraft paved the way for the development of manned aircraft (Chinese kites and hot air balloons), and the process has reversed as systems advances in manned aviation led to the integration of manned technologies into unmanned systems.

The developmental process is the same for manned and unmanned aircraft and is parallel in all respects. Structures, propulsion, flight control systems, stabilization systems, navigation systems, and the integration of all these components into flight automation systems made the nearly parallel development of manned and unmanned aircraft systems possible.

The reader is introduced to the three Ds: *dangerous, dirty,* and *dull. Dangerous* means that either someone is trying to bring down the aircraft or the life of the pilot may be at undue risk operationally. *Dirty* is where the environment may be contaminated by chemical, biological, or radiological hazards precluding human exposure. Finally, *dull* is where the task requires long hours in the air, making manned flight fatiguing, stressful, and not desirable.

Nicola Tesla invented radio control and perhaps the radio itself (before Marconi), to guide torpedoes. Lawrence Sperry developed the first practical gyro-control system, for the same reason. These inventions led to the first practical mechanical autopilot. In 1918, Sperry teamed with Glenn Curtiss to create the first practical unmanned aircraft, the Curtiss N-9 Aerial Torpedo (the first "drone" was never put into production). The Army answered with the Kettering aerial bomb, or "Bug," the first mass-produced unmanned aircraft.

The interwar period (1919–1939) produced a demand for target drones, which proved effective against antiaircraft guns during tests by U.S. and British navies, and that led to air power doctrine that favored aircraft carriers to defend the fleet.

The famous British actor Reginald Denny, model aircraft enthusiast, created an unmanned target drone for the U.S. Army, and eventually produced over 15,000 OQ-1 Radioplanes (TDD-1-Target Drone Denny- for the Navy), the most popular drone of WWII. U.S. and Royal navies continued to develop newer versions of bigger and heavier target drones to train antiaircraft crews during gunnery trials, with great success.

The first detection device on a drone was a primitive 75-pound RCA TV camera in the nose of a TDN-1 Assault Drone (flown from another manned guiding aircraft). Later versions were deployed successfully in the Pacific Theater during WWII. Eventually, the Navy and Army air forces outfitted four-engine bombers with Sperry three-axis autopilots, radio control links, and RCA TV cameras in the cockpit. They were loaded with explosives, the idea being that crews would control it during takeoff, set up the automatic functions and aim to the target, and then bail out. The tests and early missions were largely unsuccessful, resulting in abandonment of the program, but not before one such aircraft's payload detonated prematurely and killed the two pilots, one of them being President John F. Kennedy's older brother Joseph P. Kennedy.

The Germans produced the most advanced unmanned aircraft of the era with the V-1 Buzz Bomb, which terrorized England for much of the war. The device was the first successful, mass-produced, cruise-missile-type unmanned aircraft, and its design influenced many postwar unmanned aircraft designs. Over 25,000 were built, which made it the highest production unmanned combat aircraft in history, excluding modern hand-launched small unmanned aerial vehicles (UAVs). Most were ground-launched, but some were deployed from aircraft, making them the first air-launched unmanned system as well. While relatively primitive with regard to accuracy and navigability, the V-1 was mass produced, and, employing many firsts for autonomously flown aircraft, it influenced future designs and provided the historical context to fund many more sophisticated unmanned programs during the following Cold War. The U.S. Navy reverse-engineered a copy for use in the invasion of Japan and made later improvements on it to evolve into the first naval-launched, jet-powered, unmanned cruise missile.

The Germans developed other weapons systems throughout WWII. The lines between guided missile and unmanned aircraft are not always clear, and the V-1 assault drones, explosive-packed, radio-controlled bombers, and the piggyback Mistletoe (Mistel) configuration all involved forms of an airplane, which places them in the category of unmanned aircraft.

Unmanned systems development shifted from target drones and weapons delivery platforms to reconnaissance aircraft during the Cold War years, a trend that continues today, with 90% of all unmanned aircraft now being used for data gathering, law enforcement, and environmental monitoring.

The Vietnam War witnessed the development of radar decoy drones designed to deceive surface-to-air (SAM) missile radar into locking onto a drone instead of a manned aircraft, using a combination of radar reflectors and radios that mimicked electronic signatures of large aircraft such as B-52s.

In the 1960s and 1970s, the high-performance Ryan AQM-34 "Lightning Bug" reconnaissance aircraft enjoyed a long career in intelligence gathering and radar decoy roles, primarily in high-priority missions of great national importance and later on as target drones for fighter air-to-air missiles. The first helicopter UAS was also developed during the same time frame, most notably the pioneering Navy QH-50 DASH (Drone Anti-Submarine Helicopter), an aircraft developed to deploy homing torpedoes over submarines. Over 700 were built, and were also operated by other countries such as France and Japan.

All along this development history, the goal was to achieve as much autonomy from manned ground control as possible. It was not until the advent of small, lightweight digital computers, inertial navigation technology, and finally the global positioning system (GPS) satellite network that autonomous unmanned aircraft operation gained flight autonomy on par with a human-piloted vehicle. The worldwide explosion in personal computers and the digitalization of everyday items, from wrist watches to kitchen appliances, played the most significant role in unmanned aircraft autonomy.

Moving forward into the most recent era, Middle East conflicts brought the emergence of the twin-boom pusher UAV, such as the RQ-7 Shadow, and the Israel Aerospace Industries' "Scout," both used for tactical battlefield surveillance and equipped with advanced imagery capability. The Scout was accompanied by two other systems, the IAI Decoy UAV-A and the Ryan-built Mabat. They were primarily used to detect and destroy SAM batteries and provide quality eyes-on battlefield imagery for ground commanders. The lessons learned from the 1982 Bekaa Valley conflict between Israel and Syria, where the Scout UAS proved to be very effective, initiated a worldwide race to develop close-battle unmanned aircraft.

Unmanned systems continued to evolve through the Desert Storm conflict in 1991 and accelerated after the 9/11 terrorist attacks on the United States, as the military's demands for both reconnaissance and weapons delivery platforms have overcome inherent biases from the manned aviation community. Military leadership has openly embraced the technology and has committed significant resources to the acquisition and deployment of unmanned systems, increasing the number of aircraft from 30 in 2001 to over 2000 as of 2010. Amateurs and commercial UAS operators in the United States have leveraged the military experience and advantages in capability through microelectromechanical systems (MEMS), miniature power plants, and advanced radio systems to promote the transformation of a toy into a viable tool. This revolution has happened so quickly that the FAA has had great difficulty controlling the proliferation of commercial UAS manufacturers and operators.

The question of whether unmanned aircraft will ever replace manned aircraft is yet unanswered, and the challenge to that concept is the vulnerability of a fully autonomous system to interference or jamming by a hostile force or element. Another challenge is the chain of responsibility in the military weapons delivery arena, which suggests that a fully autonomous system would not be allowed to employ lethal force.

Chapter 2

In nearly all instances, the purpose of operating a UAS is to gather data or information through the process of remote sensing. Numerous definitions are available, but one well-articulated characterization, generalized enough to fit most situations, is offered by Lillesand and Kiefer (2000, p. 1): "Remote sensing is the science and art of obtaining information about an object, area, or phenomenon through the analysis of data acquired by a device that is not in contact with the object, area, or phenomenon under investigation." Most often, specific frequencies of the electromagnetic spectrum are remotely sensed; the determination of which frequencies are obtained is predicated upon the nature of the research or mission and the type of remotely sensed data required—that is, the application of that information.

Applications for which UAS missions are commonly flown are numerous, diverse, and increasing in number. These include photogrammetric analysis (e.g., surveying and volumetric assessments), infrastructure inspections, precision agriculture, natural resource management, aerial photography and cinematography, environmental monitoring and remediation (e.g., tracking hazardous and radioactive materials), journalism and newsgathering, and search and rescue operations. In the hands of first responders and emergency managers, UASs have become tools in the service of the common good.

At the 2004 ISPRS (International Society for Photogrammetry and Remote Sensing) Congress held in Istanbul, the organization passed a resolution in which UAS were recognized as an emerging, cost-effective, and novel platform uniquely suited for rapidly obtaining remotely sensed data (Everaerts 2008). The Congress also noted the capacity for unmanned aircraft to operate at great distances, for long periods, and under conditions so hazardous that these would endanger human operators—what has elsewhere in this book been referred to as dull, dirty, and dangerous or 3D missions. Consequently, ISPRS anticipated a rapid growth in the breadth and number of UAS applications and resolved to "inventory … current and technologically feasible miniature sensors … [to inventory] possible future civil applications … [and to report these findings to the] global community" (Everaerts 2008, 1187). Four years later, Everaerts (2008, 1190) reported in one of the peer-reviewed journals published by the ISPRS that "many remote sensing applications have benefited from the use of UAVs. In most cases, this was due to the cost of the mission, the need for rapid response or the fact that observations need to be carried out in an environment that may be harmful or dangerous to an aircrew." The author further proffered that the proliferation of UASs in such a short time was facilitated by the increased availability of extremely capable and cost-effective commercial-off-the-shelf platforms and predicted that UASs would become the preferred platform for remote sensing applications. It would appear, another decade removed, that this prediction was not excessively optimistic.

Chapter 3

This chapter introduces the reader to the components that make up an "unmanned aircraft system." Whether military or civilian, the system typically consists of the human element, the command and control element, the aircraft itself, the communication data link, the payload, and the launch and recovery element.

The systems described may have different names, such as remotely piloted aircraft, or remotely piloted vehicle, in addition to "unmanned aircraft system." Whatever the name, they are commonly categorized by weight and in the future perhaps by risk-based criteria. The physical and operational characteristics of both fixed-wing and vertical takeoff and landing UAS platforms are described.

The elements of the command and control component consist of an autopilot, a programmable lost-link procedure, and some degree of autonomy in navigation. The ground control station (CS) can be land- or sea-based, and can range in size from a handheld transmitter to a workstation that can accommodate multiple personnel, such as a pilot and sensor operator.

The communication data link connects the ground control station to the autopilot on the aircraft and can be line-of-sight or beyond-line-of-sight (BLOS), with varying choices for operating frequencies, subject to Federal Communications Commission (FCC) regulations. The BLOS option requires a satellite link, which invokes the drawback of latency or delay of transmission of the signal.

Payloads usually refer to an instrument on board the aircraft that provides imaging, video, communications, sensing devices, radar, or other instruments that provide the operator with information about a particular area of interest or phenomenon that requires observation. Electro-optical, thermal infrared, multi- and hyper-spectral cameras, and laser range finders offer capabilities to record and retain images and data that are useful in a wide and ever-growing variety of applications in the civilian and scientific communities, in addition to multiple military uses.

The launch and recovery element ranges from hand-launched small UAS to catapult-launched larger aircraft, with many others in between that take off and land very much like a conventional aircraft.

The human element is the most important, consisting of pilot, sensor operator, and supporting ground crew. The pilot in command is still in charge of the safe operation of the aircraft. The human interaction (human factors) component of this element is covered in detail in Chapter 11.

Chapter 4

This chapter takes the reader through a comprehensive review of the key characteristics of the payloads commonly found on UAS of varying sizes and lift capacities. The primary reason that UAS are flown at all is to collect some form of data. Data can be generated *in situ* (in place) or remotely with sensors, and the types of sensors that can accomplish this task is the focus of this chapter.

There are two types of remote sensing sensors: spot sensors and imaging sensors. The former measures a single location, such as identifying surface melt water on sea ice, whereas the latter records multiple data points utilizing a number of different methods.

The most common remote sensors are cameras, which can be visible-spectrum cameras, near-infrared cameras, long-wave infrared cameras, and hyperspectral imagers. Each has its own unique capability of collecting and storing data, and the choice for the user is to match the type of camera or imager with the mission and the type of platform to be employed to carry it.

Other devices for data collection include LIDAR (light detection and ranging) and synthetic aperture radar (SAR). LIDAR is useful for measuring distance to the ground from the aircraft, and SAR radars are often used for mapping and have the advantage of being able to "see" through clouds, smoke, and dust. SARs are also able to pick up surface features, and they have high spatial resolution.

Remote sensing technologies support many applications in academic research and science. They also offer much potential in emergency response and search and rescue operations.

Other uses of the technology include creation of background imagery for mapping, 3D point cloud/modeling, and measuring vegetation health, which has proved to be a significant tool in precision agriculture and thermal mapping, as well as forestry and vegetation management. Since the turn of the century, the industries that use remote sensing tools have witnessed an explosion of digital technology that has been the foundation for the growth of the UAS market as a whole, and of the remote sensing component in particular. However, the field of UAS remote sensing is in its adolescence and developers are still exploring its possibilities.

Chapter 5

Chapter 5 provides a broad overview of the complexities of the aviation regulatory system within which manufacturers, developers, users, and operators of unmanned systems must navigate to stay on the compliance side of the law. The reader is introduced to a brief history of the U.S. regulatory system in general, the FAA's "tool box" in particular, and the methods by which the FAA enforces those regulations, policies, orders, and guidance documents.

Regulations applying to unmanned systems cannot be read in isolation from international regulations and standards, so additional attention is given to the International Civil Aviation Organization (ICAO) and other European government aviation organizations and commissions that have also developed or are in the process of developing comprehensive regulations for unmanned aircraft in sovereign as well as international airspace.

A brief exposition of the differences and similarities between regulations and consensus-based industry standards is offered, with further exploration of how the standards process has and will impact UAS/RPA operations and airworthiness requirements in the United States and globally.

How the regulatory process works in the United States, from proposal and crafting of regulations, to implementation, and then compliance and enforcement policies, is outlined to inform the reader and the "prudent entrepreneur" on how to access, understand, and comply with the relevant regulations and policies, with the ultimate goal being to avoid the enforcement phase of the process.

Finally, a way forward is proposed, with suggestions to the user community as well as the regulators on possible methods to reduce the pain of the "sausage making" process in the development of reasonable, logical, and enforceable rules and regulations for unmanned aircraft. The rule-making process is ongoing as this book goes to press, and there may be some basis for predicting how the final UAS rule may read, but it is recommended that industry and user groups be involved from beginning to end so that all voices are heard and the best outcome that maintains the high level of safety in the U.S. national aviation system can be achieved.

Chapter 6

While remotely piloted aircraft (RPA) are often referred to as (unmanned), it is clear that the humans involved and their actions are critical elements to safe operations. The goal of human factors is to provide operators and support personnel with the necessary knowledge, skills, and abilities and to achieve the overall goal of safe, effective, and efficient operations.

The aim of this chapter is to provide a broad discussion of human factors concepts and their application to the field of RPA operations. The chapter provides the reader with an overview and in-depth analysis of the core challenges to all who are interested in or committed to the development and evolution of unmanned systems, which is how to integrate a wide range of operators and their experiences with equipment of varying capabilities flying very different missions in often unpredictable conditions.

Many of the human factors concepts discussed are common to manned as well as unmanned systems, and appear across other professions operating in complex, dynamic settings. The scope of the problem is enormous.

The pitfalls of hindsight bias (the "belief that an event is more predictable after it becomes known than before it became known") are explained with reference to addressing the concerns arising from the transition from manned aircraft to RPA systems. Examples such as imaging technologies being used for sensing and avoiding hazards, the capabilities of human vision versus the limitations of RPA visual displays as they influence in-flight maneuvers, and differences in perception, peripheral vision, and relative motion demonstrate that the transition faces many challenges.

Charles Wickens' multiple resource theory was developed to explain human cognitive challenges of performing simultaneous and/or difficult tasks. "Attention" is defined as the management of brain resources, which can be managed in several ways, and are referred to as attention types. Those types are categorized as "selective attention" (utilizing a systematic visual scan of information sources that involves the operator's skill to select and process these information sources); "focused attention" (directing brain resources to a single task, potentially resulting in poor distraction management and loss of situation awareness); "divided attention" (performing two tasks simultaneously such as an RPA pilot observing visual displays for optical images while controlling the aircraft, made more difficult when tasks share resources); and "sustained attention" (the process of monitoring the environment for changes, which may or may not be foreseen, such as an RPA operator monitoring an automated system for failures, flight performance, and navigation, subject to degradation over time).

Human errors occur when an action results in an outcome that was not intended, even if the actions performed are intended. It is evident that well-trained and current operators still commit errors. The nature of human error involves a wide range of activities, from overall strategy and planning errors, to inadequate monitoring of information systems, to poor technical execution of a skill. Some errors are inconsequential, which means that there is no negative outcome if they are not identified and corrected. Many errors are identified and corrected, through monitoring actions of operators or other individuals. Anticipating, monitoring, identifying, and correcting errors are common tasks in the aviation industry. When trained and qualified individuals commit errors, it is seen as an unexpected event. In reality, errors are common and should be expected. Training professionals in RPA operations should include active monitoring and expectation of errors.

Negative transfer is a common error when transitioning to new systems with similar layouts. Negative transfer is using a previous, well-established skill in a new setting in which the action is incorrect.

Threat and error management strategies began evolving in the 1970s with the beginning of "cockpit resource management" (CRM) to assist crew interactions with regard to communication, crew coordination, team building, and decision making. The work of James Reason, Douglas Wiegmann, Thomas Shappell, and others further identified ways to classify errors and contributing causes, with the intent to identify trends and develop effective countermeasures. Shappell and Wiegmann's publication of the human factors analysis and classification system (HFACS) provided a framework to identify and classify errors and contributing factors.

Crew resource management may be defined as "using all available resources—information, equipment, and people—to achieve safe and effective flight operations," a concept arising out of a series of crew coordination incidents in the 1970s. Team concept principles stem from manned aviation operations, where the goal is to train pilots and other crewmembers to work together in a coordinated fashion, sharing information, decision making, and actions. In RPA operations, there are likely to be others outside the operations team that would provide input or direct operations. The initial goals of CRM training involved increasing participative management and assertiveness in the crews. While a clear chain of command is important in RPA operations, participative management refers to leaders who involve other crewmembers in the decision making and actions of the team. This discussion involves the use of power distance, effective communication, ambiguity resolution, distraction management, and avoidance of negative transfer. These are a few of the topics in advanced crew coordination coursework, but are considered central issues in safe operations of unmanned systems.

Situation awareness is critical in unmanned aircraft operations. In spite of the diversity of professions involved in situation awareness training, common cognitive processes are used. Mica Endsley defined situation awareness as "an internalized mental model of the current state of the flight environment." Delta Air Lines' training programs state it is "the ability to recognize events occurring to you and around you then reacting correctly to those events."

The task of vigilance requires individuals to direct their attention to information (e.g., aircraft position, instrument readings, airspace conflicts, and structures), which has the most immediate impact on safety. Vigilance not only is managing the appropriate type of attention but also requires knowledge of the operations to identify and correctly assess the risks involved. Vigilance is commonly degraded by distractions. When using focused attention to provide inputs to the control systems, even momentary distractions can lead to significant problems. Distraction management strategies are designed to improve vigilance of tasks.

When working in complex systems, correctly identifying and understanding a developing situation ("diagnosis") can be challenging. Achieving the correct diagnosis involves many skills. Gary Klein describes the process as creating a story to explain the findings, using experience to shape the explanation. One challenge to make a correct diagnosis can arise from the clarity of the information presented to the operator. Incomplete, inaccurate, misleading, or conflicting information can impair this process. As aviation professionals, training emphasizes the need to consider abnormal readings that cannot be explained.

Risk analysis asks, "What is the likelihood and consequence of failure of each option I am considering?" "Vicarious learning" is defined as an individual whose prior poor choices and safety risks have not resulted in negative outcomes and, as a result, the individual does not have a realistic assessment of the risks they are taking.

Individuals who exhibit appropriate vigilance, correctly identify the problems occurring, and have selected an appropriate course with regard to risk assessment may still fail in that they do not act to correct the problem appropriately. Individuals who commit errors may not be willing to identify and correct the error, as this will bring attention and possibly blame and liability.

The goal when designing workstations (control modules, information displays, etc.) is to create a system the operators can use efficiently and effectively (human–machine interface). Often, the control systems are analogs of previously designed control systems, such as video gaming, aircraft controls, or computer keyboards.

In man–machine interfacing, compatibility refers to the consistency of the information and control systems to the operator's expectations. When transitioning to a new workstation, attention to compatibility can identify aspects that need modification or additional training to achieve compatibility. Conceptual compatibility is the effective use of symbols, colors, sounds, or other indicators to convey information. Spatial compatibility refers to the organization of the information and control systems. Are the information systems located where the operator would expect them? Movement compatibility refers to the direction and sensitivity of the movement. The direction and sensitivity of controls can lead to operator error. Controlling a small, radio-controlled RPA through direct visual observation is a good example of movement compatibility challenges. Systems that are poorly compatible will increase demands on the users. Human–machine interactions can be challenging. Proficiency, currency, and developing correct mental models of the systems you are using cannot be overemphasized.

Chapter 7

In this chapter, the reader is introduced to system safety tools aimed at promoting the safe integration of UAS into the National Airspace System (NAS). As organizational safety measures and procedures are increasingly being prioritized at the "system" or management level to ensure appropriate implementation, this chapter applies some of the time-tested principles of safety management system (SMS) theory to a UAS setting and provides the reader with practical tools that can be applied to any UAS operation as a system safety enhancement.

Although not a step-by-step how-to guide for SMS implementation, this chapter examines the hazard analysis process, which is a key foundational step in any SMS program, whereby potential operational hazards are identified and quantified. Closely associated with that is the "change analysis" process, which investigates organizational changes in light of their impact on safety and informs the hazard analysis process. The reader can then use the risk assessment tool to begin to assess risks in their own organization. The chapter continues with a look at flight test cards and their role in the safety process, as well as a look at how these tools can enhance the airworthiness process. Finally, the chapter concludes with accident investigation considerations specific to UAS.

Chapter 8

This chapter is a comprehensive treatment of the export control regulations related to unmanned aircraft systems in the United States. The chapter starts with a glossary of important terms related to export control. The discussion continues with an outline of the sources of export controls in the United States and describes in detail the U.S. policies that underlie the export control regulations. Also covered in this chapter is the multitude of statutes, executive orders, and regulations that make up the U.S. export control system. Many examples are given of products that are covered by the export control system and would potentially require an export license to transfer, ship, or move outside of the United States, and a full list of "exports" as defined by International Trade in Arms Regulations (ITAR) is provided.

The Export Administration Regulations (EAR) and ITAR regulations are very complicated, and the unwary faces significant sanctions for noncompliance. The chapter explores the purpose of export control, the underlying purpose or justification for enacting export control regulations, the purpose of the export control license, who must obtain an export control license, and which federal agency issues such licenses, and describes the Arms Export Control Act (AECA) and ITAR regulations. The authors describe how these regulations apply to UAS and associated technologies, and also explain the differences between ITAR and EAR regulations, which agency controls exports under the latter regulation, and what is meant by the term "export" under ITAR regulations.

The reader is provided with information to be able to describe in detail EAR regulations, to list and describe those UAS technologies that have been removed by the Departments of State and Commerce from EAR and ITAR control, why export controls are described as "strict liability" laws, and to describe the Missile Technology Control Regime (MTCR) and corresponding annex in detail.

The chapter also explains the ITAR categories, including those categories and sections that refer to UAS, describes what is meant by a compliance program, and explains why it is important to have one if there is potential EAR and ITAR application.

Chapter 9

This chapter introduces the reader to the assumptions and processes that go into the selection and execution of the design of an unmanned aircraft system. The design process is outlined, emphasizing the philosophy that the design should be driven by the mission to be accomplished by the aircraft and the system. A brief history of UAS development, from military to public to commercial operations in the United States, beginning in the early twentieth century and continuing through the present, provides context for the evolution of mission-driven UAS designs. Raymer's "design wheel" illustrates the iterative nature of the process where preceding trade studies sets requirements, concepts are derived from requirements, the design analysis may generate new concepts, and the cycle may repeat.

The chapter covers design tools, airframe materials and components, propulsion systems (battery-powered electric motors, gas turbine, and internal combustion engines), flight control systems (autopilots, wireless remote, stability augmentation, auto takeoff, and landing), control stations (GCS, handheld controllers, CS interface), payloads (cameras, meteorological sensors EO/IR sensors, lidar, SAR), and communications, command, and control (C3, wireless data links).

The ASTM International F38 Committee on Unmanned Systems is developing engineering standards supporting UAS type certification. RTCA Special Committee 203 was established to develop standards for sense-and-avoid and C3 criteria and SC 228 is developing minimum operational performance standards for detect-and-avoid equipment and C2 command and control data links.

Hardware-in-the-loop (HIL) testing is widely used for the UAS design verification of the aircraft control system, sensor/payload integration, and the CS interface. After systems verification, a simulated environment might be used to prepare for mission validation, or for pilot/operator training, where the tasks of an actual mission might be simulated.

Chapter 10

In comparison to the design choices available to those who engineer and create manned aircraft, a much larger number of configuration options inhere to the creation of unmanned aircraft, particularly small UAS or sUAS. Should the vehicle be a single- or multirotor or fixed-wing aircraft? If a fixed-wing design, should it be a flying wing, conventional configuration, or canard? Should the tail be a cruciform, T-tail, inverted-T, a Y-tail, V-tail, or inverted V? Should the structure be traditional monocoque or composite? Is the power plant electric or liquid fuel? If the latter, then options may include two-stroke, four-stroke, heavy fuel engines (HFE), Wankle, and gas turbine. A variety of factors, practical, economic, and exogenous, influence the ultimate form taken by an unmanned aircraft, but the paramount determining design factor must be the nature of the mission, which is, in turn, often a function of the type and amount of data to be collected. In fact, it is not an exaggeration to say that the successful culmination of any mission will be best served by operating a UAS optimized to the particular task at hand.

UASs currently operate in a protean environment. The future is similarly unsettled, but one commonality between the two states is that the mutable and evolving nature shared by the two will continue to drive UAS airframe and powerplant design. With the integration of increasingly larger UASs into the NAS will come a new set of missions as design drivers. The development of detect-and-avoid systems, the evolution of regulations, and the refinement of BLOS enabling technologies will promote the operation of UAS for pipeline and power line patrol, railway surveillance, the international operation of unmanned freighters, and domestic package delivery. Some speculate that, inevitably, unmanned transports will carry passengers to their destinations. These new applications will define missions that will continue to evolve UAS into larger, faster, safer, and more capable aircraft.

Chapter 11

An electrical system is essential to the operation of all controllable unmanned aircraft (UA). Even the least complex, RPA, relies upon electrical power to receive, process, and distribute input signals to achieve command and control and often for propulsion. The intricacy and sophistication of the onboard systems of large unmanned aircraft are comparable to those of large, turbine-powered manned aircraft—and some make the case that, due to payload and command and control requirements, large UASs are more complex than comparably sized manned aircraft. The electrical system interconnects all subsystems and, therefore, is the systemic nexus common to all UASs.

The evolution of UAS capabilities and the expansion of mission profiles will result in larger, more sophisticated UAS coming into service. This circumstance will impact component and subsystem design, which, in turn, will drive the evolution of larger, more complex electrical systems and associated components. Brushless generators and starter/generators have already been developed for UAS applications. A continuation of this trend will see the replacement of hydraulic system components and fluid lines with electrical actuators and wires. This will require more efficient and possibly higher-capacity generators, larger electrical distribution system components (to handle heavier current flows), and greater complexity throughout the entire interconnecting subsystem nexus.

Chapter 12

Chapter 12 takes a look at UAS communication systems. This chapter starts with a practical way to explain UAS communication to the novice and moves into a more detailed discussion. Items discussed in this chapter include electromagnetic wave propagation in free space, and basic communication systems and its elements, which includes modulation and frequency hopping.

A comprehensive outline of a basic communication system and its elements follows. The elements should consist of modulation devices, transmitters, the channel for the wireless link (antenna directivity, antenna gain, and polarization issues), the receiver (signal-to-noise ratio, receiver sensitivity, dispreading the signal), and demodulation design strategies. System design requires the establishment of bandwidth requirements and effective link design. A summary of design principles is provided, along with a discussion of bandwidth requirement and the associated problems for EMI (electromagnetic interference) internal interference, jamming, and multipath (the different pathways a signal can take to a receiver and how some of these signals arrive at the receiver after bouncing around).

Chapter 13

The command and control (C2) components comprise the UAS subsystem that enables controllable and autonomous flight. An essential C2 element is human, and the term operator "in the loop" implies direct control of the vehicle through pilot intervention, and operator "on the loop" indicates that the operator is monitoring autonomous flight. Other components most generally comprising the C2 subsystem are the ground station (GS) and associated software and electronics, antennae, ground-based and airborne transceivers, the autopilot to enable autonomous flight, air data and GPS systems, MEMS gyros, accelerometers and magnetometers for navigation and vehicle control, the interconnecting circuits and data buses, and onboard intelligences for computing and data processing. Larger UASs may also include auto-takeoff and auto-landing systems. Although UAS command and control may occur via other methods (e.g., light transmission or through the umbilical cable of a tethered vehicle), C2 is most commonly accomplished via two-way transmissions that communicate commands to the UAS through an uplink and telemetry from the aircraft to the GS through the downlink. The medium through which this communication is accomplished is the radio frequency (RF) portion of the electromagnetic spectrum.

The proliferation of UASs has paralleled improvements in binary computing devices and associated electronics. The miniaturization of these devices and their incorporation into microcircuits was enabled by developments in materials sciences, quantum physics/solid-state theory, and semiconductor manufacturing techniques. As C2 systems continue to evolve and mature to become more compact, less expensive, and more capable, it is likely that the requisite level of human intervention will correspondingly decrease, just as occurred in manned flight where transport crews have shrunk from five (pilot, copilot, navigator, radio operator, and flight engineer) to two as the result of improved technologies. If predictions that unmanned transports will one day carry passengers to their destinations prove true, the ability to do so will be dependent not only upon the evolution of C2 systems to become nearly 100% reliable, but also the integration of detect-and-avoid components into the command and control subsystem.

Chapter 14

The integration of subsystems into unmanned aircraft is what allows the aircraft to carry out its intended mission. Payload subsystems come in a variety of sizes and configurations, from extremely complex, with heavy computational capabilities, to relatively simple designs. Each system tries to meet the objectives of the operator. The myriad of applications of UAS with their respective payloads provides great opportunity for researchers in many fields to implement and exploit the capabilities of UAS.

The integration of subsystems onto UAS ideally uses a systems engineering approach to design. Payloads typically involve many components that must be integrated together to carry out the desired mission. This approach necessitates multiple steps on the way to designing a complete UAS subsystem. Those steps typically consist of concept development and trade studies, preliminary design review, critical design review, fabrication, system testing, and flight testing.

This chapter provides guidelines that have been proven to be successful for UAS payload development in a number of settings. As regulations and airspace operations continue to develop in this rapidly evolving industry, increasingly specialized regulations and requirements will be imposed, and these new rules will in all likelihood become standards for legal operation and development.

Chapter 15

One of the foundational tenets of flight is the responsibility of the pilot to see and safely avoid other aircraft. From the dawn of manned flight, pilots have been expected to maintain vigilance so as to remain "well clear" from other aircraft. The advent of UAS has changed how we approach technology's role in supporting or replacing functions once solely performed by human pilots.

The systems that allow a means of compliance with "see and avoid" and remaining "well clear" are called detect and avoid (DAA) systems. The legal definition of "well clear"—a quantified specification of the time, distance, or both that must separate aircraft—provides

UAS a means to not only meet, but likely exceed the ability of aircraft with onboard pilots to avoid a potential collision. This same definition supports the introduction of similar systems into manned aircraft, allowing pilots the benefit of increased situation awareness as well as reliable alerting for aircraft that may present a collision risk.

For UAS, the challenge is demonstrating compliance with the rules of the air—ensuring that the aircraft, even when not under direct pilot control, is capable of maneuvering sufficiently so as to avoid a potential collision hazard. This challenge requires execution of two main tasks: (1) avoid a collision and (2) remain "well clear."

The task description for DAA consists of two major functions: self-separation (SS), the ability of the system to remain "well clear" of other aircraft, and collision avoidance (CA), the ability of the system to prevent a collision with another aircraft.

To support the core functions of SS and CA, it is necessary to identify and sequence the component subfunctions that support a safe and reliable SS or CA maneuver. The FAA SAA Workshop identified eight subfunctions for SAA. The U.S. Department of Defense (DoD) SAA Science and Research Panel (SARP) provided its own decomposition, which included the addition of three new subfunctions. These 11 subfunctions, in time-linear order, are: detect target, track target, fuse target tracks, "orient" tasks, identify object, evaluate threat, prioritize threat, decide tasks, declare/alert, determine maneuver, command maneuver, execute maneuver, and return to course.

The role of the pilot (pilot in-the-loop, pilot on-the-loop, pilot-independent), and the role of air traffic control are also examined as a foundation for a description of the DAA system components (sensors, avoidance algorithms, and displays).

In 2013, the DoD SARP was tasked with defining a means of compliance with "well clear" for UAS. As a foundation for their approach, the SARP arrived at five guiding principles for UAS to comply with "well clear":

1. "Well clear" is a separation standard for UAS.
2. DAA systems need a quantitative standard, while humans may judge "well clear" subjectively.
3. "Well clear" is defined based on minimizing collision risk, but informed by operational considerations and compatibility with existing manned aircraft CA systems.
4. "Well clear" is defined in the horizontal dimension based upon time and distance.
5. "Well clear" is defined in the vertical dimension based on distance.

The result of this effort was the creation of a model that accounted for both time and distance from other aircraft, and provided a sufficient vertical (altitude) separation to support appropriate avoidance maneuvering (including a minimum time to closest point of approach of 35 s, a minimum horizontal distance of 4000 ft, and a minimum vertical distance of 700 ft from other aircraft).

Chapter 16

This chapter takes the reader through the colorful history of the UAS program at the Mesa County, Colorado Sheriff's Office (MCSO). Mesa County was one of the first state or local public safety agencies in the United States to acquire and deploy unmanned systems in a variety of missions. The "democratization" of aviation allowed MCSO to acquire an aviation asset that the agency could not otherwise have afforded (manned aircraft being considerably more expensive to acquire and operate than small UAS). Public concerns over privacy and weaponization of the aircraft were successfully dealt with by maintaining strict transparency standards and explaining the sound reasons for the employment of this technology to the public. They did this by hosting public meetings and media briefings, and the creation of an informational website.

MCSO's process of evaluating the costs and benefits of obtaining, training on, and using small UASs to perform functions such as accident reconstruction, search and rescue, surveillance of building fires in support of firefighting units, and data collection and retention is described in detail. The scenarios were illustrated by narrated case studies, including color images, of several actual UAS missions.

The MCSO experience demonstrates that small UASs are an affordable and valuable tool in the hands of public safety organizations and can be used in a way that does not inflame the public with fears of privacy invasion, warrantless searches, and other potential abuses of a highly efficient and safe technology. The limits of the technology, such as missions that are better served by a larger, manned aircraft, are also described, but the clear advantages of small UASs are in cost savings and time-sensitive deployments, as well as operating in denied, dangerous, or "dirty" environments.

In the application of standards and use of UAS, it is recommended that public safety organizations approach UAS with solid research, professionalism, and pragmatism. Humans have an amazing capability to create tools to extend beyond our physical limitations. UAS can provide significant expansion beyond those physical limitations for the benefit of all mankind.

Chapter 17

The current UAS industry, social and cultural contexts, and operational environments are dynamic, and specific predictions are sometimes difficult to formulate with any degree of confidence. However, the continuation of certain established trends is probable and, consequently, is a more reliable predictor of the future of the UAS enterprise than more inconstant, speculative, uncertain aspects of the industry. Driven by robust technological advancement, the market for UAS manufacturing and support services that began as a minor segment of the industry has become a major component of global commerce in the relatively short span of 15–20 years. All projections give credence to the expectation that this strong growth will continue. And just as technology drives the market, economics, in turn, will encourage the continued improvement of technology in terms of better manufacturing techniques, stronger, lighter, more suitable, and facilely applied materials and smaller, faster, more capable, and less expensive microprocessors, avionics, autopilots, and related UAS electronics. As regulations continue to evolve in concert with technology, larger unmanned aircraft with more complex and rigorous mission profiles will emerge. At the confluence of these trends will be an expanding job market for those involved in the UAS segment of the aerospace industry.

With a longer time horizon, the development of alternative energy sources that include fuel cells and hydrogen fuel pellets would afford another means of improving UAS performance. Several companies, including Airbus, Boeing (Phantom Works), and

AeroVironment (in conjunction with NASA), are developing solar-powered UASs with expected mission durations of 5 years or more. Sometimes termed atmospheric satellites, these UAS platforms are being developed to act as weather and earth observatories and telecommunications relay stations. Improvements in battery chemistries and construction techniques will improve both range and endurance of electrically powered UASs, as well as the useful load of such platforms. In-flight refueling to recharge the batteries of electric UASs is also being explored as a means of extending flight parameters.

As one peers further into the future, accurate predictions become more difficult. A continued acceleration of the pace of technological advancement in this area will not be surprising. Neither will be an increasing diversity of UAS mission profiles and proliferation of the technology. As has often been stated, the future is unlimited and unmanned.

References

Everaerts, J. The use of unmanned aerial vehicles (UAVs) for remote sensing and mapping. *The International Archives of the Photogrammetry, Remote Sensing and Spatial Information Sciences*, Vol. XXXVII. Part B1, 1187–1192, 2008.

Lillesand, T. and R. Kiefer. *Remote Sensing and Image Interpretation*. New York: John Wiley & Sons, 2000.

Index

D

E

K

L

M

V

W